Lecture Notes in Physics

Vol. 68: Y. V. Venkatesh, Energy Methods in Time-Varying System Stability and Instability Analyses. XII, 256 pages. 1977.

Vol. 69: K. Rohlfs, Lectures on Density Wave Theory. VI, 184 pages. 1977.

Vol. 70: Wave Propagation and Underwater Acoustics. Edited by J. Keller and J. Papadakis. VIII. 287 pages. 1977.

Vol. 71: Problems of Stellar Convection. Proceedings 1976. Edited by E. A. Spiegel and J. P. Zahn. VIII, 363 pages. 1977.

Vol. 72: Les instabilités hydrodynamiques en convection libre forcée et mixte. Edité par J. C. Legros et J. K. Platten. X, 202 pages. 1978.

Vol. 73: Invariant Wave Equations. Proceedings 1977. Edited by G. Velo and A. S. Wightman. VI, 416 pages. 1978.

Vol. 74: P. Collet and J.-P. Eckmann, A Renormalization Group Analysis of the Hierarchical Model in Statistical Mechanics. IV, 199 pages. 1978.

Vol. 75: Structure and Mechanisms of Turbulence I. Proceedings 1977. Edited by H. Fiedler. XX, 295 pages. 1978.

Vol. 76: Structure and Mechanisms of Turbulence II. Proceedings 1977. Edited by H. Fiedler. XX, 406 pages. 1978.

Vol. 77: Topics in Quantum Field Theory and Gauge Theories. Proceedings, Salamanca 1977. Edited by J. A. de Azcárraga. X, 378 pages 1978.

Vol. 78: Böhm, The Rigged Hilbert Space and Quantum Mechanics. IX, 70 pages. 1978.

Vol. 79: Group Theoretical Methods in Physics. Proceedings, 1977. Edited by P. Kramer and A. Rieckers. XVIII, 546 pages. 1978.

Vol. 80: Mathematical Problems in Theoretical Physics. Proceedings, 1977. Edited by G. Dell'Antonio, S. Doplicher and G. Jona-Lasinio. VI, 438 pages. 1978.

Vol. 81: MacGregor, The Nature of the Elementary Particle. XXII, 482 pages. 1978.

Vol. 82: Few Body Systems and Nuclear Forces I. Proceedings, 1978. Edited by H. Zingl, M. Haftel and H. Zankel. XIX, 442 pages. 1978.

Vol, 83: Experimental Methods in Heavy Ion Physics. Edited by K. Bethge. V, 251 pages. 1978.

Vol. 84: Stochastic Processes in Nonequilibrium Systems, Proceedings, 1978. Edited by L. Garrido, P. Seglar and P. J. Shepherd. XI, 355 pages. 1978

Vol. 85: Applied Inverse Problems. Edited by P. C. Sabatier. V, 425 pages. 1978.

Vol. 86: Few Body Systems and Electromagnetic Interaction. Proceedings 1978. Edited by C. Ciofi degli Atti and E. De Sanctis. VI, 352 pages. 1978.

Vol. 87: Few Body Systems and Nuclear Forces II, Proceedings, 1978. Edited by H. Zingl, M. Haftel, and H. Zankel. X, 545 pages. 1978.

Vol. 88: K. Hutter and A. A. F. van de Ven, Field Matter Interactions in Thermoelastic Solids. VIII, 231 pages. 1978.

Vol. 89: Microscopic Optical Potentials, Proceedings, 1978. Edited by H. V. von Geramb. XI, 481 pages. 1979.

Vol. 90: Sixth International Conference on Numerical Methods in Fluid Dynamics. Proceedings, 1978. Edited by H. Cabannes, M. Holt and V. Rusanov. VIII, 620 pages. 1979.

Vol. 91: Computing Methods in Applied Sciences and Engineering, 1977, II. Proceedings, 1977. Edited by R. Glowinski and J. L. Lions. VI, 359 pages. 1979.

Vol. 92: Nuclear Interactions. Proceedings, 1978. Edited by B. A. Robson. XXIV, 507 pages. 1979.

Vol. 93: Stochastic Behavior in Classical and Quantum Hamiltonian Systems. Proceedings, 1977. Edited by G. Casati and J. Ford. VI, 375 pages. 1979.

Vol. 94: Group Theoretical Methods in Physics. Proceedings, 1978. Edited by W. Beiglböck, A. Böhm and E. Takasugi. XIII, 540 pages. 1979.

Vol. 95: Quasi One-Dimensional Conductors I. Proceedings, 1978. Edited by S. Barišić, A. Bjeliš, J. R. Cooper and B. Leontič. X, 371 pages. 1979.

Vol. 96: Quasi One-Dimensional Conductors II. Proceedings 1978. Edited by S. Barišić, A. Bjeliš, J. R. Cooper and B. Leontić. XII, 461 pages. 1979.

Vol. 97: Hughston, Twistors and Particles. VIII, 153 pages. 1979.

Vol. 98: Nonlinear Problems in Theoretical Physics. Proceedings, 1978. Edited by A. F. Rañada. X, 216 pages. 1979.

Vol. 99: M. Drieschner, Voraussage – Wahrscheinlichkeit – Objekt. XI, 308 Seiten. 1979.

Vol. 101: A. Martin-Löf, Statistical Mechanics and the Foundations of Thermodynamics. V, 120 pages. 1979.

Vol. 102: H. Hora, Nonlinear Plasma Dynamics at Laser Irradiation. VIII, 242 pages. 1979.

Vol. 103: P. A. Martin, Modèles en Mécanique Statistique des Processus Irréversibles. IV, 134 pages. 1979.

Vol. 104: Dynamical Critical Phenomena and Related Topics. Proceedings, 1979. Edited by Ch. P. Enz. XII, 390 pages. 1979.

Lecture Notes in Physics

104

Dynamical Critical Phenomena and Related Topics

Proceedings of the International Conference, Held at the University of Geneva, Switzerland, April 2–6, 1979

Edited by Charles P. Enz

Springer-Verlag Berlin Heidelberg GmbH

Editor

Charles P. Enz
Département de Physique Théorique
Université de Genève
24, quai Ernest-Ansermet
CH-1211 Genève 4

ISBN 978-3-540-09523-1 ISBN 978-3-540-35124-5 (eBook)
DOI 10.1007/978-3-540-35124-5

Originally published by Springer-Verlag Berlin Heidelberg New York in 1979

INTRODUCTION

It is a great satisfaction to present this volume to the scientific community. For, it contains reports which, I believe, stand out in quality as well as actuality. Therefore, my first word is addressed to the authors of these reports thanking them for their enthousiastic collaboration as invited speakers to the "International Conference on Dynamical Critical Phenomena and Related Topics", held at the University of Geneva, Switzerland, from 2 to 6 April 1979, in short, the Critical Dynamics Conference - CDC 79. I also thank the invited speakers for having submitted their manuscripts quickly so that the remarkable freshness of the results reported in these proceedings is not impared by delays of publication.

It was the intention, that the program for CDC 79 should cover dynamical phenomena not only in the usual critical domain of closed, equilibrium systems but also around instabilities of open, non-equilibrium systems. This "opening" towards a new and fastdeveloping field contained some risks, as was pointed out by several experts asked for their opinion about the interest of such a conference back in March 1978. In order to arrive at a program which was internally "at equilibrium", the borderline was drawn by limitation to physical systems, thus excluding endeavours into chemistry, or even further, into life and social sciences.

The vicinity in time of several meetings of a similar kind presented a certain danger for CDC 79, as signaled by a number of the experts contacted for advice. These meetings are listed below because some of them are referred to in the contributions to CDC 79 and have produced proceedings of interest :

- The "International Symposium on Nonlinear Nonequilibrium Statistical Mechanics" held in Kyoto, Japan, on July 10-14, 1978, and chaired by H. Mori;
- the 17th International Solvay Conference on Physics on "Order and Fluctuations in Equilibrium and Non-Equilibrium Statistical Mechanics" held in Brussels, Belgium, on November 20-23, 1978, and directed by I. Prigogine;
- the 1979 Midwinter Solid State Research Conference on "Non-Linear and Non-Equilibrium Phenomena in Condensed Matter" held in Laguna Beach, California, on January 15-19, 1979, and chaired by G. Ahlers and B. Huberman;
- the "6th International Seminar on Phase Transitions and Critical Phenomena in Solids and Liquids" held in Trieste, Italy, on March 26-28, 1979, and chaired by C. Di Castro;
- the "International Symposium on Synergetics" held in Elmau Castle, Bavaria, on April 30 - May 5, 1979, and directed by H. Haken.

If these meetings had at all been dangerous to CDC 79, it may have been with regard to participation but certainly not with regard to quality. In fact, while many of the experts in the field unfortunately did not attend, the CDC 79 community turned out to be a motivated, coherent, enthousiastic and surprisingly young group, as was evident from the quality of the discussions.

These discussions would actually have merited publication, as would have the contributed papers. However, reasonable size and quick publication of the proceedings were considered of primary importance, and I believe that the final product fully justifies this reasoning. Still, in order to retain some information about the contributed papers, the list of contents of the Abstract-Handbook (which contained the abstracts of all contributions and was handed to each participant at CDC 79) is reproduced here with only minor modifications, in place of a simple list of contents. In this *list of contributions* all invited lectures, which alone are printed in this volume, figure with their paper numbers while the contributed papers are only listed by title and have no page numbers attached.

All contributions are grouped into the *seven topics* which formed the underlying structure of the program. Making use of the *address list of participants* included as an appendix, it will be possible to obtain more ample information directly from the authors of the contributed papers. In this way it is hoped that these proceedings will offer maximum information at minimum cost both in volume and in time delay.

The first topic of the program was a review of the theoretical methods and was meant as an introduction to the conference. In his comparison of mode-coupling theory and the dynamical renormalization group method, Prof. Gunton clearly showed the extent and limits of the equivalence between the two approaches. Prof. Janssen gave a review of the elegant field-theoretic method, which had been introduced into critical dynamics by himself and by De Dominicis, and also pointed to new applications, some of which were discussed in contributed papers. The particularities and difficulties of critical dynamics in the ordered phase and the role of Goldstone modes were discussed by Prof. Szépfalusy.

Although not truely separated from the first topic, the real space dynamical renormalization group methods have assumed a quite distinct importance in present-day research, mainly because of their applicability to low-dimensional systems. While Prof. Suzuki's lecture, as well as the other contributions, gave an impressive view of the problems still ahead, the beautiful report by Prof. Mazenko took up the question at a more fundamental level and was full of promise.

In view of the very recent break-through in the understanding of critical damping of second sound in liquid Helium, a full day was devoted to this substance. The latest measurements of the second sound damping and other effects of critical dynamics obtained with powerful light scattering techniques were the subject of an illuminating report by Prof. Greytak. Dr. Peliti reviewed the theory by De Dominicis and himself, which for the first time had exhibited the discrepancy between theory and experiment. The final explanation of this discrepancy by Prof. Ferrell and his collaborators Dr. Bhattacharjee and Dr. Dohm was reported for the first time at CDC 79 by these three authors. The abundant experimental data obtained near the tricritical point of He^3-He^4 mixtures were reviewed by Prof. Meyer while the theoretical aspect was also contained in Prof. Gunton's and Dr. Peliti's talks.

In view of the still controversal question of whether the central peak is an intrinsic feature of critical dynamics near structural phase transitions, a panel was organized on this subject. The very latest experimental and theoretical developments together with an extensive historic overview was presented by Prof. Müller. A second panel discussion devoted to the problems left open by this conference was directed by Prof. Mazenko.

Two other important topics of critical dynamics could only be touched, namely low-dimensional systems and systems with random parameters. Concerning the first topic, Prof. Sjölander reported on exact results obtained for the one-dimensional Heisenberg chain, which was supplemented by some results for two dimensions by Dr. Reiter. Unfortunately, it had not been possible to have a report on the exciting new results for two-dimensional two-component systems (Helium films). But instead, there was an expert report by Prof. De Dominicis on the sophisticated theoretical methods developed for systems with randomness.

Much emphasis was also given to the experimental aspect in the following topic concerning hydrodynamic instabilities. Here, Prof. Bergé and his group reported their imaginative light scattering work, most of which was communicated for the first time at CDC 79. Bergé also discussed the transition to turbulence, emphasizing the importance of the aspect ratio of the system. The formidable theoretical problems related with both hydrodynamic instabilities and turbulence were discussed respectively by Prof. Velarde and Dr. Fournier, the latter reporting also on some attempts to apply renormalization group methods to turbulence.

The last topic was devoted to critical dynamics far away from equilibrium. Here again CDC 79 witnessed a first report, namely the break-through achieved by Prof. Kawasaki and his collaborator Dr. Onuki in the problem of a shear flow of a liquid near its critical point. It was a remarkable coincidence that the experimental verification of one of Kawasaki's predictions was also first reported at CDC 79 by Beysens and Gbadamassi from Prof. Bergé's group.

Prof. Kawasaki also reviewed other systems which are far away from equilibrium and which may be analyzed by his new theory. One was non-equilibrium superconductivity which, due to its importance, would have deserved an independent account. Instead, the classic far-away-from-equilibrium system , namely the laser, was brilliantly reviewed by Prof. Arecchi, and many contributed papers on non-equilibrium problems were to follow.

Having mentioned some missing topics, it is worth adding that the program of CDC 79 was probably a major cause for a stimulating conference. This program grew out of a collaboration with Michel Droz for which I am most grateful to him. I also thank Michel and my other associates, as well as the secretaries, accountants and printers for their unassuming but efficient presence whenever it was needed.

But if CDC 79 succeeded with only a local staff of three, it was also due to its third member, Mrs. Christine Baly. She was not only the helpful and efficient Conference secretary but contributed significantly to the shaping of the social program : the excursion to CERN and to the famous Bodmer collection of ancient manuscripts and art objects, as well as the high-spirited banquet in the castle of the historic Madame de Staël at Coppet with the authentic Swiss folklore.

A particular word of thanks is due to the authorities of the Canton and the City of Geneva who so graciously offered the cocktail at the elegant Palais Eynard but, more indirectly, are also at the origin of the excellent facilities at the new Sciences II building.

Obviously, without the sponsors, there would have been no CDC 79. Therefore, I here reiterate my thanks to them in the name of the participants, above all to the Swiss National Science Foundation, but also to the Fond General of Geneva University, to the International Union of Pure and Applied Physics and to the Faculté des Sciences, the Section de Physique and the Département de Physique Théorique of Geneva University. I think, the present proceedings are the best proof for the good use made of their funds.

Charles P. Enz

Geneva, May 6, 1979

TABLE OF CONTENTS

LIST OF CONTRIBUTED PAPERS NOT PUBLISHED IN THIS VOLUME

Mode-Coupling and the Dynamical Renormalization Group

Relaxation Processes of Isotropic Spin-Fields Below T_c
R. BAUSCH, H.K. JANSSEN, Y.YAMAZAKI

The Dynamic Renormalization Group in the Large n Limit
P. SZEPFALUSY, T. TEL

Nonlinear Critical Slowing Down in the Glauber-Ising Model
R. BAUSCH, E. EISENRIEGLER, H.K. JANSSEN

Critical Dynamics and Crossover due to Elastic Coupling in Systems with Marginal Dimensionalities
G. MEISSNER, R. PIRC

Crossover Scaling Function of the Spin-Diffusion Coefficient for an Isotropic Ferromagnet
L. SASVARI

Real Space Dynamical Renormalization Group Methods

Master Equation Approach to Critical Dynamics
U. DEKER, F. HAAKE

A Simple Approach to Critical and Non-Critical Dynamics Based on Real Space Renormalization
W. KINZEL

Real Space Renormalization for Dynamic Gaussian Systems
Y. YAMAZAKI

Real Space Renormalization Group for the Kinetic Ising Model in $1+\varepsilon$ Dimensions
M. DROZ

MODE COUPLING THEORY IN RELATION TO THE DYNAMICAL RENORMALIZATION GROUP METHOD

J.D. Gunton
Physics Department, Temple University
Philadelphia, Pa. 19122

Mode Coupling Theory in Relation to the Dynamical Renormalization Group Method

J. D. Gunton
Physics Department
Temple University
Philadelphia, Pa. 19122

A review is given of the relationship between the mode coupling and renormalization group theories of critical dynamics. A brief discussion is given of the origin of the semiphenomenological nonlinear Langevin equations of motion on which both theories are usually based. A summary of the basic approximations and limitations of the mode coupling theory is presented. It is noted that although the renormalization group approach represents a more fundamental theory of critical dynamics, the theories are intimately related. In particular it is shown that to lowest order in $\epsilon = d_c - d$, where for $d > d_c$ conventional theory is valid, the renormalization group equations for several different models of critical dynamics are related by a simple transformation to those obtained from mode coupling theory. A model of the critical dynamics of fluids is discussed in some detail to compare explicitly the two theories.

1. Introduction

The development of the modern theory of dynamic critical phenomena provides an interesting example of the evolutionary nature of scientific progress. It is perhaps worthwhile to summarize here the basic physical ideas that have led to what we now consider a successful dynamical theory. At least six major contributions stand out in retrospect. The first was the concept that the relaxation time to approach equilibrium diverges as one approaches a critical point. This was clearly formulated in the so-called conventional theory of critical slowing down developed independently by Van Hove[1] and Landau and Khalatnikov[2]. This theory assumes that the kinetic coefficient for the order parameter remains finite at the critical point. The anomaly in the relaxation rate is then attributed completely to the divergence in the order parameter susceptibility, since in general the relaxation rate is given by the ratio of the transport coefficient to the static susceptibility. Thus the origin of the dynamic anomaly is assumed to be completely thermodynamic in origin. This theory is invalid for real systems, but has been useful in providing some qualitative understanding of critical dynamics. In addition, of course, it is now known that conventional theory is valid for large enough dimensionality, d. This has provided the basis for the recent successful ϵ-expansion in dynamics, where $\epsilon = d_c - d$, with d_c being the dimensionality such that for $d > d_c$ conventional theory is valid.

The second major theoretical advance came with the work of Fixman[3] who proposed mechanisms by which the kinetic coefficients themselves could diverge at a critical point. These ideas were more fully developed by Kawasaki[4,5] and Kadanoff and Swift[6] in what is now known as mode coupling theory. In general this work involved the basic idea that nonlinear couplings between the slow dynamical modes (the conserved variables and the order parameter) can lead to divergences in kinetic coefficients. In particular the nonlinear couplings specifically considered in these theories were of a nondissipative nature.

A third, closely related advance was the development of phenomenological nonlinear equations of motion in which the earlier intuitive mode coupling approach was put on a firmer theoretical basis, in such a way that detailed calculations of correlation and response functions, for example, could be performed. The pioneer work in this field is that of Kawasaki[4,5] who derived these phenomenological nonlinear Langevin equations from exact, generalized nonlinear Langevin equations with memory. It would seem in retrospect that this third development is one of the major theoretical advances in the field, since these equations now serve as a basis for both mode coupling and renormalization group calculations. The next significant idea came with the advent of the renormalization group in dynamics as developed originally by Halperin, Hohenberg and Ma[7,8], following the fundamental work of Wilson[9] in equilibrium critical phenomena. The idea that emerged from this original work was that there exists a second source of nonlinear couplings between the slow modes (gross variables) which are of a dissipative nature and which can cause kinetic coefficients for relaxational models to vanish at the critical point. These dissipative couplings can also be of importance in determining critical exponents and scaling functions. It should be noted that such couplings are present in the phenomenological nonlinear equations of motion developed in the mode coupling theory, but has been overlooked in the original mode coupling approximations for the Ginzburg-Landau free energy functional.

A fifth significant development was the notion of universality classes, whose origin lies in equilibrium critical phenomena. Its extension to dynamics was developed in the mode coupling theory where, for example, it is clear that the Poisson brackets which determine the non-dissipative, "streaming" terms in the equations of motion are important in determining to which universality class a given system belongs. However it is now quite clear that the natural theoretical framework for discussing dynamic universality classes is the renormalization group. Further discussion of this point is given in the excellent review article by Hohenberg and Halperin[8]. Finally, we end this summary of the main developments in the evolution of our understanding of critical dynamics by noting the very useful idea of dynamical scaling, as first suggested by Ferrell and collaborators and developed by Halperin and Hohenberg. A complete discussion of this work is found in reference 8. Although we do not treat this idea here, it will be dealt with at this meeting with particular reference to the λ-transition in ^{4}He. This has long posed a challenge to our theory of critical dynamics[8], but recent work (to be discussed at this meeting) by Ferrell and Dohm seems to represent a break through in this area. A discussion of the current experimental status and its comparison with the recent theory is given by Greytak in this volume.

We now turn to the purpose of this present paper, which is to discuss the mode coupling theory in relation to the renormalization group. Before discussing this subject in greater detail in later sections, we summarize our main conclusions here. To begin with, it is now quite clear that the renormalization group is an improvement over its mode coupling predecessor. However, in fairness to the mode coupling work, it is important to note that the renormalization group involves the same equations of motion and ideas as developed in mode coupling. At present the major improvement in the theory lies in the development of the ϵ-expansion in dynamics. There are several specific areas in which one can see the improvement in the theory. First, the ϵ-expansion provides a more accurate treatment of nonlinear couplings, at least for small ϵ. It also provides a systematic way to prove scaling relations and to calculate exponents and scaling functions, again at least for small ϵ. Second, the renormalization group has provided a refinement or clarification of the nonlinear Langevin equations of motion. Two good examples of this are the relaxational (kinetic Ising) models and the model of ^3He - ^4He tricritical dynamics. In both cases the improved model equations result from an explicit treatment of nonlinear dissipative couplings not considered in earlier mode coupling work. Third, as noted earlier the renormalization group provides a natural way in which to attack the problem of identifying universality classes. According to this concept, all members of a given universality class have essentially the same critical dynamics. Finally, and perhaps most promising for future theory, the development of the renormalization group has allowed the possibility of developing new and perhaps powerful methods for studying critical dynamics, particularly in low dimensions. One promising new advance has been the extension of real space renormalization group methods to critical dynamics. This is reviewed by Mazenko and by Suzuki in this volume.

Although the results are still quite preliminary and even somewhat controversial this does seem to provide a promising new direction of research for what is still one of the major areas of statistical physics.

The outline of this paper is the following. In section II we briefly review the origin of the semiphenomenological equations of motion which form the basis of both the mode coupling and renormalization group theories. In retrospect as noted above it would appear that one of the major accomplishments of mode couplint theory was the derivation of these semimacroscopic equations from the microscopic Liouville equations. In section III we discuss the approximations made by Kawasaki in developing a solution of these equations. These include a quadratic approximation for the free energy as well as a "self-consistent" perturbation theory. In sections IV and V we discuss the relationship between the two theories in more detail. Namely, we show that to lowest order in ϵ the renormalization group equations for several models of critical dynamics can be obtained from the corresponding mode coupling equations by a simple transformation[10,11]. Thus to order ϵ the two theories agree exactly. In section VI we discuss one example, the simple fluid (or binary fluid) to second order in ϵ, in order to compare the original very successful mode coupling calculation for this system with the exact second order result. Finally, in section VII we briefly discuss the tricritical dynamics of He^3 - He^4 where the renormalization group work led to improved model[13] equations of motion over those of the earlier mode coupling theories[14,15].

II. Semiphenomenological Nonlinear Langevin Equations

One of the significant accomplishments in the development of the theory of critical dynamics has been the derivation from the exact microscopic equations of nonlinear Langevin or, equivalently, Fokker-Planck equations of motion. The pioneer work in this field is that of Kawasaki, who relied heavily on the projection operator formalism developed by Mori[16] and by Zwanzig[17]. This projection operator formalism was in turn motivated by the work of Green[18], who showed that the probability distribution function for hydrodynamic variables satisfies a Fokker-Planck equation. These nonlinear equations of motion are the fundamental starting point of both the mode coupling and renormalization group theories. This is in fact one reason for the close relationship between these two theories. However, it is to be noted that these equations are not exact but involve three basic approximations which seem physically plausible and which we discuss below. These equations are hence called semiphenomenological; one justification for their validity lies a posteriori in comparison between the resultant theory and experiment. Further work in this field will certainly involve refinement of these equations of motion.

Two particularly simple derivations of these nonlinear Langevin equations have been given, one by Mori and Fujisaka[19] and another by Kawasaki[20]. Other discussions of these equations or the corresponding Fokker-Planck equations also exist in the literature[4, 21-23]. A very nice summary of these equations as well as a useful discussion of some of their general properties is given by Ma and Mazenko[24]. In this section we shall present a short version of Kawasaki's derivation which quite naturally involves the introduction of the Zwanzig projection operator. The first point to note here is that in general we are not interested in the complete, highly complicated microscopic time evolution of a system. Rather, we are usually interested in time scales characteristic of its slowly varying dynamical quantities, such as those associated with hydrodynamic modes. Therefore Zwanzig and independently Mori introduced a projection operator formalism which would extract from the microscopic dynamics the equations of motion appropriate for these slowly changing quantities, the so-called gross variables, which we denote by $\{A_1, \ldots, A_n\}$ or $\{A\}$. The correct choice of these variables is crucial to obtaining useful equations of motion. Near a critical point the set would include the order parameter, even if it were not a conserved quantity, due to its anomalous relaxation time. To derive the nonlinear equations appropriate near a critical point, we first introduce a projection operator P, such that for a dynamical variable X(t),

$$PX(t) = \sum_j (X, \psi_j(\{A\}))\, \phi_j(\{A\}), \tag{2.1}$$

where (A,B) denotes an appropriate inner product of two arbitrary dynamical variables. For simplicity we only consider the classical mechanics case where $\psi_m(\{A\}) = \phi_m(\{A\})$. The choice of $\{\phi_m(\{A\})\}$ then determines the projection operator. The separation of X(t) into its "slowly" varying and "more rapidly" varying parts is then given by $X = PX + QX$, where $Q = 1 - P$ and QX represents the rapidly varying part of X.

In order to discuss the particular problem posed by critical dynamics within the context of this projection operator formalism, we next introduce the operator identity of Kawasaki,

$$\frac{d}{dt} e^{it\mathcal{L}} = e^{it\mathcal{L}} i\mathcal{L}_0 + \int_0^t ds\, e^{i(t-s)\mathcal{L}} i\mathcal{L}_0 e^{is(\mathcal{L}-\mathcal{L}_0)} i(\mathcal{L}-\mathcal{L}_0) + e^{it(\mathcal{L}-\mathcal{L}_0)} i(\mathcal{L}-\mathcal{L}_0) \quad (2.2)$$

for an arbitrary operator $\mathcal{L}_0$, where the time development of a dynamical variable X(t) is given by the Liouville operator as

$$X(t) = e^{it\mathcal{L}} X. \quad (2.3)$$

Here X denotes the initial value of X(t) at t = 0. In order to obtain a generalized Langevin equation for the gross variables we next choose $\mathcal{L}_0 = P\mathcal{L}$ and apply (2.2) to $A_i(t)$. This yields the formally exact equations

$$\frac{d}{dt} A_i(t) = \sum_j (i\mathcal{L}A_i, \Psi_j(\{A\}))\, \phi_j(\{A(t)\}) - \sum_j \int_0^t ds\, (f_i(s), f_j)\, \phi_j(\{A(t-s)\}) + f_i(t) \quad (2.4)$$

where

$$f_i(t) = e^{itQ\mathcal{L}} iQ\mathcal{L}A_i, \quad (2.5)$$

$$f_j = Q i\mathcal{L} \phi_j(\{A\}). \quad (2.6)$$

The first term in (2.4) is the "adiabatic" term; the second expresses the damping of this adiabatic motion by the frictional effects arising from the "random" forces $f_i(t)$ which satisfy the orthogonality conditions

$$(f_i(t), \Psi_j) = 0, \quad (2.7)$$

$$(\phi_j, f_k) = 0. \quad (2.8)$$

This friction term in general involves memory effects, which brings us to the core of the problem of critical dynamics. Namely, so far what we have done is formally exact for any projection operator in classical mechanics which satisfies (2.1). The issue, however, is what constitutes a useful projection operator for critical dynamics? An answer can be given, if we decide that it would be nice to be able to essentially ignore memory effects, i.e. to make a Markovian approximation for the kernel $(f_i(s), f_j)$. This is tantamount to the assertion that the $\{f_i(t)\}$ are really random. Once we decide that this is a sensible goal, we can immediately see why critical dynamics differs from the usual hydrodynamics which we are accustomed to handle. For example, at first sight a natural choice to make for the ϕ_j is to choose $\phi_j(\{A\}) = \chi_i^{-1/2} A_i$. If we then choose the gross variables to be orthogonal to each other and define the inner product to be the Kubo canonical correlation[20] (turning, momentarily to the quantum mechanical case) we obtain from (2.4) the generalized linear Langevin equation originally derived by Mori:

$$\frac{d}{dt} A_i(t) = \sum_j (i\mathcal{L} A_i, A_j)\chi_j^{-1} A_j(t) - \sum_j \int_0^t (f_i(s), f_j)\chi_j^{-1} A_j(t-s) + f_i(t). \tag{2.9}$$

The problem with this choice near a critical point, however, is that the $\{f_i(t)\}$ are not "really random". In this case the forces $\{f\}$ are orthogonal only to linear functions of the gross variables, via (2.7) and (2.8). Thus these forces then contain products of the gross variables and hence are not random[4,5,20,22]. Correspondingly, there is a memory effect and one cannot therefore make a Markovian approximation, i.e. $(f_i(s), f_j) = 2\,\Upsilon_{ij}\,\delta(s)$.

Therefore, if we wish to find an approximately Markovian theory, we must choose a different projection operator. This rather naturally leads to nonlinear interactions between the slow modes, as we now see. Following Kawasaki we thus go to the opposite extreme from the above choice of the Mori projection operator and introduce the Zwanzig projection operator,[22] i.e. we choose the set $\{\phi_j\}$ to include all the suitably symmetrized polynomials of $\{A\}$. We also choose the inner product (A,B) to be the equilibrium average $\langle A\, B^*\rangle$ (or the Kubo canonical correlation in quantum mechanics) and introduce the completeness condition

$$\sum_j \phi_j(\{A\})\,\phi_j^*(\{A'\}) = \delta(A-A')\,P_e^{-1}(\{A\}) \tag{2.10}$$

where $A = A(x)$, $A' = A(x')$ and x denotes the representative point of the system in phase space. The equilibrium probability distribution function for the gross variables is $P_e(\{A\})$, while $\delta(A - A')$ denotes the product of the delta functions $\delta(A_i - A_i')$. Then from (2.4) we obtain

$$\frac{d}{dt} A_i(t) = v_i(\{A(t)\}) + \sum_j \int_0^t ds\, P_e^{-1}(\{A(t-s)\}) \frac{\partial}{\partial A_j^*(t-s)} K_{ij}^{\circ}(s; A(t-s)) P_e(A(t-s)) + f_i(t), \tag{2.11}$$

where the streaming term is given by

$$v_i(\{a\}) = \langle i\mathcal{L}A_i;\quad \{a\}\rangle / P_e(\{a\}). \tag{2.12}$$

The random forces are given by

$$f_i(t) = e^{itQ\mathcal{L}}\; Q\, i\mathcal{L} A_i \tag{2.13}$$

and the kernel

$$K^{\circ}_{ij}(t;\{a\}) = \langle f_i(t)\, f^{*}_j(0);\quad \{a\}\rangle / P_e(\{a\}) \tag{2.14}$$

where

$$\langle X;\{a\}\rangle = \langle X\,\delta(A-a)\rangle. \tag{2.15}$$

Thus by choosing the Zwanzig projection operator we are led to nonlinear equations of motion. The nonlinear coupling between the slow modes $\{A(t)\}$ arises from two sources, the reversible mode-coupling term $v_i(\{A\})$ and the irreversible term arising from the equilibrium probability function through the kernel K°_{ij}. On the other hand, what we have gained at the expense of these nonlinear couplings is a representation in which the forces $\{f(t)\}$ should be random. This would result from the orthogonality of the $\{f(t)\}$ with the $\{\phi_j(\{A\})\}$, provided that the set A contains all of the slowly varying dynamical variables of the system. Thus a proper choice of the gross variables is crucial to a successful theory of critical dynamics.

To obtain the usual phenomenological equations of motion from the exact generalized Langevin equations (2.11) we need to make three basic, plausible assumptions. The first is the Markovian approximation, which was the original motivation for choosing the Zwanzig projection operator. Namely,

$$K^{\circ}_{ij}(t;\{a\}) \cong 2L^{\circ}_{ij}(\{a\})\,\delta(t) \tag{2.16}$$

where the $L^{\circ}_{ij}(\{a\})$ are the "bare" kinetic coefficients. The nonlinear couplings renormalize these bare coefficients and lead to the physical transport coefficients L_{ij} which are the object of experimental and theoretical interest. The second approximation is to assume that these bare coefficients are independent of the gross variables, i.e.

$$L^{\circ}_{ij}(\{a\}) \cong L^{\circ}_{ij}\;. \tag{2.17}$$

The final assumption is that the noise is Gaussian distributed. These assumptions then approximate (2.11) by the phenomenological equations of motion

$$\frac{d}{dt}A_i(t) = v_i(\{A(t)\}) - \sum_j L^{\circ}_{ij}\,\frac{\partial F(\{A(t)\})}{\partial A^{*}_j} + f_i(t) \tag{2.18}$$

where

$$F(\{A\}) = -\ln P_e(\{A\}). \tag{2.19}$$

These equations are the starting point for both the mode coupling and renormalization group theories.

Next, we note that the "mode coupling" or "streaming" terms $v_i(\{A\})$ can be given explicitly in terms of certain Poisson brackets or commutators of the gross variables $\{A\}$. These terms can be written in general in the form[5,24]

$$v_i(\{A\}) = \lambda \sum_j \left\{ \frac{\partial}{\partial A_j} Q_{ij}(\{A\}) - Q_{ij}(\{A\}) \frac{\partial F(\{A\})}{\partial A_j} \right\} \tag{2.20}$$

where λ is a constant and $Q_{ij} = -Q_{ij}$ are variables obtained from the Poisson brackets or commutators of A_i. Since these depend on the system of interest, the Poisson brackets are one set of quantities which determine the universality classes in critical dynamics. In general the $v_i(\{A\})$ involve nonlinear interactions between the slow modes of the system (hence the name "mode coupling") and were shown by Kawasaki[4,5] and by Kadanoff and Swift[25,26] to give rise to anomalies in transport coefficients near the critical point. We should also note that in many mode coupling calculations a simplified version of the above model equations is used. Namely, in $v_i(\{A\})$ one retains only the linear and quadratic terms in the gross variables[5], i.e.

$$v_j(\{a\}) = i\omega_j a_j - i \sum_{\ell,m} V_{j\ell m} (a_\ell a_m - \langle a_\ell a_m \rangle). \tag{2.21}$$

The ω_j and $V_{j\ell m}$ constitute characteristic frequencies and mode coupling coefficients , respectively. We do not give explicit expressions for them here but they involve in general the Poisson brackets and susceptibilities for the gross variables.

Finally, we end this section with a relatively simple example of these nonlinear Langevin equations, namely the isotropic Heisenberg ferromagnet[4,24]. In this case the appropriate gross variables are the set of Fourier components $\{\vec{S}_{\vec{k}}\}$ of the magnetization density $\vec{s}(\vec{x},t)$ with $k<\Lambda$, where Λ is the upper cut-off wave number. The Ginzburg-Landau free energy in zero magnetic field is

$$F = \frac{1}{2} \int d^d x \left[(\nabla \vec{S})^2 + r_0 \vec{S}^2 + \frac{1}{2} u_0 (\vec{S}^2)^2 \right] \tag{2.22}$$

where $\vec{S}$ is a three-component vector field and $r_0 = a(T-T_c)$. The "mode coupling" term follows from the standard spin commutation relations which leads to

$$Q_{kk',\alpha\beta} = \sum_{\gamma=1}^{3} \epsilon_{\alpha\beta\gamma} S^{\gamma}_{k+k'} . \tag{2.23}$$

Thus one finds

$$v_i(\{\vec{S}\}) = -\lambda V^{-\frac{1}{2}} \sum_{\beta,\lambda} \sum_{q<\Lambda} \epsilon_{\alpha\beta\gamma} S^{\beta}_{k+q} \frac{\partial F(\{\vec{S}\})}{\partial S^{\gamma}_{q}} . \quad (2.24)$$

If we write the bare kinetic coefficient for the conserved order parameter as $L(q) = q^2 \Gamma$ and use the Ginzburg-Landau free energy (2.22), our nonlinear Langevin equations (2.19) can be written as

$$\frac{\partial \vec{S}}{\partial t} = \lambda \vec{S} \times \vec{H} - \Gamma \nabla^2 \vec{H} + \vec{f} \quad (2.25)$$

where the local magnetic field is

$$\vec{H} = - \delta F / \delta \vec{S} . \quad (2.26)$$

The Gaussian noise term satisfies

$$\langle f_{\alpha k}(t) f_{\beta k'}(t') \rangle = 2 \Gamma k^2 \delta_{\alpha\beta} \delta_{-k,k'} \delta(t-t'). \quad (2.27)$$

In this language, the conventional theory consists of three approximations: first, the Onsager coefficient Γ remains finite at the critical point. Second, the mode coupling term is ignored ($\lambda = 0$). Third, a quadratic approximation is made for the free energy functional, such that in Fourier space

$$\vec{H}(\vec{k}) = -\partial F / \partial \vec{S}_{-\vec{k}} \cong \chi_k^{-1} \vec{S}_k \quad (2.28)$$

where $\chi^{-1}_{k=0}$ is the thermodynamic susceptibility. In the mode-coupling theory only the latter approximation is made. Finally, in the renormalization group analysis the full Ginzburg-Landau free energy functional is used. The presence of the quartic term $u_0(\vec{S}^2)^2$ then leads to additional nonlinear coupling in the Langevin equations beyond that included in the mode coupling theory[24].

III. Mode Coupling Theory

In this section we summarize briefly the self-consistent mode coupling theory as developed by Kawasaki[4]. Other approximate calculational schemes similar to this have been developed, but we will not discuss them here[25-28]. The basic starting point in the mode coupling theory is the set of phenomenological equations given in (2.18). These, however, are extremely difficult to solve, due to their nonlinearity which arises both from the reversible, streaming term $v_i(\{a\})$ as well as from the irreversible term through the free energy $F(\{a\})$. Furthermore, it is worth noting that at the time the theorists were attempting to solve these equations, the ϵ-expansion did not exist. One was faced, therefore, with the additional difficulty that there is no smallness parameter in three dimensions in these nonlinear Langevin equations. The approach which was developed, primarily by Kawasaki, consisted of two parts. The first was to consider the local free energy $F(\{a\})$ in these equations as known, in the same spirit as conventional theory. He then made the Gaussian approximation

$$F(\{a\}) = \sum_j |a_j|^2 / \langle |a_j|^2 \rangle \qquad (3.1)$$

where a_j stands for the gross variable whose corresponding phase function is $A_j(x)$. We also choose variables such that $\langle a_j\, a_l^* \rangle = 0$ for $j \neq l$. It is important to note two points here. The first is that this Gaussian approximation involves the correct equilibrium susceptibility $\chi_j = \langle |a_j|^2 \rangle$ rather than a "bare" susceptibility such as r_0^{-1} which occurs for example in the original Ginzburg-Landau Hamiltonian (2.22). Thus in a sense this approximation includes some of the effects of the quartic term which occurs in the Ginzburg-Landau Hamiltonian. However, as we will see later it does not adequately describe all of the effects of u_0. The second point is that due to this approximation all of the nonlinearity in the Langevin equations arises from the reversible streaming term $v_i(\{a\})$. The renormalization group approach, on the other hand, does not make a Gaussian approximation but rather begins with the full Ginzburg-Landau Hamiltonian with bare coupling constants. In particular this means that there is an additional source of nonlinearity in these equations which arises, for example, from the quartic term which involves the coupling constant u_0. It should also be noted that in many of the mode coupling calculations the Ornstein-Zernike approximation was used for the order parameter susceptibility, i.e. $\chi_k = A(k^2 + \kappa^2)^{-1}$ where κ is the inverse correlation length and k is the wave number. This approximation was made only for calculational convenience.

The second part of the mode coupling calculation involves a perturbation solution of the mode coupling equations. To discuss this it is convenient to convert our Langevin equations into equations for the propagators

$$G_j(t_1 - t_2) = \Theta(t_1 - t_2) \langle a_j(t_1)\, a_j^+(t_2) \rangle / \langle a_j\, a_j^+ \rangle \qquad (3.2)$$

$$\Theta(t) = \begin{cases} 1 & t \geq 0 \\ 0 & t < 0 \end{cases}$$

where we use the notation that j now stands for both the wave number as well as the label for the type of gross variable being considered. The superscript $+$ denotes Hermitian conjugate. It is straightforward to show[4] that from (2.19) we obtain a Dyson equation

$$G_j(t) = G_j^{\circ}(t) + \int_0^t dt_1 \int_0^{t_1} dt_2\, G_j(t-t_1)\Sigma_j(t_1-t_2)\, G_j^{\circ}(t_2), \qquad (3.3)$$

where $\Sigma_j(t)$ is the proper self-energy and $G_j^{\circ}(t)$ is the bare propagator which can be obtained from the Langevin equation by dropping all of the nonlinear terms. To solve the Dyson equation Kawasaki developed a perturbation expansion in powers of the nonlinear terms which arise from $v_i(\{a\})$. Since there is no smallness parameter in this expansion it is not obvious a priori as to whether this is a useful procedure. The simplest nontrivial approximation in this scheme is to ignore all "vertex" renormalizations which leads to the following set of coupled equations[4] for the Fourier transform of $G_j(t)$:

$$G_j(\omega) = [-i\omega + i\omega_j + \gamma_j - \Sigma_j(\omega)]^{-1} \qquad (3.4)$$

where the self energy is given by

$$\Sigma_j(\omega) = -2\int_0^{\infty} dt \sum_{\ell,m} \frac{\chi_\ell \chi_m}{\chi_j} |V_{j\ell m}|^2\, G_\ell(t) G_m(t)\, e^{i\omega t} \qquad (3.5)$$

The ω_j are the characteristic frequencies and the $V_{j\ell m}$ are the mode coupling coefficients which arise from the explicit evaluation of the streaming term $v_i(a)$[4,5] as in equation (2.21). The γ_j are proportional to the bare kinetic coefficients L_j°, where we restrict ourselves to models in which $L_{ij}^{\circ} = L_j^{\circ}\delta_{ij}$.

These equations are the "self-consistent" mode coupling equations originally derived and successfully applied to a variety of models of critical dynamics by Kawasaki. The most notable success of this theory is for the critical point of simple fluids and binary fluid mixtures. We will return to this particular example in more detail in section V. We now turn to the question of the relationship between this self-consistent mode coupling theory and the more recent renormalization group analyses.

IV. Solution of the Mode Coupling Equations to Lowest Order in ϵ

In order to discuss the relationship of the two theories we first continue our discussion of the nonlinear integral equations which we obtained in the preceding section in the self-consistent mode coupling scheme. Due to the complexity of these equations not much progress has been made so far in solving them in three dimensions. However, in light of the renormalization group work it is natural to analyze these equations for $\epsilon \ll 1$ where $d = d_c - \epsilon$ (for $d > d_c$ we know that the mode coupling term can be ignored in the Langevin equation and that conventional theory is valid). To do this we make the Markoffian approximation

$$G_j(t) = e^{-(\Gamma_j + i\omega_j)t} \tag{4.1}$$

where

$$\Gamma_j(\omega) = \gamma_j - \mathrm{Re}\,\Sigma_j(\omega) \tag{4.2}$$

and

$$\Gamma_j = \Gamma_j(\omega = 0). \tag{4.3}$$

This yields from (3.4), (3.5) and (4.1)

$$\Gamma_j = \gamma_j + 2\sum_{l,m} \frac{\chi_l \chi_m}{\chi_j} |V_{jlm}|^2 \frac{1}{(\Gamma_l + \Gamma_m)} \tag{4.4}$$

as our set of coupled equations, where we have dropped negligible terms involving ω_l and ω_m. Next, we note that for small ϵ wave numbers much greater than q give the dominant contribution to the second term of (4.4) where q is the wave number associated with the mode j. We therefore take the $q \to 0$ limit of the summand and introduce a lower cut-off in the wave number sum which is proportional to q. The arbitrary proportionality constant in this cutoff can be shown to be irrelevant for small ϵ. Next, we change the sum into an integral over wavenumber and perform the angular integral to lowest order in ϵ. Finally, we then take $\partial\Gamma_j/\partial q$ which converts the integral equations into simple partial differential equations which are then easily solved to give Γ_j.

To illustrate these general remarks, we consider the particular example of the model equations for the isotropic ferromagnetic Heisenberg system discussed in section II. For simplicity we consider only the zero magnetic field, $T = T_c$ situation for which there is only one independent correlation function which we choose as $G_{\vec{q}}(t) = \langle S^z_{\vec{q}}(t) S^z_{-\vec{q}}(0)\rangle / \chi_{\vec{q}}$, $t > 0$. Then (4.4) becomes

$$L(q) = L_0 + \frac{A}{q^4}\int_0^\Lambda dk \int_0^\pi \frac{d\theta}{2\pi} \frac{\sin^{d-2}\theta\, k^{d-1}\{\chi_{\vec{k}}\chi_{\vec{k}-\vec{q}}(\chi^{-1}_{\vec{k}} - \chi^{-1}_{\vec{q}-\vec{k}})^2\}}{\chi_{\vec{q}}[k^4 L(k) + (\vec{q}-\vec{k})^4 L(|\vec{q}-\vec{k}|)]} \tag{4.5}$$

where $A = 2^{-(d-1)} \pi^{-d/2}/\Gamma(d/2)$; v_0 is the volume of a unit cell and $\Gamma(q) = q^4 L(q)$. Since $d_c = 6$, we have $\chi_k \cong \chi_0 k^{-2}$ at $T = T_c$ for $\epsilon \ll 1$. Therefore our approximations discussed above result in the equation

$$L(q) = L_0 + g^2 \int_q^\Lambda dk / k^{1+\epsilon} L(k), \tag{4.6}$$

where $L_0 = L(\Lambda)$ and $g^2 = \chi_0 / 192\pi^3$. Next, we obtain from (4.6) the partial differential equation

$$\frac{\partial L(q)}{\partial q} = \frac{-g^2}{q^{1+\epsilon} L(q)}, \qquad q \ll 1. \tag{4.7}$$

Thus the renormalized kinetic coefficient is

$$L(q) = \left[L^2(q_1) + \frac{2g^2}{\epsilon}\left(\frac{1}{q^\epsilon} - \frac{1}{q_1^\epsilon}\right)\right]^{1/2} \tag{4.8}$$

where q_1 is an upper limit of integration such that $q < q_1 \ll 1$. Therefore we find

$$\Gamma(q) \underset{q \to 0}{\longrightarrow} \left(\frac{2g^2}{\epsilon}\right)^{1/2} q^{4-\epsilon/2} \tag{4.9}$$

which yields $z = 4 - \frac{1}{2}\epsilon = \frac{d+2}{2}$ for the dynamical critical exponent, in agreement with the renormalization group results. Similar calculations can be done for the mode coupling equations for other models such as the planar ferromagnet, superfluid He^4, and the binary fluid to give results which agree with renormalization group calculations[10]. In each case the mode coupling integral equations can be reduced to partial differential equations for $\epsilon \ll 1$.

With these preliminary remarks to indicate how an ϵ -expansion calculation can be carried out in the mode-coupling scheme, we now consider the relationship between the two theories to lowest order in ϵ.

V. Relationship Between the Mode Coupling and Renormalization Group Theories

In order to compare the two theories it is convenient to summarize the renormalization group approach in the context of Fokker-Planck model equations which correspond to the nonlinear Langevin equations introduced earlier. These take the form[5,11]

$$\frac{\partial P}{\partial t} = \mathcal{L} P \tag{5.1}$$

where $\mathcal{P}(\{a\}, t)$ is the probability distribution function for the set of gross variables $\{a_i(\vec{k})\}$ with wave number $\vec{k}$. The stochastic operator $\mathcal{L}$ is

$$\mathcal{L} = \sum_{i,j} \sum_{\vec{k}} \frac{\partial}{\partial a_i(\vec{k})} k^{p_{ij}} L_{ij} \left\{ \frac{\partial}{\partial a_j^*(\vec{k})} + \frac{\partial F(\{a\})}{\partial a_j^*(\vec{k})} \right\} - \sum_i \sum_{\vec{k}} \frac{\partial}{\partial a_i(\vec{k})} v_{i\vec{k}}(\{a\}) \quad (5.2)$$

where we have explicitly displayed the $\vec{k}$ dependence. We have also written the kinetic coefficient $L^{\circ}_{ij}(\vec{k})$ as $k^{p_{ij}} L_{ij}$, where p_{ij} is two or zero depending on whether or not the variables $a_i(\vec{k})$ and $a_j(\vec{k})$ are conserved. As is well known, the renormalization group as applied to dynamics consists of two operations. The first consists of integrating out those Fourier components $a_i(\vec{k})$ in (5.1) with $b^{-1}\Lambda < k < \Lambda$, where Λ is the upper cutoff in the wave number sums and $b > 1$. The second is the set of scale transformations

$$\begin{aligned} \vec{k} &\to b\vec{k} \\ a_i(\vec{k}) &\to b^{x_i} a_i(b\vec{k}) \\ t &\to b^z t \end{aligned} \quad (5.3)$$

where $x_i = (2-\eta)/2$ when a_i is the order parameter. These two operations yield a renormalization group transformation on the set of <u>scaled</u> variables which parametrize $\mathcal{L}$. To obtain the relationship between these scaled variables and the physical transport coefficients, we first note that the renormalization group transformation relates the original stochastic operator $\mathcal{L}^1$ and the new stochastic operator $\mathcal{L}^2$ through the equation

$$\mathcal{L}^2(\{b^{x_i} a_i(b\vec{k}), \xi, V, \{L^2_{ij}\}, \{\lambda^2_i\}) = b^{-z} \mathcal{L}^1(\{a_i(\vec{k})\}, \xi/b, V/b^d, \{b^{z-p_{ij}-x_i-x_j} L^2_{ij}\}), \quad (5.4)$$

where μ_i is the exponent which characterizes the transformation of the streaming term, i.e.

$$v^2_{i\vec{q}}(\{b^{x_i} a_j(b\vec{k}), \xi, V, \lambda^2_i) = b^{-\mu_i} v^1_{i,b\vec{q}}(\{a_j(\vec{k})\}, \xi/b, V/b^d, \lambda^2_i) \quad (5.5)$$

where the superscripts 1 and 2 denote the functions before and after the elimination of the short wave length fluctuations, respectively. It is to be noted that the form given in (5.4) is correct to first order in ϵ. In higher order memory effects would have to be included.

Upon iterating the renormalization transformation ℓ times we can write

$$\lambda_{i,\ell} = \lambda^1_i, \qquad \lambda_{i,\ell+1} = b^{z-\mu_i-x_i} \lambda^2_i \quad (5.6)$$

and

$$L_{ij,\ell} = L^1_{ij} \quad , \quad L_{ij,\ell+1} = b^{z-p_{ij}-x_i-x_j} L^2_{ij} \; .$$

If we denote by $\{\mu\}$ the set of coupling constants $\{L_{ij}\}$, the mode coupling coefficients $\{\lambda_i\}$ and the thermodynamic variables, then formally we can write the renormalization group equations as

$$\mu_{\ell+1} = R_b \mu_\ell \tag{5.7}$$

where R_b represents the renormalization group operations given above. For the purpose of comparing this with the mode coupling theory, it is useful to give a differential formulation of (5.7). Let $k = k_0 b^\ell$ for some constant k_0 and choose (b-1) to be infinitesimal. Also, define scaled variables $\hat{L}_{ij}(k) = L_{ij,\ell}$ and $\hat{\lambda}_i(k) = \lambda_{i,\ell}$ so that $\{\mu_\ell\} \rightarrow \{\hat{\mu}(k)\}$. Then (5.7) becomes

$$d\hat{\mu}/d\tau = G\hat{\mu} \tag{5.8}$$

where $\tau = -\ln k$ and G is the infinitesimal generator of the renormalization group transformation.

We are now in a position to compare the two theories, at least to order ϵ . First recall that the mode coupling approach yields equations for the <u>physical</u> transport coefficients $L_{ij}(k)$ and the mode coupling coefficients λ_i , for small values of the wave number k. The nonlinear coupled equations (3.4) and (3.5) or the simplified partial differential equations such as (4.7) express the effect of long-wavelength ($k \rightarrow 0$) fluctuations on the transport properties of the system. On the other hand, the renormalization group yields equations for the scaled variables $\hat{L}_{ij}(k)$ and $\hat{\lambda}_i(k)$, in the limit $k \rightarrow \infty$, where k again describes the effects of long wavelength fluctuations, as $\ell \rightarrow \infty$. It is to be noted that the reason one deals with <u>scaled</u> variables in the renormalization group approach is due to the desire to find a scale invariant theory i.e., a fixed point solution of the group equations. Mode coupling, on the other hand, is not formulated to yield a scale invariant theory. The two approaches could be formally related, then, if one could link the scaled variables to the physical variables by suitable transformations which effectively "unscale" the scaled coefficients and which at the same time take into account the different limiting values of k which are involved in the two cases. Thus we write

$$\hat{L}_{ij}(k) = A_{ij}\, k^{z-z_{ij}} L_{ij}(k_m^2/k) \tag{5.9}$$

and

$$\hat{\lambda}_i(k) = A_i\, k^{z-w_i} \lambda_i \; , \tag{5.10}$$

where

$$z_{ij} = \rho_{ij} + x_i + x_j$$
$$w_i = \mu_i + x_i \quad . \tag{5.11}$$

The coefficient z_{ij} is just the conventional dynamical critical exponent while w_i is the scaling exponent for the mode-coupling term involving λ_i. The various constants in (5. 9) and (5.10) arise from the arbitrariness in choosing a scale for k and to the invariance of the renormalization group equations to multiplication of the scaled coefficients by arbitrary constants. It is also important to note that the different k-dependencies of $\hat{L}_{ij}$ and L_{ij} in (5.9) arises from the fact that $\hat{L}_{ij}(k)$ corresponds to iterating the renormalization equations $\ln(k/k_0)/\ln b$ times, whereas $L_{ij}(k)$ contains the contribution of fluctuations whose wave numbers are greater than k. To obtain differential equations for the physical transport coefficients one substitutes (5.9) and (5.10) into the renormalization group equations (5.8). Explicit calculation for several models shows that these differential equations are exactly the same as obtained from the mode coupling equations (3.4) and (3.5), to first order in ϵ.

As a simple example of these ideas we return to the isotropic Heisenberg model discussed earlier. The renormalization group equations for this model were first obtained by Ma and Mazenko[24] and can be written as (with $\Lambda = 1$)

$$\lambda_{\ell+1} = b^{z-4+\epsilon/2} \lambda_\ell \tag{5.12}$$

$$L_{\ell+1} = b^{z-4}\left(L_\ell + \frac{\chi_0 \lambda_0^2}{192\pi^3} \frac{\ln b}{L_\ell}\right), \tag{5.13}$$

where $\chi_k = \chi_0 k^{-2}$ at criticality. The differential form of these equations is

$$k\frac{d}{dk}\hat{\lambda}(k) = (z-4+\epsilon/2)\hat{\lambda}(k) \tag{5.14}$$

$$k\frac{d}{dk}\hat{L}(k) = (z-4)\hat{L}(k) + \frac{\chi_0}{192\pi^3}\frac{\hat{\lambda}(k)^2}{\hat{L}(k)} \quad . \tag{5.15}$$

Introducing (5.9) and (5.10) into (5.14) and (5.15) yields

$$\frac{d}{dk}L(k) = -\frac{\chi_0 \lambda_0^2}{192\pi^3}\frac{1}{k^{1+\epsilon}L(k)} \tag{5.16}$$

which is the mode coupling equation (4.7) with $\lambda_0 = 1$. The scaled $\hat{\lambda}(k)$ is just given by A $k^{2-4+\epsilon/2}\lambda_0$. Similar results can be obtained for the other models mentioned earlier.

Finally, we note that it is not really surprising that there is such a close relationship between the mode coupling and renormalization group theories. In both cases one attempts to eliminate the short wavelength fluctuations in the dynamical equations of motion. In the original mode coupling theory, one does this in one step, whereas in the renormalization group approach one integrates out these fluctuations in small steps. To first order in ϵ, the only difference in these theories is that one seeks a scale invariant theory whereas the other does not. In particular, this means that the Gaussian approximation and the neglect of vertex renormalization is valid to first order in ϵ. In higher order in ϵ, however, this ceases to be true.

VI. Critical Dynamics of a Simple Fluid

Since the self-consistent mode coupling theory is based in part on an approximate perturbation theory in which "vertex corrections" are ignored, it is worth considering an example in which the nature of this approximation becomes clear. For this reason we consider a model for the critical dynamics of a simple fluid in which a renormalization group analysis to second order in ϵ by Siggia et. al.[12] clarifies the nature of the self-consistent scheme. Since the mode coupling theory has been successful in explaining the experimental situation for fluids, it seems a particularly good system to review here.

The phenomenological equations of section II as applied to the fluid are given by

$$\frac{\partial\psi}{\partial t} = \lambda_0 \nabla^2 \delta F/\delta\psi - g_0 \nabla\psi \cdot \delta F/\delta\vec{j} + \theta \tag{6.1}$$

$$\partial\vec{j}/\partial t = T\cdot[\bar{\eta}_0 \nabla^2 \delta F/\delta\vec{j}] + g_0(\nabla\psi)\,\delta F/\delta\psi + \vec{\zeta} \tag{6.2}$$

where the Ginzburg-Landau like Hamiltonian is

$$F = \int d^dx \left\{\frac{1}{2} r_0 \psi^2 + \frac{1}{2}(\nabla\psi)^2 + u_0 \psi^4 + \frac{1}{2}\vec{j}^{\,2}\right\} \tag{6.3}$$

and the noise terms θ and ζ are Gaussian distributed and satisfy

$$\langle\theta(x,t)\,\theta(x'\,t')\rangle = -2\lambda_0 \nabla^2\, \delta(x-x')\,\delta(t-t') \tag{6.4}$$

$$\langle\zeta_\alpha(x,t)\,\zeta_\beta(x',t')\rangle = -2\bar{\eta}_0 \nabla^2\, \delta(x-x')\,\delta(t-t')\,\delta_{\alpha\beta}\ . \tag{6.5}$$

This is model H first introduced by Halperin et. al.[12,29]. The order parameter ψ is a linear combination of the energy and mass densities and $\vec{j}$ is the transverse momentum density for a d-dimensional incompressible fluid. The operator T picks out the transverse part of the vector in brackets ($T_k^{\alpha\beta} = \delta_{\alpha\beta} - k_\alpha k_\beta / k^2$). The coefficients λ_0 and $\bar{\eta}_0$ are the "bare" thermal conductivity and shear viscosity, respectively, while the mode coupling coefficient g_0 is introduced for calculational convenience. A good discussion of the relevance of this model to the dynamics of a real fluid is given by Siggia et al.[12]. This model contains both dissipative and reversible interactions, through the coupling constants u_0 and g_0 respectively. There is a dimensionless bare coupling constant f_0

$$f_0 = K_d g_0^2 \Lambda^{d-4} / \lambda_0 \bar{\eta}_0 \tag{6.6}$$

where $K_d = (2\pi)^{-d} 2\pi^{d/2}/\Gamma(d/2)$ and Λ is the ultra violet cutoff. The ϵ expansion shows that f_0 is a relevant variable for $d < d_c = 4$, with its fixed point value being of order ϵ.

We first summarize the mode coupling theory for this system which is based on a first order expansion of (6.1) - (6.2) in powers of g_0. Although this is an uncontrolled expansion, since there is no smallness parameter in three dimensions, this self-consistent calculation yields results in quite good agreement with experiment. The coupled equations are given by

$$\lambda(\vec{k}) = \lambda_0 + g_0^2 k^{-2} \int \frac{d^dp}{(2\pi)^d} \frac{\chi_\psi(\vec{p}+\vec{k}/2)\,(\vec{k}\cdot T_{\vec{p}}\cdot\vec{k})}{\bar{\eta}(\vec{p}-\vec{k}/2)(\vec{p}-\vec{k}/2)^2 + \lambda(\vec{p}+\vec{k}/2)\chi_\psi^{-1}(\vec{p}+\vec{k}/2)(\vec{p}+\vec{k}/2)^2} \tag{6.7}$$

$$\bar{\eta}(\vec{k}) = \bar{\eta}_0 + \frac{g_0^2 k^{-2}}{(d-1)} \int \frac{d^dp}{(2\pi)^d} \frac{\chi_\psi(\vec{p}-\vec{k}/2)\left[\chi_\psi^{-1}(\vec{p}+\vec{k}/2)-\chi_\psi^{-1}(\vec{p}-\vec{k}/2)\right](\vec{p}\cdot T_k\cdot\vec{p})}{\lambda(\vec{p}+\vec{k}/2)\chi_\psi^{-1}(\vec{p}+\vec{k}/2)(\vec{p}+\vec{k}/2)^2 + \lambda(\vec{p}-\vec{k}/2)\chi_\psi^{-1}(\vec{p}-\vec{k}/2)(\vec{p}-\vec{k}/2)^2} \, . \tag{6.8}$$

Although no one has solved these equations exactly to date, a number of approximate solutions have been given and used in the analysis of experimental measurements. The quantities with which we will concern ourselves most here are the exponents x_λ and $x_{\bar{\eta}}$ which characterize the divergences in the thermal conductivity and shear viscosity, respectively, i.e.

$$\lambda(T) \sim \xi^{x_\lambda} \tag{6.9}$$

$$\bar{\eta}(T) \sim \xi^{x_{\bar{\eta}}} \, . \tag{6.10}$$

In addition we will be interested in a universal amplitude ratio R,

$$R = (k_B T_c)^{-1} \lambda(T) \bar{\eta}(T) \chi_\psi^{-1}(T) \xi^{d-2} \tag{6.11}$$

in units in which $g_0 = 1$ and the free energy has units of $k_B T$. This equation can also be written in terms of the diffusion constant

$$D = \lambda \chi_\psi^{-1} \tag{6.12}$$

as

$$D = R k_B T_c / \bar{\eta} \xi^{d-2} . \tag{6.13}$$

Now the exact second order calculation yields

$$x_\lambda = \frac{18}{19} \epsilon \left[1 - 0.033\epsilon + O(\epsilon^2)\right] \simeq 0.916 \tag{6.14}$$

$$x_{\bar{\eta}} = \frac{1}{19} \epsilon \left[1 + 0.238\epsilon + O(\epsilon^2)\right] \simeq 0.065 \tag{6.15}$$

which satisfies the scaling relation

$$x_\lambda + x_{\bar{\eta}} = \epsilon - \eta . \tag{6.16}$$

In addition one obtains

$$R = K_d \frac{19}{24} \epsilon^{-1} \left[1 + 0.06\epsilon + O(\epsilon^2)\right] . \tag{6.17}$$

To obtain the second order results for the self-consistent mode coupling calculation of Kawasaki, one has to omit certain diagrams which correspond to vertex corrections in the exact calculation. In addition, since the original self-consistent calculation of Kawasaki used the Ornstein-Zernike approximation for the static susceptibility (for calculational convenience) one has to set $\eta = 0$ in the scaling relation (6.16). This then yields the values

$$x_\lambda = \frac{18}{19} \epsilon \left[1 - 0.013\epsilon + O(\epsilon^2)\right] \simeq 0.935 \tag{6.18}$$

$$x_{\bar{\eta}} = \frac{1}{19} \epsilon \left[1 + 0.242\epsilon + O(\epsilon^2)\right] \simeq 0.065 \tag{6.19}$$

$$R = K_d \frac{19}{24} \epsilon^{-1} \left[1 - 0.023 \epsilon + O(\epsilon^2) \right] \tag{6.20}$$

Thus one can see that the mode coupling results which are exact in first order are quite good even in 2nd order. Furthermore, the Ornstein-Zernike approximation was only introduced by Kawasaki to simplify calculations. One could improve the self-consistent scheme by using[12,29] the correct η , see also all recent work by Ohta[30].

Finally we remark that, as is discussed in Siggia et. al.[12] it appears that the reason for the excellent agreement between Kawasaki's theory and experiment lies in the existence of what amounts to a smallness parameter, $x_\eta / x_\lambda \simeq 0.05$ in the corrections to the mode coupling calculation that arise in the renormalization group analysis. Although this parameter does not lead to any systematic expansion, it does appear in most of the higher-order diagrams.

VII. Tricritical Dynamics of He^3-He^4 Mixtures

We conclude this paper with some brief remarks about the tricritical dynamics of He^3-He^4 mixtures. Our purpose is only to indicate that in this particular case using the Ginzburg-Landau free energy functional rather than making a quadratic approximation to it leads to significantly different results for the transport coefficients. A full discussion of this subject is given by Siggia and Nelson[13]. Other speakers at this conference will no doubt have more to say about this subject.

The phenomenological model which has been studied by ϵ-expansion techniques is an example of the nonlinear Langevin equations discussed in section II. Specifically these equations are

$$\frac{\partial \psi}{\partial t} = -2\Gamma_0 \frac{\delta F}{\delta \psi^*} - i g_{1,0} \frac{\delta F}{\delta q} - i g_{2,0} \psi \frac{\delta F}{\delta q} + \zeta \tag{7.1}$$

$$\frac{\partial c}{\partial t} = \lambda_0 \nabla^2 \frac{\delta F}{\delta c} + L_0 \nabla^2 \frac{\delta F}{\delta c} + 2 g_{2,0} \,\mathrm{Im}\left(\psi^* \frac{\delta F}{\delta \psi^*} \right) + \theta \tag{7.2}$$

$$\frac{\partial q}{\partial t} = K_0 \nabla^2 \frac{\delta F}{\delta q} + L_0 \nabla^2 \frac{\delta F}{\delta c} + 2 g_{1,0} \,\mathrm{Im}\left(\psi^* \frac{\delta F}{\delta \psi^*} \right) + \varphi \tag{7.3}$$

where

$$F = \int d^d x \left\{ \frac{1}{2} r_0 |\psi|^2 + \frac{1}{2} |\nabla \psi|^2 + \tilde{u}_0 |\psi|^4 + \frac{1}{2} \chi_0^{-1} c^2 + \gamma_0 c |\psi|^2 + \frac{1}{2} C_0^{-1} q^2 \right\}, \tag{7.4}$$

Using the notation of Siggia and Nelson,[13] the random forces satisfy the usual relations

$$\langle \zeta(x,t)\zeta(x',t')\rangle = 4\,\mathrm{Re}\,\Gamma_0\,\delta(x-x')\,\delta(t-t') \tag{7.4}$$

$$\langle \theta(x,t)\theta(x',t')\rangle = -2\lambda_0\nabla^2\delta(x-x')\,\delta(t-t') \tag{7.5}$$

$$\langle \varphi(x,t)\theta(x',t')\rangle = -2L_0\nabla^2\delta(x-x')\,\delta(t-t') \tag{7.6}$$

$$\langle \varphi(x,t)\varphi(x',t')\rangle = -2K_0\nabla^2\delta(x-x')\,\delta(t-t') \tag{7.7}$$

with the other force correlations being zero. The various quantities in these equations are the following: ψ is the complex order parameter for the superfluid transition. The variable c is the local ^{3}He mass concentration and q is a linear combination of the local entropy and local concentration. The quantities Γ_0, λ_0, L_0 and K_0 denote kinetic coefficients, while the $g_{1,0}$ and $g_{2,0}$ denote mode coupling coefficients.

The original mode coupling theories[14,15] were based on this model except that only quadratic terms in (7.1) - (7.3) were kept, with the equilibrium inverse susceptibilities in place of the bare inverse susceptibilities. This is a situation, however, in which the dissipative coupling between the order parameter ψ and c which was ignored in the original mode coupling theories turns out to be quite important. Indeed the first order ϵ-expansion results for the model defined by (7.1) through (7.7) differs from those obtained in the earlier mode coupling results, precisely because of this coupling. The point here is that this is a good example of a situation in which the renormalization group has led to an improved model, precisely through an explicit display of the dissipative nonlinear coupling term. It is also true, of course, that a mode coupling calculation of the model (7.1) - (7.7) would agree to order ϵ with the renormalization group analysis.

VIII. Conclusion

It should be clear from the preceding discussion that the renormalization group has led to an improvement in the theory of critical dynamics. It should not be overlooked, however, that the earlier mode coupling theory was nevertheless quite successful. Indeed it seems to this author that these recent developments provide strong evidence for the great intuition exhibited by the theorists responsible for the formulation of mode coupling theory.

References

1. L. van Hove, Phys. Rev. 93, 1374 (1954)
2. L. D. Landau and I. M. Khalatnikov, Dokl. Acad. Nauk SSSR 96, 469 (1954)
3. M. Fixman, J. Chem. Phys. 36, 310 (1962)
4. K. Kawasaki, Ann. Phys. 61, 1 (1970)
5. K. Kawasaki, in Critical Phenomena, Proceedings of the International School of Physics Enrico Fermi, Course L1, edited by M. S. Green (Academic Press, N. Y.) 1971
6. L. P. Kadanoff and J. Swift, Phys. Rev. 166, 89 (1968)
7. B. I. Halperin, P. C. Hohenberg and S. Ma, Phys. Rev. Lett. 29, 1548 (1972)
8. P. C. Hohenberg and B. I. Halperin, Rev. Mod. Phys. 49, 435 (1977)
9. K. G. Wilson and J. Kogut, Phys. Rep. 12C, 75 (1974)
10. J. D. Gunton and K. Kawasaki, J. Phys. A8, L 9 (1975)
11. K. Kawasaki and J. D. Gunton, Phys. Rev. B13, 4658 (1976)
12. E. D. Siggia, B. I. Halperin and P. C. Hohenberg, Phys. Rev. B13, 2110 (1976)
13. E. D. Siggia and D. R. Nelson, Phys. Rev. B15, 1427 (1977)
14. K. Kawasaki and J. D. Gunton, Phys. Rev. Lett. 29, 1661 (1972)
15. M. K. Grover and J. Swift, J. Low Temp. Phys. 11, 751 (1973)
16. H. Mori, Prog. Theor. Phys. (Kyoto) 33, 423 (1965)
17. R. Zwanzig, Phys. Rev. 124, 983 (1961)
18. M. S. Green, J. Chem. Phys. 20, 1281 (1952)
19. H. Mori and H. Fujisaka, Prog. Theor. Phys. 49, 764 (973)
20. K. Kawasaki, J. Phys. A 6, 1289 (1973)
21. K. Kawasaki, Rendiconti S.I.F. Course LI edited by M. S. Green (New York: Academic Press) p. 342 (1971)
22. R. Zwanzig, Statistical Mechanics, edited by S. A. Rice, K. F. Freed and J. C. Light (University of Chicago Press) p. 241 (1972)
23. K. Kawasaki, ibid, p. 259
24. S. Ma and G. F. Mazenko, Phys. Rev. B11, 4077 (1975)
25. L. P. Kadanoff and J. Swift, Phys. Rev. 165, 310 (1968)
26. L. P. Kadanoff and J. Swift, Phys. Rev. 166, 89 (1968)
27. R. Perl and R. A. Ferrell, Phys. Rev. Lett. 29, 51 (1972)
28. R. Perl and R. A. Ferrell, Phys. Rev. A 6, 2358 (1972)
29. B. I. Halperin, P. C. Hohenberg and E. D. Siggia, Phys. Rev. Lett. 32, 1289 (1974)
30. T. Ohta, J. Phys. C10, 791 (1977); T. Ohta and K. Kawasaki, Prog. Theor. Phys. 55, 1384 (1976)

FIELD-THEORETIC METHOD APPLIED TO CRITICAL DYNAMICS

H.K. Janssen

Institut für Theoretische Physik
Universität Düsseldorf
W.-Germany

FIELD-THEORETIC METHOD APPLIED TO CRITICAL DYNAMICS

H.K. JANSSEN
Institut für Theoretische Physik
Universität Düsseldorf
W.-Germany

Introduction

Since the discovery of Wilson's renormalization group theory of critical phenomena based on the successive elimination of short wavelength fluctuations also the field theoretic formulation of renormalization group ideas was developed [1,2] to deal with critical systems. The main idea of the field theoretic method rests on the fact that measuring the critical fluctuations on a convenient (long wave length) scale one has to chose a scale on which the interatomic distances go to zero in the critical limit. In this limit the fluctuations of the physical variables are now given by fields defined over a spatial continuum. Then the correlation functions of the fluctuations of an interacting system entail ultraviolet divergencies which can be eliminated by suitable multiplicative and additive renormalizations of the bare physical quantities. The renormalizations introduce instead of the internal length scale (the interatomic scale which goes to zero) an external length scale suitable to measure the long wave length phenomena. The independence of the original bare theory from this external length scale leads to the renormalization group which exhibits the scaling properties of the critical system.
In the area of dynamical critical phenomena (for a review see [3]) field theory was first developed by De Dominicis et al. [4] for pure relaxational models. As independently shown by De Dominicis [5] and Janssen [6] more general Markovian processes can be easily described in terms of a path integral which involves besides the physical variables a conjugate set of variables. The perturbation theory based on this path integral leads to the Martin, Siggia, Rose theory [7]. The path integral proves in particular as a convenient tool to develop the renormalized field theory of critical dynamics as shown by Bausch, Janssen, Wagner [8] and later by De Dominicis, Peliti [9] (for related field theoretic concepts see Kawasaki, Gunton [10]).
In this lecture we shall develop the field theory of dynamical critical phenomena by considering a model system with a continuous symmetry group. In chapter I the path integral formulation is given. Chapter II shows the renormalization of the model and leads to the critical properties. In chapter III we calculate by a combination of renormalization group and renormalized perturbation theory the dynamical susceptibility of a relaxational system below T_c at the coexistence curve.

I. Path Integral Description of Statistical Dynamics

Let us start from a set of stochastic variables $\psi(t) = (\ldots, \psi_i(t), \ldots)$ and underlying equations of motion

$$\frac{d}{dt}\psi(t) = V(\psi(t)) + \zeta(t) \tag{I.1}$$

$\zeta(t)$ denotes a set of random forces assumed to be Gaussian distributed with white noise and probability density in the time interval $T_1 \leq t \leq T_2$:

$$w(\{\zeta\}) \sim \exp\left(-\frac{1}{4}\int_{T_1}^{T_2} dt\, \zeta(t)\, L^{-1}\, \zeta(t)\right) \tag{I.2}$$

The deterministic velocities $V(\psi)$ can be written in the form (see for instance Kawasaki [11] and the literature cited there)

$$V_i = \sum_j \left(-R_{ij}\frac{\partial \mathcal{H}}{\partial \psi_j} + \frac{\partial}{\partial \psi_j} R_{ij}\right) \tag{I.3}$$

$$R_{ij} = M_{ij} + L_{ij}\;,\quad M_{ij} = -M_{ji}\;,\quad L_{ij} = L_{ji} \tag{I.4}$$

The matrix M defines the (reversible) mode coupling and the matrix L (for simplicity we take it to be independent of ψ) the(irreversible) damping related to the fluctuations of the random forces ζ (I.2) by the Nyquisttheorem (2.fluctuation-dissipation theorem). In (I.3) $\mathcal{H}$ denotes the "Hamiltonian" given by the equilibrium distribution:

$$W(\psi) \sim \exp(-\mathcal{H}(\psi)) \tag{I.5}$$

Remark: the formulation (I.1-5) of course includes the classical microscopic equations of motion as the limit L $\rightarrow$ 0 with a suitably chosen M . . The random forces ζ thermalize the motion of the system and lead ultimately to the thermal equilibrium distribution exp $(-(k_B T)^{-1}\cdot H)$

Our aim is to calculate correlation functions from (I.1). Instead of solving (I.1) for ψ (t) in terms of the random force ζ and averaging then over ζ with the weight (I.2) we proceed by eliminating ζ in favour of ψ. This can be achieved (see Graham [12]) by introduction of a path probability density $\mathcal{W}(\{\psi\})$ for the stochastic variables ψ via $\left(d\{\zeta\} = \prod_{T_1}^{T_2}\prod_i d\zeta_i(t)\;,\ \text{etc.}\right)$

$$w(\{\xi\})\, d\{\xi\} = \mathcal{W}(\{\psi\})\, d\{\psi\} \tag{I.6}$$

To obtain $\mathcal{W}(\{\psi\})$ in a convenient form we first introduce a set of new variables $\tilde{\psi}(t)$ by a functional Fourier transformation

$$w(\{\xi\}) \sim \int d\{i\tilde{\psi}\} \exp\left(\int_{T_1}^{T_2} dt\, [\tilde{\psi}(t) L \tilde{\psi}(t) - \tilde{\psi}(t)\xi(t)]\right) \tag{I.7}$$

As shown by Graham the Jacobian of the transformation from ξ to ψ via the equation of motion (I.1) is

$$\frac{d\{\xi\}}{d\{\psi\}} = \exp\left(-\frac{1}{2}\int_{T_1}^{T_2} dt\, \frac{\partial V(\psi(t))}{\partial \psi(t)}\right) \tag{I.8}$$

hence

$$\mathcal{W}(\{\psi\}) = \text{const.} \int d\{i\tilde{\psi}\} \exp\left(-\mathcal{J}_{T_1}^{T_2}(\{\tilde{\psi},\psi\})\right) \tag{I.9}$$

with the action integral

$$\mathcal{J}_{T_1}^{T_2}(\{\tilde{\psi},\psi\}) = \int_{T_1}^{T_2} dt\, \mathcal{L}(\tilde{\psi}(t), \psi(t), \dot{\psi}(t)) \tag{I.10}$$

and the Lagrangian

$$\mathcal{L}(\tilde{\psi},\psi,\dot{\psi}) = -\tilde{\psi} L \tilde{\psi} + \tilde{\psi}(\dot{\psi} - V(\psi)) + \frac{1}{2}\frac{\partial V(\psi)}{\partial \psi} \tag{I.11}$$

This form of the path integral was given independently by De Dominicis [5] and Janssen [6] in order to treat critical dynamics. For more details see Bausch, Janssen, Wagner [8] and De Dominicis, Peliti [9]. We may now calculate correlation functions by a path integral $(T_1 < t_i < T_2)$

$$\langle \psi(t_1) \ldots \psi(t_n) \rangle = \text{const.} \int d\{i\tilde{\psi},\psi\}\, \psi(t_1) \ldots \psi(t_n) \cdot \exp\left(-\mathcal{J}_{T_1}^{T_2}(\{\tilde{\psi},\psi\}) - \mathcal{H}(\psi(T_1))\right)$$

$$= \text{const.} \int d\{i\tilde{\psi},\psi\}\, \psi(t_1) \ldots \psi(t_n) \exp\left(-\mathcal{J}(\{\tilde{\psi},\psi\})\right) \tag{I.12}$$

where we defined $\mathcal{J} =: \mathcal{J}_{-\infty}^{\infty}$ and used the ergodicity of the stochastic process. The correlation functions can be obtained as well from an additive external term $- F(t)\cdot\psi$ in the Lagrangian (I.11) by taking successive derivatives with respect to F. We add external time dependent forces $\tilde{F}(t)$ to the deterministic velocities $V(\psi)$. Then a second additional term $- \tilde{F}(t)\cdot\tilde{\psi}$ arises in the Lagrangian (I.11). Expanding the correlations (I.12) with respect to such an external force $\tilde{F}$ we generate response functions:

$$\begin{aligned}&\langle \psi(t_1)\ldots\psi(t_n)\tilde{\psi}(\tilde{t}_1)\ldots\tilde{\psi}(\tilde{t}_m)\rangle = \\ &= \text{const.}\int d\{i\tilde{\psi},\psi\}\,\psi(t_1)\ldots\psi(t_n)\cdot\tilde{\psi}(\tilde{t}_1)\ldots\tilde{\psi}(\tilde{t}_m)\cdot\exp\left(-\mathcal{J}(\{\tilde{\psi},\psi\})\right)\end{aligned} \tag{I.13}$$

This is why we call the variables $\tilde{\psi}$ response fields. The generating functional for the cumulants of all correlation and response functions is now given by the logarithm of a path integral

$$\begin{aligned}W(\{F,\tilde{F}\}) = \ln\Big(\int d\{i\tilde{\psi},\psi\}\exp\Big(-\mathcal{J}(\{\tilde{\psi},\psi\}) + \int_{-\infty}^{\infty} dt\,[\psi(t)F(t) \; + \\ + \tilde{\psi}(t)\tilde{F}(t)]\Big)\Big)\end{aligned} \tag{I.14}$$

$$\langle \psi(t_1)\ldots\psi(t_n)\tilde{\psi}(\tilde{t}_1)\ldots\tilde{\psi}(\tilde{t}_m)\rangle^{(\text{cum.})} = \left.\frac{\delta^{n+m}\,W(\{F,\tilde{F}\})}{\delta F(t_1)\ldots\delta F(t_n)\,\delta\tilde{F}(\tilde{t}_1)\ldots\delta\tilde{F}(\tilde{t}_m)}\right|_{F=\tilde{F}=0} \tag{I.15}$$

We may calculate the cumulants as the connected parts in a diagramatical perturbation expansion in terms of the non-linear couplings in $\tilde{\psi}\cdot V(\psi)$. The Gaussian form of the unperturbed weight $\exp(-\mathcal{J}_0)$ ($\mathcal{J}_0$ arises from linear terms in $V(\psi)$) leads immediatly to the Wick theorem. Thus the elements of the diagrams are (besides the usual numerical factors) contractions (lines) or propagators:

$$t)\!\!-\!\!\!<\!\!-\!\!(t' = \langle\psi(t)\tilde{\psi}(t')\rangle_0 = \Theta(t-t')\cdot G_0(t-t') \quad , \; G_0(0) = 1$$

$$t)\!\!-\!\!-\!\!(t' = \langle\psi(t)\psi(t')\rangle_0 = \mathcal{U}_0(t-t') = [\Theta(t-t')G_0(t-t') + \Theta(t'-t)G_0(t'-t)]\cdot\langle\psi\psi\rangle_0$$

and vertices:

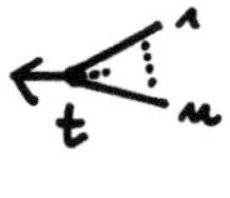

= coefficient of $\tilde{\psi}(t)\cdot\psi(t)^m$ in $\tilde{\psi}(t)\cdot V_{\text{nonlinear}}(\psi(t))$

The role of the Jacobian is to compensate selfloops of the response propagators

 $\hat{=} \; \langle \tilde{y} y \rangle_0 \cdot \dfrac{\partial V}{\partial y} = \dfrac{1}{2} \cdot \dfrac{\partial V}{\partial y}$

 $\hat{=} \; -\dfrac{1}{2} \dfrac{\partial V}{\partial y}$

$\Bigg\}$  $+$  $= 0$

Thus the Jacobian can be set to one by defining $\Theta(0) = 0$ which excludes self loops. Causality is then easily demonstrated (see [8])

$$\langle y(t_1) \ldots y(t_n) \, \tilde{y}(\tilde{t}_1) \ldots \tilde{y}(\tilde{t}_m) \rangle = 0 \qquad \text{if one } \tilde{t}_j > \text{all } t_i \tag{I.16}$$

This perturbation theory now generated by the path integral is the same as the Martin-Siggia-Rose perturbation theory [7] .
The path integral has another important property: symmetry against time inversion (detailed balance). Assume that we have chosen the variables y_i to have definite time parities $\varepsilon_i = \pm 1$. We now define a time inversion transformation $\mathcal{T}$:

$$\mathcal{T} y_i(t) = \varepsilon_i \, y_i(-t) \tag{I.17}$$

$$\mathcal{T} \tilde{y}_i(t) = -\varepsilon_i \left(\tilde{y}_i(-t) - \frac{\partial \mathcal{H}(y(-t))}{\partial y_i(-t)} \right) \tag{I.18}$$

Assuming that the matrix R in (I.3) obeys Onsager-Casimir reciprocity relations (which follow from microreversibility)

$$R_{ij}(\varepsilon y) = \varepsilon_i \varepsilon_j R_{ji}(y) \tag{I.19}$$

and the invariance of the Hamiltonian

$$\mathcal{T} \mathcal{H}(y) = \mathcal{H}(\varepsilon y) = \mathcal{H}(y) \tag{I.2o}$$

One can easily prove from (I.3, 4, 1o, 11, 17-2o) $(\mathcal{H}_t =: \mathcal{H}(y(t)))$

$$\mathcal{T}\left(\exp\left(-\mathcal{J}_{T_1}^{T_2} - \mathcal{H}_{T_1}\right)\right) = \exp\left(-\mathcal{J}_{-T_2}^{-T_1} - \mathcal{H}_{-T_2}\right) \tag{I.21}$$

which expresses time inversion symmetry. From (I.12) we conclude that inside the correlation and response functions we can set

$$\mathcal{T}\mathcal{J} = \mathcal{J} \tag{I.22}$$

Once causality is established a general relation is easily obtained. From (I.16,17, 18,22) we have ($\tilde{\psi} = \tilde{\psi}(0)$)

$$\begin{aligned}\Theta(t)\langle\psi_i(t)\cdot(\tilde{\psi}_j - \frac{\delta\mathcal{H}}{\delta\psi_j})\rangle &= -\varepsilon_i\varepsilon_j\,\Theta(t)\langle\mathcal{T}(\psi_i(-t)\tilde{\psi})\rangle = \\ &= -\varepsilon_i\varepsilon_j\,\Theta(t)\langle\psi_i(-t)\tilde{\psi}\rangle = 0\end{aligned} \tag{I.23}$$

and we find

$$\Theta(t)\langle\psi(t)\frac{\partial\mathcal{H}}{\partial\psi}\rangle = \langle\psi(t)\tilde{\psi}\rangle \tag{I.24}$$

This relation can immediatly be generalized by taking instead of $\psi(t)$ an arbitrary functional of $\psi(t)$ and $\tilde{\psi}(t)$ with only positive time arguments.
A further relation follows from:

$$\langle\psi(t)\frac{\delta\mathcal{J}}{\delta\tilde{\psi}(0)}\rangle = \langle\frac{\delta\psi(t)}{\delta\tilde{\psi}(0)}\rangle = 0 \tag{I.25}$$

which is easily shown from (I.13) by partial functional integration. With (I. 3,1o, 11) using the generalized form of (I.24) we get from (I.25) if $t>0$:

$$\langle\psi_i(t)\left(\dot{\psi}_j - \sum_k(\tilde{\psi}_k - \frac{1}{2}\frac{\partial}{\partial\psi_k})R_{kj}\right)\rangle = 0 \tag{I.26}$$

The last term in this equation again cancels response propagator self loops in a diagramatical expansion. Excluding such self loops we therefore can omit this compensating term and obtain the correlation response theorem

$$-\Theta(t)\frac{d}{dt}\langle\psi(t)\psi\rangle = \langle\psi(t)\cdot\tilde{\psi}R\rangle \tag{I.27}$$

Again this relation can be generalized in the way mentioned below (I.24). Whereas the right hand side of (I.24) is the linear response to an external force added linearly to the equation of motion (I.1), the right hand side of (I.27) is the linear response to an external force coupled linearly to ψ in the Hamiltonian, $\mathcal{H}\to\mathcal{H} - H(t)\cdot\psi$ .This is easily seen from the definition of the deterministic velocity V (I.3), the Lagrangian $\mathcal{L}$ (I.11) and the action integral (I.1o) by calculating $\psi(t)$ via (I.12) taking up to linear terms in H(t). It is easy to see that in general the various insertions of the variable

$$\Psi = \tilde{\psi}\cdot R(\psi) \tag{I.28}$$

into the mean value of some quantity A

$$\langle A\, \Psi(t_1)\, \Psi(t_2) \ldots \rangle = \frac{\delta \langle A \rangle}{\delta H(t_1)\, \delta H(t_2) \ldots} \tag{I.29}$$

yield all (nonlinear) responses to the so called "physical" force H(t). Relations of the type (I.24,27) have also been derived from the Fokker-Planck equation by Deker and Haake [13].

At the end of this chapter let us generalize the action integral to general non-markovian processes based on classical Hamiltonian microdynamics (Janssen [14]) Let us choose the microvariables $\psi = (\varphi, \chi)$. The set φ will be retained and the set χ will be eliminated. The set χ is chosen such that the Hamiltonian splits into independent parts:

$$\mathcal{H}(\varphi, \chi) = \mathcal{H}^{(\varphi)}(\varphi) + \mathcal{H}^{(\chi)}(\chi) \tag{I.3o}$$

This can always be achieved by a (nonlinear) transformation $\chi' = F(\chi, \varphi)$. As remarked following below (I.5) our formulation (I. 1-5) includes microdynamics. With (I.3o) also the time inversion transformation (I. 17,18) holds for each set φ or χ separately. If now the variables $\chi, \tilde{\chi}$ are eliminated via

$$\exp(-\mathcal{J}(\{\tilde{\varphi}, \varphi\})) = \int d\{\chi, i\tilde{\chi}\} \exp(-\mathcal{J}(\{\tilde{\varphi}, \tilde{\chi}, \varphi, \chi\})) \tag{I.31}$$

an action integral arises which is no longer local in time. This action integral is only restricted by causality and time inversion invariance (and other symmetries possibly). One is therefore lead to the general form (including counter terms for self loops)

$$\begin{aligned} \mathcal{J}(\{\tilde{\varphi}, \varphi\}) = & \int_{-\infty}^{\infty} dt \left(\tilde{\varphi}(t) \cdot (\dot{\varphi}(t) - v(\varphi(t))) + \frac{1}{2} \frac{\partial v(\varphi(t))}{\partial \varphi(t)} \right) \\ & - \iint_{-\infty}^{\infty} dt\, dt' \left(\tilde{\varphi}(t) - \frac{1}{2} \frac{\partial}{\partial \varphi(t)} \right) \left(\tilde{\varphi}(t') - \frac{\partial \mathcal{H}(\varphi(t'))}{\partial \varphi(t')} + \frac{1}{2} \frac{\partial}{\partial \varphi(t')} \right) \cdot \\ & \cdot L(t, t'; \{\tilde{\varphi}(t''), \varphi(t'') \mid t \geq t'' \geq t'\}) \end{aligned} \tag{I.32}$$

where $$v_i(\varphi) = \sum_j \left(\frac{\partial \mathcal{H}}{\partial \varphi_j} - \frac{\partial}{\partial \varphi_j} \right) M_{ji}(\varphi) \, , \quad M_{ij} = - M_{ji} \tag{I.33}$$

with time inversion behavior

$$M(\varphi)_{ij} = \varepsilon_i \varepsilon_j\, M(\varepsilon \varphi)_{ji} \tag{I.34}$$

$$L(t,t';\{\tilde{\varphi},\varphi\})_{ij} = \varepsilon_i\varepsilon_j\, L(-t',-t;\{-\varepsilon(\tilde{\varphi}-\frac{\partial\mathcal{H}}{\partial\varphi}),\varepsilon\varphi\})_{ji} \qquad (I.34)$$

The correlation response theorems follow as:

$$\theta(t)\langle\varphi(t)\frac{\partial\mathcal{H}}{\partial\varphi}\rangle = \langle\varphi(t)\tilde{\varphi}\rangle$$

$$-\theta(t)\frac{d}{dt}\langle\varphi(t)\varphi\rangle = \langle\varphi(t)\left[\tilde{\varphi}M(\varphi) + \int_0^t dt'\,\tilde{\varphi}(t')\,L(t',0;\{\tilde{\varphi},\varphi\})\right]\rangle \qquad (I.35)$$

The general form (I.32) (with time inversion symmetry under the transformations (I. 17,18) for φ) is now sufficient to yield exp $(-\mathcal{H})$ as the equilibrium distribution.

II. Renormalization and the Renormalization Group Equation

1. The Model

Near to the critical point the long wavelength and low frequency behaviour of correlation and response functions is dominated by the motion of slow variables like the order parameter itself and, e.g., densities of conserved quantities. As a model we shall consider systems in which only internal continuous symmetries provide us with such quantities. In these cases the infinitesimal generators of the symmetry are conserved due to the Noether theorem. To simplify our model we exclude the energy from the relevant slow variables. This is justified if the order parameter system is energetically strongly coupled to another system which works as a local heat reservoir (e.g., a spin system coupled strongly to the lattice phonon system).The order parameter itself is not conserved in our model. For simplicity we shall further assume that the order parameter transforms according to an irreducible representation of the internal symmetry group which has only one fourth order invariant and the direct product of which reduces only once (and then antisymmetrically) to the regular representation of the group (remember that the group generators transform according to the regular representation). Without these assumptions we have to work with many but similar couplings and we must resort to the Clebsch-Gordan coefficients of the group. A familiar example of such models are the O(n)-symmetric or the SSS-model [15] which includes, e.g., the symmetric antiferromagnet and the planar model. As in these models we shall assume that the group generators have negative time parity and that the order parameter has either positive or negative time parity. (This is in contrast to the multicomponent Bose system [16] where the group generators and the real part of the order parameter have positive, the imaginary part however negative time parity. Therefore the couplings in the multicomponent Bose system are different from the ones in our model).

The relevant variables $\varphi(t)$ of our model are now

1) the order parameter field $s(\underline{x},t) = \{s_\alpha(\underline{x},t) \mid \alpha = 1,\dots,N\}$ which is nonconserved
2) the generator field $m(\underline{x},t) = \{m_i(\underline{x},t) \mid i = 1,\dots,P\}$ which is conserved and has negative time parity.

All these fields are taken to be real. Therefore N is equal to the dimension for a real and twice the dimension for a complex representation of the symmetry group, and P is equal to the number of parameters of the continuous symmetry group. The infinitesimal generators (real antisymmetric matrices $I_i = \{I_{i,\alpha\beta}\}$, i=1,...,P) obey the commutation rules $[I_i, I_j] = c_{ijk} I_k$ where c_{ijk} are the structure constants of the group.

The equilibrium fluctuations are governed by a Hamiltonian of Ginzburg-Landau type

$$\mathcal{H} = \int d^d x \{ \tfrac{1}{2} Z_\tau \cdot \tau \cdot s(\underline{x})^2 + \tfrac{1}{2} Z_s \cdot (\nabla s(\underline{x}))^2 + + \tfrac{1}{4!} \mu^\varepsilon \cdot Z_{\tilde{u}} \cdot \tilde{u} \cdot (s(\underline{x})^2)^2 + \tfrac{1}{2} Z_m \cdot m(\underline{x})^2 \} \tag{II.2}$$

Because of the symmetry (including time reversal) of the model other relevant couplings are not admitted. We have written the Hamiltonian in renormalized form going over from bare quantities to the renormalized ones by means of Z-factors ($Z_i = Z_i(u)$)

$$s_b = Z_s^{1/2} \cdot s , \qquad m_b = Z_m^{1/2} \cdot m , \tag{II.3}$$
$$Z_s \cdot \tau_b = Z_\tau \cdot \tau , \qquad Z_s^2 \cdot g_b = \mu^\varepsilon \cdot Z_{\tilde{u}} \cdot \tilde{u} , \qquad \varepsilon = 4 - d$$

where we have introduced an external length scale μ and τ is a linear measure of temperature $\tau \sim T - T_c$. An (infinite) critical $\tau_b^{(c)}$ has not been taken into account because in dimensional regularization this term is formally set to zero. In a loop expansion about the "free" Hamiltonian

$$\mathcal{H}_0 = \int d^d x \{ \tfrac{\tau}{2} \cdot s^2 + \tfrac{1}{2} \cdot (\nabla s)^2 + \tfrac{1}{2} m^2 \} \tag{II.4}$$

the one-line-irreducible diagrams (the so called vertex- or Γ-functions) if evaluated by analytic continuation in dimension d exhibit poles in $\varepsilon = 4 - d$. The Z-factors are determined to cancel exactly these poles and this enables us to establish the ε-expansion which is indispensable to avoid infrared singularities of the critical theory $\tau = 0$. Therefore the Z-factors are determined uniquely in the form

$$Z = 1 + \sum_{k=1}^{\infty} \frac{a_k(\tilde{u})}{\varepsilon^k} \tag{II.5}$$

This procedure is the so called minimal renormalization scheme of t'Hooft and Veltman [17]. For more details see Amit [2] or Brezin, Le Guillou, Zinn-Justin [1]. Because the fluctuations of m are Gaussian we find immediatly

$$Z_m = 1 \tag{II.6}$$

Since H is dimensionless we conclude from (II.2) that the (naive) dimensions of the fields are given by

$$s \sim \mu^{\frac{d-2}{2}} , \qquad m \sim \mu^{d/2} \tag{II.7}$$

Let us now consider equations of motion (I.1) for the Fourier-transformed fields $s(q,t)$ and $m(q,t)$

$$\frac{d}{dt} s(q,t) = L_s(q) \cdot (\tau + q^2) \cdot s(q,t) \quad + \text{nonlinear terms} \tag{II.8}$$

$$\frac{d}{dt} m(q,t) = L_m(q) \cdot m(q,t) \quad + \text{ nonlinear terms} \tag{II.9}$$

$L_s(q)$ and $L_m(q)$ are wave vector dependent Onsager-coefficients. An expansion of $L_m(q)$ in powers of q has to start with q^2 because the conservation property of m(q,t) demands

$$\lim_{q \to 0} \frac{d}{dt} m(q,t) = 0 \tag{II.1o}$$

whereas an expansion of $L_s(q)$ can start with a constant. Therefore we have

$$L_s(q) = \lambda_s \cdot (1 + O(q^2)) \quad , \quad L_m(q) = \lambda_m q^2 \cdot (1 + O(q^2)) \tag{II.11}$$

If μ is a suitable scale for τ or q

$$\tau \sim \mu^2 \quad , \quad q \sim \mu \tag{II.12}$$

we obtain from (II.8,9,11) the scale of (inverse) time or frequencies

$$\omega \sim \lambda \mu^2 \tag{II.13}$$

where λ is an appropriate damping constant of scale λ_s and λ_m. We note another consequence of m-conservation. From (II.1o) it also follows that each term on the right hand side of (II.9) carries at least one power of q. Rewriting the equation of motion in Lagrangian form (I.11) we see that each $\tilde{m}(q,t)$ (except the term containing a time derivative) is decorated by at least one power of q.
From the time derivative term we obtain the dimension of the response fields $\tilde{s}(\underline{x},t)$ and $\tilde{m}(\underline{x},t)$ using the dimension of $s(\underline{x},t)$ and $m(\underline{x},t)$ from (II.7) and taking into account that the action $\mathcal{J}$ is dimensionless

$$\tilde{s} \sim \mu^{\frac{d+2}{2}} \quad , \quad \tilde{m} \sim \mu^{d/2} \tag{II.14}$$

We are now ready to reduce the general form of the action to its relevant part.
At first let us consider an expansion of the memory kernel $L\,(t,\underline{x};t',\underline{x}';\{\tilde{s},s,\tilde{m},m\})$ in monomials of the fields. Because of translational symmetry the coefficients of the monomials depend only on differences of the time and space coordinates of the fields. Consider now the Fourier transforms of these coefficients. They should be analytic functions of frequencies and wave vectors if all slow variables are retained (e.g., in a system with quenched impurities their density must be treated as a slow variable, averaging over the impurity distribution leads to nonanalyticities in the frequency variables, see De Dominicis [18]). Therefore we can expand in powers of frequencies ω and wave vectors q. Each power of ω or q^2 lowers the μ-dimension of the (constant) coefficients of this expansion by 2 (see II.12,13). Recalling that on-

ly coupling coefficients with positive μ-dimension are relevant couplings and that each field m(q) is decorated with at least one power of q we recognize from (II.7,14) that the memory reduces to a Markovian form (local in time) with relevant terms: (without self loop counter terms)

$$-\tilde{s}\cdot\lambda_s\cdot\left(\tilde{s}-\frac{\delta\mathcal{H}}{\delta s}\right)-(\nabla\tilde{m})\cdot\lambda_m\cdot\left(\nabla\left(\tilde{m}-\frac{\delta\mathcal{H}}{\delta m}\right)\right) \tag{II.15}$$

up to contributions which have the same structure as the term $\tilde{\varphi}\cdot\vartheta(\varphi)$ in the action integral (I.32)

The velocity $\vartheta(\varphi)$ is determined by the antisymmetric mode coupling matrix M (φ) in (I.39). Expanding this matrix in monomials of the fields and taking into account time inversion and internal symmetry one sees easily that only one relevant term with coupling constant of positive μ-dimension survives; it is proportional to s and couples the s and m modes. Employing the infinitesimal generators $I_i=\{I_{i,\alpha\beta}\,|\,i=1,..P,\ \alpha,\beta=1,..N\}$ we obtain except irrelevant terms

$$\begin{aligned}\tilde{\varphi}\,M(\varphi)\,\frac{\delta\mathcal{H}}{\delta\varphi}&=f\cdot\sum_{i,\alpha,\beta}I_{i,\alpha\beta}\cdot s_\beta\left\{\tilde{s}_\alpha\frac{\delta\mathcal{H}}{\delta m_i}-\tilde{m}_i\frac{\delta\mathcal{H}}{\delta s_\alpha}\right\}\\&=f\cdot\left\{\tilde{s}\cdot(m\cdot I)\cdot s-s\cdot(\tilde{m}\cdot I)\cdot\nabla^2 s\right\}\end{aligned} \tag{II.16}$$

We now combine (II. 15 and 16) to extract the relevant part of the action integral (in renormalized form without self loop counter terms)

$$\begin{aligned}\mathcal{J}=\int dt\,d^dx\Big\{&-\tilde{Z}_s Z_s Z_\lambda Z_g\cdot\lambda g\cdot\tilde{s}^2-\tilde{Z}_m Z_m Z_\lambda Z_g^{-1}\cdot\lambda g^{-1}\cdot(\nabla\tilde{m})^2+\\&+\tilde{s}\left[\tilde{Z}_s Z_s\cdot\dot{s}+Z_\lambda Z_g\cdot\lambda g\cdot\left(Z_\tau\tau-Z_s\cdot\nabla^2+\tfrac{1}{3!}Z_{\tilde{u}}\cdot\tilde{u}\mu^{\varepsilon}\cdot s^2\right)s\right]+\\&+\tilde{m}\left[\tilde{Z}_m Z_m\cdot\dot{m}-Z_\lambda Z_g^{-1}\cdot\lambda g^{-1}\cdot\nabla^2 m\right]+\\&+Z_\lambda Z_\vartheta\cdot\lambda\vartheta\mu^{\varepsilon/2}\cdot\left[\tilde{Z}_s\cdot\tilde{s}(mI)s-\tilde{Z}_m\cdot s(\tilde{m}I)\nabla^2 s\right]\Big\}\end{aligned} \tag{II.17}$$

The connection between the bare and the renormalized quantities is given by (II.3) and

$$\begin{aligned}&\tilde{s}_b=\tilde{Z}_s Z_s^{1/2}\cdot\tilde{s}\ ,\quad \tilde{m}_b=\tilde{Z}_m Z_m^{1/2}\cdot\tilde{m}\\&\tilde{Z}_s\cdot\lambda_{s,b}=Z_\lambda Z_g\cdot\lambda g\ ,\quad \tilde{Z}_m\cdot\lambda_{m,b}=Z_\lambda Z_g^{-1}\cdot\lambda g^{-1}\\&Z_s Z_m^{1/2}\cdot f_b=Z_\lambda Z_\vartheta\cdot\lambda\vartheta\mu^{\varepsilon/2}\end{aligned} \tag{II.18}$$

2. Determination of the Z-factors

The determination of the various Z-factors follows the same recipe as in statics. The vertex functions (one-line-irreducible diagrams) are calculated in a loop expansion about the free action

$$\mathcal{J}_0 = \int dt\, d^dx \left\{ -\lambda\varrho\cdot\tilde{s}^2 + \tilde{s}\cdot\left[\dot{s} + \lambda\varrho\cdot(\tau - \nabla^2)s\right] - \lambda\varrho^{-1}\cdot(\nabla\tilde{m})^2 + \tilde{m}\cdot\left[\dot{m} - \lambda\varrho^{-1}\cdot\nabla^2 m\right] \right\} \qquad \text{(II.19)}$$

by analytic continuation in dimension d and the arising poles in $\varepsilon = 4\text{-}d$ are cancelled by suitably chosen Z-factors. If the Z-factors have been determined up to a given order of the loop expansion in the next order only a finite number of vertex functions (the so called primitiv divergent) shows poles proportional to ε^{-1} which then determine the Z-factors to that order. Because the primitively divergent vertex functions

$$\Gamma_{\tilde{s}\tilde{s}},\ \Gamma_{\tilde{m}\tilde{m}},\ \Gamma_{\tilde{s}s},\ \Gamma_{\tilde{m}m},\ \Gamma_{\tilde{s}sss},\ \Gamma_{\tilde{s}sm},\ \Gamma_{\tilde{m}ss} \qquad \text{(II.2o)}$$

correspond to the relevant terms in J there are normally enough Z-factors to cancel all poles. This fact is called renormalizability of the theory. Proving the renormalizability of our action we have to regard all symmetries of the theory and in particular time inversion symmetry. The latterleaves only one coupling constant and one associated Z-factor for the two mode coupling terms. On the other hand, time inversion symmetry results in a vast number of correlation response theorems. These theorems lead to connections of dynamic and static quantities calculated with the Hamiltonian $\mathcal{H}$; in particular, we can show that the static renormalizations Z_s, Z_m, Z_τ, Z_u work in dynamics as well and we need only one mode coupling renormalization Z_v. For a detailed proof of the renormalizability by means of relations between different primitive divergencies see De Dominicis, Peliti [9] .
The superficial degree of divergence of the Fourier transformed vertex functions is given by their naive or μ-dimension which is equal to the dimension of the coefficients of the corresponding monomials in the action (II.17). Due to the conservation property of the field m, however, from any vertex function containing a $\tilde{m}(q)$ can be extracted at least one factor q. Thus the superficial degree of divergence is reduced and the vertex functions $\Gamma_{\tilde{m},\tilde{m}}$, $\Gamma_{\tilde{m}m}$, $\Gamma_{\tilde{m}ss}$ exhibit only logarithmic divergencies (degree = 0) at d=4. Further, $\partial/\partial\omega\, \Gamma_{\tilde{m}m}$ has no primitive divergencies and therefore requires no subtraction. We conclude that in minimal renormalization

$$\tilde{Z}_m Z_m = 1 \qquad \text{(II.21)}$$

On the other hand, (I.24) in renormalized form leads to

$$\Theta(t) \langle m(x,t)\, m(0,0)\rangle = \tilde{Z}_m \cdot \langle m(x,t)\, \tilde{m}(0,0)\rangle \qquad \text{(II.22)}$$

As the two mean values are renormalized they are free of poles in ε and therefore

$$\tilde{Z}_m = 1 \tag{II.23}$$

which leads back to (II.6).

The conservation property also lowers the superficial degree of divergence of some irreducible vertex function with composite field insertions. For instance $\Gamma_{\tilde{m}s;(\tilde{m}s)}$, which is calculated from all irreducible diagrams with a $\tilde{m}$ and a s leg and one additional interaction vertex $\tilde{m}s$, has μ-dimension zero but its divergence is lowered by one because an over-all q-factor splits off. Therefore, $\Gamma_{\tilde{m}s;(\tilde{m}s)}$ has no primitive divergencies, a fact to be exploited below.

Consider the fluctuation response theorem (I.27) (renormalized)

$$-\theta(t)\frac{d}{dt}\langle s(x,t)s(0,0)\rangle = \langle s(x,t)\,\Psi_s(0,0)\rangle \tag{II.24}$$

where we have defined the composite field (using (II.21)

$$\Psi_s = Z_\lambda Z_s\cdot\lambda g\cdot\tilde{s} + \frac{Z_\lambda Z_v}{Z_s Z_m}\cdot\lambda v\mu^{\varepsilon/2}\cdot s(\tilde{m}I) \tag{II.25}$$

Insertions of Ψ_s yield all responses to a physical field coupled to s (see I.29). Relation (II.24) shows that Ψ_s is fully renormalized and does not give rise to new divergencies or poles in ε. Writing now in irreducible parts

$$\langle m\tilde{s}\,\Psi_s\rangle = \langle m\tilde{m}\rangle\cdot\langle\tilde{s}s\rangle\cdot\Big(\Gamma_{\tilde{m}ss}\cdot\langle s\,\Psi_s\rangle + \frac{Z_\lambda Z_v}{Z_s Z_m}\cdot\lambda v\mu^{\varepsilon/2}\cdot\Gamma_{\tilde{m}s;(\tilde{m}s)}\cdot I\Big) \tag{II.26}$$

we get (for $v \neq 0$)

$$Z_\lambda Z_v = Z_s Z_m \tag{II.27}$$

because divergencies do not arise in the various parts of (II.26). From (II.18) we find

$$f_b = Z_m^{1/2}\cdot\lambda v\mu^{\varepsilon/2} \tag{II.28}$$

and recognize that in our case ($Z_m=1$) the mode coupling constant is not renormalized. To recapitulate: via the correlation response theorem the conservation property of the generator field m provides us with two Ward identities (II.21,27) which will be of great use for the derivation of exact results, especially of critical exponents. It only remains to calculate the three Z-factors Z_λ , Z_ϱ and $\tilde{Z}_s$.

3. The Renormalization Group Equation

Now we are in position to derive the renormalization group equation. First let us recall that the parameter μ in the Hamiltonian (II.2) and the action (II.17) has been introduced by the renormalizations (II.3,18). Written in terms of bare quantities $\mathcal{H}$ and $\mathcal{J}$ are independent of μ and so are all bare Green functions and vertex functions. Let us consider a vertex function $\Gamma_{\varphi_1,\varphi_2,\ldots}$ with external legs corresponding to the fields $\varphi_1,\varphi_2,\ldots$. These fields may be some combination of $s,\tilde{s},m,\tilde{m}$. The independence of the bare theory from the parameter μ leads to

$$0=\mu\frac{\partial}{\partial\mu}\Gamma^{\text{bare}}_{\varphi_1,\varphi_2,\ldots}(\tau_b,g_b,f_b,\lambda_{s,b},\lambda_{m,b}) = \tag{II.29}$$

$$=\mu\frac{\partial}{\partial\mu}\left(Z_{\varphi_1}^{-1/2}Z_{\varphi_2}^{-1/2}\cdots\Gamma_{\varphi_1,\varphi_2,\ldots}(\tau,\tilde{u},v,\lambda,\varrho,\mu)\right)\Big|_{\substack{\text{constant bare}\\ \text{quantities}}}$$

Introducing the functions

$$\beta_{\tilde{u}}=\mu\frac{\partial}{\partial\mu}\tilde{u}\big|_b,\quad \beta_v=\mu\frac{\partial}{\partial\mu}v\big|_b,\quad \beta_\varrho=\mu\frac{\partial}{\partial\mu}\varrho\big|_b$$

$$\tau\cdot\kappa=\mu\frac{\partial}{\partial\mu}\tau\big|_b,\quad \lambda\cdot\zeta=\mu\frac{\partial}{\partial\mu}\lambda\big|_b \tag{II.3o}$$

$$\gamma_{\varphi_i}=\mu\frac{\partial}{\partial\mu}\ln Z_{\varphi_i}\big|_b,\quad (Z_{\tilde{s}}=\tilde{Z}_s^2Z_s,\ Z_{\tilde{m}}=\tilde{Z}_m^2Z_m)$$

we finally obtain the renormalization group equation:

$$\Big(\mu\frac{\partial}{\partial\mu}+\beta_{\tilde{u}}\frac{\partial}{\partial\tilde{u}}+\beta_v\frac{\partial}{\partial v}+\beta_\varrho\frac{\partial}{\partial\varrho}+\tau\cdot\kappa\frac{\partial}{\partial\tau}+ \tag{II.31}$$

$$+\lambda\zeta\frac{\partial}{\partial\lambda}-\frac{1}{2}\sum_i\gamma_{\varphi_i}\Big)\Gamma_{\ldots\varphi_i\ldots}(\{q,\omega\},\tau,\tilde{u},v,\varrho,\lambda,\mu)=0$$

The functions $\beta_{\tilde{u}},\kappa,\gamma_s$ depend on u only whereas the other functions depend on u,v and ϱ , the dimensionless renormalized parameters. They can be computed directly from the Z-factors by use of the renormalizations (II.3,18) and they have no poles in ε . As a consequence of (II.6,23,28) we get the exact relations

$$\gamma_m=\gamma_{\tilde{m}}=0 \tag{II.32}$$

$$\beta_v=-v\cdot(\varepsilon/2+\zeta)$$

The renormalization group equation can be integrated by the method of characteristics which are defined by

$$\begin{aligned} \ell \frac{dx(\ell)}{d\ell} &= \beta_x(\tilde{u}(\ell), v(\ell), \rho(\ell)) \quad , \quad x = \tilde{u}, v, \rho \\ \ell \frac{d \ln \lambda(\ell)}{d\ell} &= \zeta(\tilde{u}(\ell), v(\ell), \rho(\ell)) \\ \ell \frac{d \ln \tau(\ell)}{d\ell} &= \kappa(\tilde{u}(\ell)) \end{aligned} \tag{II.33}$$

If the initial (l=1) dimensionless parameters u,v,ρ belong to the region of attraction (for l$\rightarrow$0) of a "infrared" stable fixed point u_*, v_*, ρ_*, this fixed point determines the behaviour of all Γ-functions in the scaling limit $|q/\mu|, |\omega/\lambda\mu^2| \ll 1$ $|\tau/\mu^2| \ll 1$. For any $\Gamma = \Gamma_{\ldots\varphi_i\ldots}$ we define a function

$$\sigma(\Gamma) = \frac{1}{2} \sum_i \gamma_{\varphi_i} \tag{II.34}$$

and a function d (Γ) as the μ-dimension of Γ, hence

$$\begin{aligned} &\Gamma(\{q,\omega\}, \tau, \tilde{u}, v, \rho, \lambda, \mu) = \\ &\quad = \lambda \cdot \mu^{d(\Gamma)} \cdot \phi\left(\left\{\frac{q}{\mu}, \frac{\omega}{\lambda\mu^2}\right\}, \frac{\tau}{\mu^2}, \tilde{u}, v, \rho\right) \end{aligned} \tag{II.35}$$

with ϕ dimensionless. Asymptotically then reads the solution of (II.31)

$$\begin{aligned} \Gamma(\{q,\omega\}, \tau)_{as} &= \ell^{\,d(\Gamma) - \sigma(\Gamma)_* + \zeta_*} \cdot \lambda \mu^{d(\Gamma)} \cdot \\ &\quad \cdot \phi\left(\left\{\frac{q}{\ell\cdot\mu}, \frac{\omega}{\ell^z \cdot \lambda\mu^2}\right\}, \frac{\tau}{\ell^{1/\nu} \cdot \mu^2}, \tilde{u}_*, v_*, \rho_*\right) \end{aligned} \tag{II.36}$$

which displays the scaling behaviour with the usual critical exponents

$$z = 2 + \zeta_* \quad , \qquad \nu = (2 - \kappa_*)^{-1} \quad , \qquad \eta = \gamma_{s\,*} \tag{II.37}$$

If $v_* \neq 0$ we see from (II.32) that $\zeta_* = -\varepsilon/2$ and obtain the well known dynamical critical exponent

$$z = 2 - \varepsilon/2 = d/2 \tag{II.38}$$

From $G_{ss} = |\Gamma_{\tilde{s}s}|^{-2} \cdot \Gamma_{\tilde{s}\tilde{s}}$ we find the particular scaling form of the dynamic structure factor setting $\ell = q/\mu$

$$G_{ss}(q,\omega,\tau) = q^{\eta - 2 - z} \cdot G(q\cdot\xi, \omega \cdot q^{-z}) \tag{II.39}$$

with the correlation length $\xi = \tau^{-\nu}$

It is possible that $\varrho_* = 0$ (weak scaling fixed point of De Dominicis, Peliti [9]). From the definition of ϱ in the action (II.17) it is obvious that in this case ϱ is a dangerous irrelevant variable and corrections to scaling play an important role. With a correction exponent

$$\bar{\omega}_\varrho = \frac{\partial}{\partial\varrho}\beta_\varrho\Big|_* > 0 \tag{II.40}$$

ϱ varies asymptotically as

$$\varrho = \varrho_* + \varrho_0\cdot \ell^{\bar{\omega}_\varrho} = \varrho_0\cdot\left(\frac{q}{\mu}\right)^{\bar{\omega}_\varrho} \tag{II.41}$$

where ϱ_0 is a nonuniversal quantity. Thus two different time scales emerge the orders of magnitude being given by characteristic frequencies

$$\omega_s \sim \lambda\varrho\cdot\mu^2 \sim q^{2+\bar{\omega}_\varrho}\cdot\Omega_s(q\cdot\xi) \tag{II.42}$$

$$\omega_m \sim \lambda\varrho^{-1}\cdot\mu^2 \sim q^{2-\bar{\omega}_\varrho}\cdot\Omega_m(q\cdot\xi)$$

To demonstrate what kinds of fixed points arise in our model we give the results of a one-loop calculation of the dynamic quantities:

$$\beta_\varrho = \varrho\cdot\left[\left(1-\frac{4}{1+\varrho^2}\cdot\frac{P}{N}\right)\cdot\left(\frac{v}{8\pi}\right)^2 + O(\tilde{u}^2, \tilde{u}v^2, v^4)\right] \tag{II.43}$$

$$\zeta = -\left(1+\frac{4}{1+\varrho^2}\cdot\frac{P}{N}\right)\cdot\left(\frac{v}{8\pi}\right)^2 + O(\tilde{u}^2, \tilde{u}v^2, v^4)$$

From $\zeta_* = -\frac{\varepsilon}{2}$ and $\beta_{\varrho *} = 0$ we find

1. <u>N < 4P:</u>

$$\varrho_*^2 = 4P/N - 1 \quad , \quad \left(\frac{v_*}{4\pi}\right)^2 = \varepsilon \tag{II.44}$$

2. <u>N > 4P:</u>

$$\varrho_* = 0 \; , \; \bar{\omega}_\varrho = \frac{\varepsilon}{2}(1-4P/N)(1+4P/N)^{-1}$$
$$\left(\frac{v_*}{4\pi}\right)^2 = 2\varepsilon\cdot(1+4P/N)^{-1} \tag{II.45}$$

Therefore, if the dimension N of the representation to which the order parameter belongs is too large in comparison to the number of group parameters P (in our model $N > 4P + O(\varepsilon)$) or, equivalently, if the continuous symmetry is not high enough only weak scaling holds with a dynamical critical exponent $z_s = d/2 + \bar{\omega}_\varrho > d/2$. The relevance of weak scaling to the superfluid transition in He has been discussed first by De Dominicis, Peliti [9] and later by Dohm et al. [19] .

III. Dynamic Susceptibility at the Coexistence Line

To demonstrate how the renormalization group and a suitable perturbation theory combine to yield universal scaling functions we paraphrase some results of Horner and Schäfer [20,21] concerning the Goldstone mode singularities at the coexistence curve.

For simplicity we neglect mode coupling (v=0 in (II.17)) whereupon our model coincides with model A of Halperin, Hohenberg, Ma [3]. The field m decouples from the order parameter and may be eliminated. So we can employ $\varrho=1$ and $Z_g=1$ in our formulas. The composite field Ψ_s (II.25) essentially now equals the field $\tilde{s}$, and the correlation response theorem leads to the Ward identity

$$Z_\lambda = 1 \tag{III.1}$$

Therefore we need only one dynamical renormalization $\tilde{Z}_s$ and from (II.18,30) we obtain

$$\zeta = \mu \frac{\partial}{\partial\mu} \ln \tilde{Z}_s \Big|_b \tag{III.2}$$

$$\gamma_{\tilde{s}} = \mu \frac{\partial}{\partial\mu} \ln (\tilde{Z}_s^2 \cdot Z_s)\Big|_b = 2\zeta + \gamma_s$$

Moreover, we can include a constant external magnetic field H or, equivalently, consider vertex functions depending on a constant magnetization $M = \langle s \rangle$. As in the static case [1] we obtain the renormalization group equation for a vertex function with n s-legs and $\tilde{n}$ $\tilde{s}$-legs

$$\Big(\mu\frac{\partial}{\partial\mu} + \beta(\tilde{u})\frac{\partial}{\partial\tilde{u}} + \tau\cdot\kappa(\tilde{u})\frac{\partial}{\partial\tau} + \zeta(\tilde{u})\Big(\lambda\frac{\partial}{\partial\lambda} - \tilde{n}\Big) + \\ - \gamma(\tilde{u})\Big(M\frac{\partial}{\partial M} + n\Big)\Big)\Gamma^{(\tilde{n},n)}(\{q,\omega\},\tau,M,\tilde{u},\lambda,\mu) = 0 \tag{III.3}$$

In particular, from (II.36) the scaling behaviour of the dynamic susceptibility

$$\chi(\omega) = \lambda\cdot\Gamma^{(1,1)}(\omega, q=0)^{-1} \tag{III.4}$$

is found to be

$$\chi(\omega,\tau,M) = \ell^{-2+\eta}\cdot\chi(\ell^{-z}\omega, \ell^{-1/\nu}\cdot\tau, \ell^{-\beta/\nu}\cdot M) \tag{III.5}$$

If by a suitable parameter ℓ the thermodynamic variables $\tau' = \ell^{-1/\nu}\cdot\tau$, $M' = \ell^{-\beta/\nu}\cdot M$ are chosen to be outside the critical region we can do a perturbative expansion of the right hand side of (III.5). This clearly requires that the frequency is not too large, otherwise one has to do a dynamical operator product expansion [8].

As Horner and Schäfer have shown, also at the coexistence line a perturbative expan-

sion is possible, if one chooses the correct free Hamiltonian or action as the starting point of the expansion and sums up the so called vertex irreducible parts. The first condition can be satisfied by adapting renormalization conditions so as to make the Goldstone singularities tractable. The resummation of vertex irreducible parts is most easily performed by introduction of dummy fields χ and $\tilde{\chi}$.
As well as with the action $\mathcal{J}$ (II.17) without the mode couplings and the fields m, $\tilde{m}$ we can take averages with a new action

$$\mathcal{J}' = \int dt\, d^d x\, \lambda \Big\{ -\tilde{s}^2 + \tilde{s}\cdot[\lambda^{-1} \dot{s} + (r - \nabla^2) s] + \qquad \text{(III.6)}$$
$$+ \tfrac{1}{2} f \cdot \tilde{\chi} s^2 + f \cdot \chi \tilde{s} s - \tilde{\chi}\chi \qquad + \text{ counter terms } \Big\}$$

Integrating out $\tilde{\chi}, \chi$ we come back to $\mathcal{J}$. Introducing now a coupling to an external field H and defining new fluctuating variables $(\langle\varphi\rangle = \langle\tilde{\varphi}\rangle = \langle\chi\rangle = \langle\tilde{\chi}\rangle = 0)$

$$\varphi = s - M\,, \quad \varphi_{\text{longit.}} = \sigma\,, \quad \varphi_{\text{transv.}} = \pi \qquad \text{(III.7)}$$

we obtain

$$\mathcal{J}' = \int dt\, d^d x\, \lambda \cdot \Big\{ -\tilde{\varphi}^2 + \tilde{\varphi}\,[\lambda^{-1} \dot{\varphi} + (r - \nabla^2)\varphi] +$$
$$- \tilde{\chi}\chi \quad + f \cdot M \cdot (\tilde{\chi}\sigma + \tilde{\sigma}\chi) + f \cdot (\tfrac{1}{2} \tilde{\chi}\varphi^2 + \chi \cdot \tilde{\varphi}\varphi) + \qquad \text{(III.8)}$$
$$+ \text{ counter terms } \Big\}$$

Rotational invariance results in a set of Ward idendities for the new vertex functions

$$\lambda H = \Gamma'_{\tilde{\pi}\pi}(Q=0)\, M \qquad (Q = (q,\omega))$$
$$\Gamma'_{\tilde{\sigma}\sigma}(Q) - \Gamma'_{\tilde{\pi}\pi}(Q) = \Gamma'_{\tilde{\pi}\pi\pi'\pi'}(Q,0,-Q,0)\cdot M^2 \qquad \text{(III.9)}$$
$$\Gamma'_{\tilde{\pi}\pi\sigma}(Q_1, Q_2, Q_3) = \Gamma'_{\tilde{\pi}\pi\pi'\pi'}(Q_1, Q_2, Q_3, 0)\cdot M$$
$$\Gamma'_{\tilde{\chi}\sigma}(Q) = \Gamma'_{\tilde{\chi}\pi\pi}(Q, -Q, 0)\cdot M$$

Now we choose the renormalization and define thereby the counter terms at the coexistence line H=0, M$\neq$0 such that

$$\Gamma'_{\tilde{\pi}\pi}(Q=0) = \Gamma'_{\tilde{\pi}\pi\pi'\pi'}(0,0,0,0) = 0 \qquad \text{(III.1o)}$$

This ensures that the Goldstone singularities arising in the transverse propagators for Q=0 are reduced by the dummy field propagators because for Q$\to$0 now $G_{\tilde{\chi}\chi} \sim \Gamma'_{\tilde{\pi}\pi} \to 0$ and the transversal modes couple to $\chi, \tilde{\chi}$ only.

The new constants in (III.8) f and σ are determined by the two conditions (III.1o). Together with the counter terms their connection with the old parameters τ and $\breve{u}$ may be found by integrating out the dummy fields $\chi, \tilde{\chi}$. Clearly condition (III.1o) which refers to the coexistence line yields r = 0.

The new vertex functions Γ' can now be calculated in a loop expansion. The old Γ-functions are then given by in general nonlinear relations, e.g.,

$$\Gamma_{\tilde{\sigma}\sigma} = \Gamma'_{\tilde{\sigma}\sigma} - \Gamma'_{\tilde{\sigma}\chi} \cdot (\Gamma'_{\tilde{\chi}\chi})^{-1} \cdot \Gamma'_{\tilde{\chi}\sigma} \tag{III.11}$$

where the inverse $\Gamma'_{\tilde{\chi}\chi}$ -function reflects the resummation. Choosing the magnetization M' to be outside the critical region, e.g.,

$$\breve{u}_{*} \cdot M'^{2} = 3\mu^{2-\varepsilon} \quad \text{with} \quad \frac{\breve{u}_{*}}{(4\pi)^2} = \frac{3\varepsilon}{N+8} + O(\varepsilon^2) \tag{III.12}$$

and using the scaling equation (III.5) and the definition (III.4) we can relate the longitudinal susceptibility χ_ℓ for any magnetization at the coexistence line M=const.$\cdot|\tau|^{\beta}$ to the perturbational calculation of $\Gamma_{\tilde{\sigma}\sigma}$ for M'. The result is (in a one-loop calculation)

$$\begin{aligned}[\chi_\ell(\omega)\cdot M^{\delta-1}]^{-1} &= iw + \frac{2\varepsilon}{N+8}\cdot(L(w)-1) + \\ &+ \frac{1+\frac{2\varepsilon}{N+8}\cdot(L(w)-1)}{1+\frac{N-1}{N+8}\left[\left(\frac{iw}{2}\right)^{-\varepsilon/2}-1\right]-\frac{\varepsilon}{2(N+8)}(L(w)-N)}\end{aligned} \tag{III.13}$$

where we have defined

$$\begin{aligned} w &= \text{const.}\ \omega/M^{\nu z/\beta} \\ L(w) &= \left(1+\frac{2}{iw}\right)\cdot \ln\left(1+\frac{iw}{2}\right)\end{aligned} \tag{III.14}$$

The expression (III.13) has much in common with that of Mazenko [22]. The Goldstone singularity $\omega^{\varepsilon/2}$ occurs in the same form, the other terms, however, are differently arranged. As Schäfer [21] pointed out, in (III.13) the Goldstone singularities are consistenly resummed and are free of exponentiations which are somewhat arbitrary but inavoidable if $\chi_\ell(\omega)$ is only calculated to some order of ε .

Acknowledgement

I thank Dr.N.Breuer for discussion and a critical reading of this paper.

REFERENCES

1) E.BREZIN, J.C. LE GUILLOU, J.ZINN-JUSTIN; in Phase Transitions and Critical Phenomena, edited by C.DOMB and M.S.GREEN (Academic Press N.Y.) 1976, vol.VI, and references cited therein

2) D.AMIT; Field Theory, the Renormalization Group, and Critical Phenomena, Mc Graw Hill Comp., 1978

3) P.C. HOHENBERG, B.I. HALPERIN; Rev.Mod.Phys. 49, 435 (1977)

4) C. DE DOMINICIS; N.Cim.Lett. 12, 567 (1975)
C. DE DOMINICIS, E. BREZIN, J. ZINN_JUSTIN; Phys.Rev. B 12, 4945 (1975)
E. BREZIN, C. DE DOMINICIS; Phys.Rev. B 12, 4954 (1975)

5) C. DE DOMINICIS; J.Phys. (Paris) 37, Colloque C-247 (1976)

6) H.K. JANSSEN; Z.Phys. B 23, 377 (1976)

7) P.C. MARTIN, E.D. SIGGIA, H.A. ROSE; Phys.Rev. A 8, 423 (1973)

8) R. BAUSCH, H.K. JANSSEN, H. WAGNER; Z.Phys. B 24, 113 (1976)

9) C. DE DOMINICIS, L. PELITI; Phys.Rev.Lett. 38, 5o5 (1977), Phys.Rev. B 18, 353 (1978)

1o) K. KAWASAKI; Progr.Theor.Phys. 54, 1665 (1975)
J.D. GUNTON, K. KAWASAKI; Progr.Theor.Phys. 56, 61 (1976)

11) K. KAWASAKI; in Phase Transitions and Critical Phenomena, edited by C. DOMB and M.S. GREEN (Academic Press N.Y.) 1976, vol Va, and references cited therein

12) R. GRAHAM; Springer Tracts in Modern Physics 66, Berlin-Heidelberg-New York: Springer 1973

13) U. DEKER, F. HAAKE; Phys. Rev. A 11, 2o43 (1975)

14) H.K. JANSSEN; Frühjahrstagung der DPG, 1976, Freudenstadt,(Germany)

15) L. SASVARI, F. SCHWABL, P. SZEPFALUSY; Physica 81 A, 1o8 (1975)
L. SASVARI, P. SZEPFALUSY; Physica 87 A, 1 (1977)

16) B.I. HALPERIN; Phys.Rev. B 11, 178 (1975)

17) G. T'HOOFT, M. VELTMAN; Nucl.Phys. B 44, 189 (1972)

18) C. DE DOMINICIS; Phys.Rev. B 18, 4913 (1978)

19) V. DOHM; Z.Physik B 31, 327 (1978)
R.A. FERREL, V. DOHM, J.K. BHATTACHARJEE; Phys.Rev.Lett. 41, 1818 (1978)
V. DOHM; Z.Physik B 33, 79 (1979)

2o) L. SCHÄFER, H. HORNER; Z.Phys. B 29, 251 (1978)

21) L. SCHÄFER; Z.Phys. B 31, 289 (1978)

22) G.F. MAZENKO; Phys.Rev. B 14, 3933 (1976)

CRITICAL DYNAMICS BELOW T_c

P. Szépfalusy

Central Research Institute for Physics+
1525 Budapest, Hungary

and

Universität des Saarlandes - Theoretische Physik
66 Saarbrücken, Germany

SUMMARY

+ permanent address

CRITICAL DYNAMICS BELOW T_c

P. Szépfalusy

Central Research Institute for Physics [+]
1525 BUDAPEST, Hungary

and

Universität des Saarlandes - Theoretische Physik
66 SAARBRÜCKEN, Germany

Summary

Going below the critical temperature, the existence of a non-zero average value of the order parameter induces qualitatively new features in the critical behaviour in a variety of systems. A crucial role is played by the symmetry in the intrinsic space of the order parameter degrees of freedom. This lecture is focused on rotationally invariant systems, others are only briefly mentioned.

At first, properties of the isotropic multicomponent systems are described on a phenomenological basis in cases of purely dissipative systems and of systems with reversible mode coupling as well. The main feature is that the orientational fluctuations are dominating the large-distance, long-time behaviour of the system, and the parallel and longitudinal order parameter correlation functions can be expressed in terms of the correlation function for the orientational fluctuations. In the purely relaxational model such qualitative arguments predict a power-law decay in space and time of the correlation functions.

These properties are subsequently considered within the framework of the semimacroscopic theory. In this context the theoretical means to handle the problems in the ordered phase i.e. new type of building blocks replacing the usual self energies are introduced, both for the transverse and longitudinal order parameter response and correlation functions. It is shown how characteristics of the Goldstone mode can be expressed in terms of them. The longitudinal correlation function is discussed especially from the point of view of how the results of the phenomenological considerations can be justified.

Special emphasis is given to purely dynamic effects, such as the Goldstone-mode induced singularity in the transport coefficient of the parallel total magnetization in an isotropic antiferromagnet, recently investigated also experimentally. It is shown that the theory can account for the experimental findings in $RbMnF_3$, not only concerning the wave-number dependence of the transport coefficient in the hydrodynamic region but also regarding its magnitude.

+ Permanent address

1. General properties of isotropic multicomponent systems

Consider systems with a vector order parameter $\varphi(\varphi_1,\ldots,\varphi_n)$ and suppose that there exists a continuous rotational symmetry in the intrinsic space of the order parameter degrees of freedom. Below T_c an important feature is the spontaneous breaking of rotational symmetry due to the non-zero average value of the order parameter. We assume that the order parameter in equilibrium points in the direction of the first axis and denote its value by P:

$$P \equiv \langle\varphi_1\rangle \quad . \tag{1.1}$$

At first we consider purely relaxational models in which only dissipative coupling between the fluctuations is included. To analyse the dynamics of the system in the ordered phase our starting point is the recognition that there are two relaxational time scales; one for the fluctuations of the magnitude of the order parameter and an other one for the orientational fluctuations[1]. It is then natural to introduce new variables, namely a random variable ρ measuring the magnitude of the order parameter

$$\rho^2 = \sum_\alpha \varphi_\alpha^2 \tag{1.2}$$

and all the variables ϕ necessary to specify the orientation of the order parameter. Let us denote the above-mentioned relaxational times by $\tau_\rho(k)$ and $\tau_\phi(k)$, respectively, for a fluctuation of wave-number k.

In equilibrium the order parameter can equally point in any direction, thus no energy is necessary to rotate the order parameter uniformly. Consequently, if there is a long wave-length orientational fluctuation of the order parameter, the associated energy is small and the relaxation time to equilibrium is long. This means that $\tau_\phi(k)\to\infty$ if $k\to 0$. $\tau_\rho(k)$ may remain finite in the same limit or may also tend to infinity depending on whether the order parameter is conserved or not. But in any case

$$\frac{\tau_\phi(k)}{\tau_\rho(k)} \sim \frac{1}{k^2}\text{, for small k.} \tag{1.3}$$

As a consequence, the long wave-length and long time fluctuations of the order parameter are orientational fluctuations maintaining the magnitude of the order parameter constant. We will be interested in this limit and will not treat the magnitude of the order parameter as a separate variable. Patashinskii and Pokrovskii[2] exploited the consequences of fixed-length orientational fluctuations of the order parameter for static properties. Since their phenomenological considerations are based on the existence of local equilibrium, they can naturally be extended to include time dependence for large enough t-s when local equilibrium is reached. Then a simple picture arises as follows.

For small-angle fluctuations we can write

$$\varphi_\alpha = P\,\phi_\alpha + \ldots \qquad \alpha \geqslant 2 \tag{1.4}$$

$$\delta\varphi_1 = -\,P\,\frac{1}{2}\sum_\alpha \phi_\alpha^2 + \ldots \tag{1.5}$$

By introducing the correlation function

$$C_\phi(r,t) = \langle\phi_\alpha(1)\phi_\alpha(2)\rangle, \qquad \alpha \geqslant 2 \tag{1.6}$$

$$r \equiv |x_1-x_2|, \quad t \equiv t_1-t_2\,,$$

the expressions

$$C_\perp(r,t) = P^2 C_\phi(r,t) \tag{1.7}$$

and

$$C_{||}(r,t) = \frac{n-1}{2}\,P^2[C_\phi(r,t)]^2 \tag{1.8}$$

follow for the perpendicular and parallel order parameter correlation functions, respectively, in the limit of large r and t. An important ingredient in the basic physics of the problem which has been taken into account is that the interaction between the orientational fluctuations disappears in the small k, small ω limit and consequently a decoupling of fluctuations becomes exact. Moreover, the leading terms are those which contain the possible smallest number of C_ϕ factors.

Before entering the discussion of the time dependences of these correlation functions let us recall briefly the main features of the static results.

$$C_\phi(r) = A_d\left(\frac{\xi_\phi}{r}\right)^{d-2} , \tag{1.9}$$

or in momentum space,

$$C_\phi(k) = \frac{1}{\rho_s}\frac{1}{k^2} \tag{1.10}$$

A_d is a constant depending on the dimensionality of space. The coherence length for the orientational fluctuations is given in terms of ρ_s ,

$$\xi_\phi = \rho_s^{-\frac{1}{d-2}} \quad \text{(Josephson's relation)}. \tag{1.11}$$

ρ_s is specified by writing the excess free energy in the presence of an orientational fluctuation as

$$\delta F = \int d^d x \frac{1}{2} \rho_s \sum_{\alpha=2}^{n} (\nabla\phi_\alpha)^2 \tag{1.12}$$

By subsituting eq. (1.9) into eq. (1.8) one obtains[2]

$$C_{||}(r) = A_d^2 \frac{n-1}{2} P^2 \left(\frac{\xi_\phi}{r}\right)^{2(d-2)} \tag{1.13}$$

and in momentum space

$$C_{||}(k) = \frac{n-1}{2} \frac{P^2}{\rho_s{}^2} \frac{1}{k^\varepsilon} , \qquad \varepsilon = 4-d \tag{1.14}$$

The singularity of the parallel correlation function as $k \to 0$ is related to the coexistence curve singularity below T_c in isotropic systems. The static correlation functions and the problem of the coexistence curve singularity have been studied by a large number of authors. Since the discussion of these static properties is not the purpose of this lecture, we can only give a selection of references by quoting an early one[3] and some of the recent ones[4-8].

Consider now the time dependence of the correlation functions. We still continue to deal with the purely relaxational models and for the sake of definiteness let us choose the case when the order parameter is not conserved. Then the correlation function of the orientational fluctuations in the small k small ω region takes the simple form

$$C_\phi(k,\omega) = C_\phi(k)\,\frac{2\omega_\phi(k)}{\omega^2+\omega_\phi^2(k)}\ , \tag{1.15}$$

where the characteristic frequency is given by

$$\omega_\phi(k) = \frac{\Gamma_\phi}{\chi_\phi} = \Gamma_\phi \rho_s k^2\ . \tag{1.16}$$

Here χ_ϕ and Γ_ϕ are the static response function and the kinetic coefficient, respectively, connected with changing the orientation of the order parameter (in our units $C_\phi(k) = \chi_\phi(k)$). Dynamical scaling[9,10] requires

$$\Gamma_\phi \sim \xi_\phi^{\,d-z} \qquad (k\xi_\phi \ll 1) \tag{1.17}$$

where z is the dynamical critical exponent which has been calculated within the renormalisation group theory by ε and 1/n expansions (see for a review ref. 11). The characteristic behaviour of the correlation functions is most clearly exhibited in terms of the space-time variables

$$C_\phi(r,t) = C_\phi(r)\ \psi\left(\frac{t}{\tau_\phi}\left(\frac{\xi_\phi}{r}\right)^2\right) \tag{1.18}$$

where

$$\tau_\phi^{-1} = \omega_\phi(k=\xi_\phi^{-1})\ . \tag{1.19}$$

For fixed r the correlation function falls off as a power law in time:

$$C_\phi(r,t) \xrightarrow[t\to\infty]{} \frac{1}{t^{(d-2)/2}}\ . \tag{1.20}$$

According to eqs. (1.7) and (1.8) the transverse order parameter correlation function is of the same form while the longitudinal one behaves as

$$C_{\parallel}(r,t) = C_{\parallel}(r)\ \psi^2\left(\frac{t}{\tau_\phi}\ (\frac{\xi_\phi}{r})^2\right) \xrightarrow[t\to\infty]{} \frac{1}{t^{d-2}} \tag{1.21}$$

for given r. It has to be emphasized that the same coherence length ξ_ϕ and the same time constant τ_ϕ, characterising the orientational fluctuations of the order parameter, enter both the transverse and longitudinal correlation functions.

Until now we have treated the purely relaxational model with an isotropic n-component non-conserved order parameter. Such a model is, of course, an artifact since due to the rotational invariance in the component space there exist a number of conserved fields which have to be included when the dynamical properties of the system are discussed[1,12]. They are the generators for the rotation of the order parameter and can be represented by an antisymmetric tensor field $L_{\alpha\beta}$ with $\frac{n(n-1)}{2}$ components, $\alpha,\beta=1,2,\ldots,n$. Examples for such fields are the magnetisation perpendicular to the easy plane in planar magnetic systems and the three components of the total magnetisation in the isotropic antiferromagnet.

Let us continue by considering the properties of such a general n-component φ-L model. First note that there are now two dissipative characteristic frequencies having the same k-dependences to be taken into account. One of them, ω_ϕ, is connected again with the relaxational process of the orientational fluctuations of the order parameter and is again given by eq. (1.16). The other one is related to the diffusion process for the L field

$$\omega_{L\alpha\beta} = \frac{\lambda_{\alpha\beta}k^2}{\chi_L} \tag{1.22}$$

where $\lambda_{\alpha\beta}$ is the transport coefficient and χ_L is the corresponding static susceptibility. The basic new feature is, however, the appearance of reversible mode-coupling terms in the equation of motion. Below T_c in the hydrodynamic region they lead to a linear coupling beween the variables ϕ_α and $L_{1\alpha}$:

$$\dot{\phi}_\alpha = g\ \chi_L^{-1}\ L_{1\alpha} \tag{1.23}$$

$$\dot{L}_{1\alpha} = -\ g\ \chi_\phi^{-1}\phi_\alpha \tag{1.24}$$

where g is the mode coupling constant. As a result, the orientational waves of the order parameter are associated with a mode of linear dispersion[1,12],

$$\omega(k) = c\ k - \frac{i}{2}\,Dk^2\ , \tag{1.25}$$

where

$$c\ k = g\ (\chi_L\chi_\phi)^{-1/2} \tag{1.26}$$

and the damping arising from the dissipative processes is

$$Dk^2 = \omega_\phi + \omega_L\ . \tag{1.27}$$

The property that Re $\omega(k)\to 0$ if $k\to 0$ is a manifestation of Goldstone's theorem and arises from the fact that the rotation of the average order parameter as a whole does not change the energy of the system. Such a motion cannot lead either to any dissipative process, which is manifested by that $\omega_\phi\to 0$ if $k\to 0$, as already mentioned.

Note that c is given in terms of purely static quantities so, using ε-expansion results for the static transverse order parameter susceptibility[13], we can write

$$c = g\ P(1+\frac{1}{4}\ \varepsilon\ \frac{1}{n+8} + O\ (\varepsilon^2)) \tag{1.28}$$

where P should be substituted by its expression to $O(\varepsilon)$.

In the hydrodynamic region also an exact expression can be derived for the transverse order parameter correlation function which, in turn, through eqs. (1.7) and (1.8) yields the parallel order parameter correlation function, too. We give here the latter in the special case when d=3 and in the region where damping is negligible:

$$C_{||}(k,\omega) = C_{||}(k)\,\frac{1}{\omega}\,\ln\left(\frac{\omega-ck}{\omega+ck}\right)^2 \qquad (1.29)$$

An other interesting quantity to be studied is the relaxation of the $L_{\alpha\beta}$ fields for $\alpha,\beta \geq 2$ i.e. of such generator fields which are not linearly coupled to the order parameter fluctuations. In case of the antiferromagnet (n=3) there is only one such quantity, namely the longitudinal total magnetisation. This problem will be discussed in Sect. 3.

2. Semimacroscopic theory

Let us now consider the properties we have discussed in the previous section on the basis of the nonlinear semimacroscopic equations of motion used extensively in the literature to treat time-dependent critical properties (see for recent reviews refs. 11, 14, 15).

2.1 The transverse order parameter correlation function in the φ-L model

As a first example the coupling of the transverse order parameter fluctuations and the fluctuations of the $L_{1\alpha}$ field will be discussed. This coupling is manifested in the structure of the self-energies[16,17]

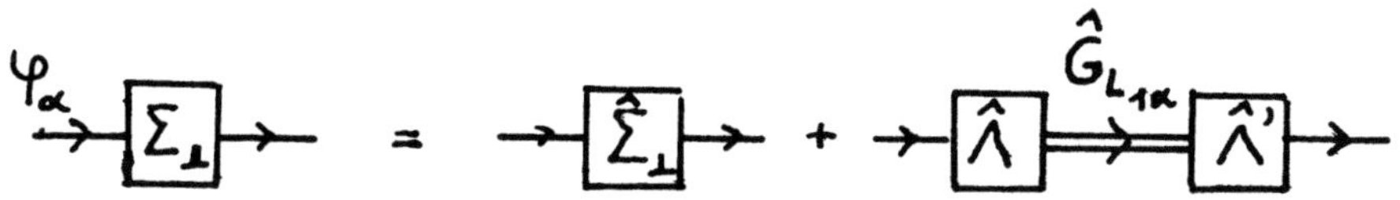

(2.1)

for the transverse order parameter propagator and

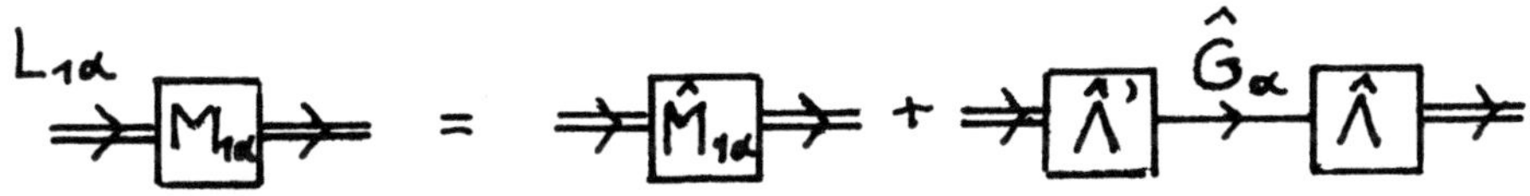

(2.2)

for the propagator of the $L_{1\alpha}$ field. Here $\Sigma_\perp$ is the self-energy of the transverse order parameter propagator and $M_{1\alpha}$ is the self-energy for the $L_{1\alpha}$ field. The blocks whose contributions are denoted by a letter with a hat contain diagrams which are double-irreducible i.e. they cannot be split into two parts by cutting a single order parameter or $L_{1\alpha}$ propagator line. $\hat{G}_\perp$ and $\hat{G}_{L_{1\alpha}}$ are transverse order parameter and $L_{1\alpha}$ propagators, respectively, containing only double-irreducible self-energies. $\hat{\Lambda}$ and $\hat{\Lambda}'$ are vertices which convert transverse order parameter fluctuations into those of $L_{1\alpha}$ and vice versa. They are non-zero only in the ordered phase and represent the coupling discussed at the end of the previous section.

These double-irreducible building blocks are non-singular, and the various physical quantities below T_c can be expressed in terms of them. Thus, for example, the kinetic coefficient for the relaxation of the orientational fluctuations Γ_ϕ can be written in the following form

$$\frac{\Gamma_\phi}{\Gamma^{(o)}} = \frac{1 - \left(\frac{\partial \hat{\Sigma}_\perp}{\partial k^2}\right)_o}{1 + \Gamma^{(o)} \left(\frac{\partial \hat{\Sigma}_\perp}{\partial i\omega}\right)_o} \tag{2.3}$$

where $\Gamma^{(o)}$ is the bare kinetic coefficient. When a perturbation expansion is applied, one has to calculate these functions instead of the original self-energies to the desired order, to ensure that the response and correlation functions fulfil the general relationships reflecting symmetry breaking. Therefore any consequent approximation involves the summation of an infinite series of selected self-energy diagrams.

On the basis of this analysis of the structure of the self-energies it has also been shown that the transverse order parameter fluctuations and those of the $L_{1\alpha}$ fields share the same excitation spectra in the ordered phase, not only in the hydrodynamic region but in the whole region where the semimacroscopic theory is applicable. The dispersion of this mode can be calculated from the equation[16,17]

$$\left(k^2 - \frac{i\omega}{\Gamma^{(o)}} - \hat{\Sigma}_\perp\right)\left(1 - \frac{i\omega}{\lambda^{(o)}k^2} - \hat{M}_{1\alpha}\right) - \hat{\Lambda}\hat{\Lambda}' = 0 \tag{2.4}$$

where $\lambda^{(o)}$ is the bare transport coefficient for the $L_{1\alpha}$ field. The velocity of the mode in the hydrodynamic region has been calculated directly by ε-expansion, in dynamics, using various approaches.

Siggia[18] and Hohenberg et al.[19] (for n=2) and Sasvari and Szépfalusy[17] (for general n) have obtained results in accord with the macroscopic expression (eqs. (1.26) and (1.28)) while the result of Mazenko et al.[20] (for n=3) is in disagreement with it. The damping D has also been calculated by the authors cited above. It can most suitably be characterised by the universal amplitude ratio

$$R_2 \equiv D/2c\xi_\phi \quad . \tag{2.5}$$

To leading order in ε

$$R_2 = \left(\frac{K_d}{\varepsilon}\,\frac{(n-1)^2}{2n-3}\right)^{1/2} \tag{2.6}$$

where K_d is the standard abbreviation

$$K_d = 2^{-d+1}\pi^{-d/2}[\Gamma(d/2)]^{-1} \quad . \tag{2.7}$$

2.2 Longitudinal order parameter correlation function

The longitudinal order parameter response and correlation funtions also have interesting structure in the ordered phase below T_c due to the symmetry breaking. The investigation of the parallel response function $G_{\parallel}$ is to be performed together with that of G_ρ, the response function for the square of the order parameter to an external field coupled to ρ[16,*] (ρ has been defined earlier by eq. (1.2)). The important point is that, in the symmetry breaking phase, diagrams of $G_{\parallel}$ appear in G_ρ as intermediate states carrying the total external momentum and frequency, and vice versa, $G_{\parallel}$ has intermediate states built up of irreducible diagrams of G_ρ. As a consequence, rearrangements of the perturbational series have to be performed similar to those which were necessary in the case of the transverse response function. To this aim let us represent the quartic interaction by

---

*Note that we use here a notation different from that of ref. 16. Namely, $G_{\parallel}$, G_ρ, $\tilde{\pi}$, $\tilde{\Lambda}$, and $\tilde{\Lambda}'$ are denoted in ref. 16 by G_L, $\chi_{\rho\rho}$, $\tilde{\chi}_{\rho\rho}$, $\tilde{\Lambda}_1$ and $\tilde{\Lambda}_2$, respectively.

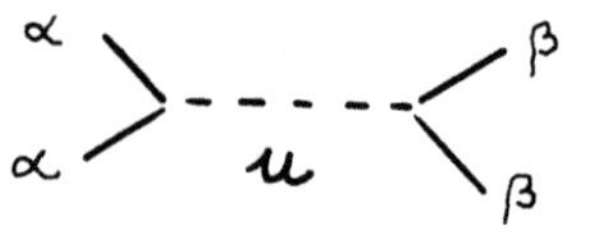

(2.8)

Those quantities, which are made up of only such diagrams which cannot be split into two parts by cutting either an order parameter propagator or an interaction u-line, are called double-irreducible blocks and will be marked by a tilde. In terms of them, the self-energies exhibit the following structure:

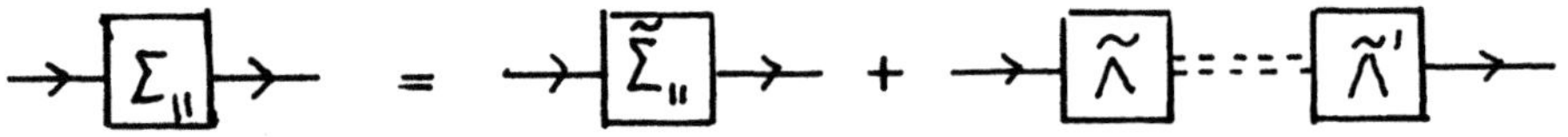

(2.9)

(2.10)

with

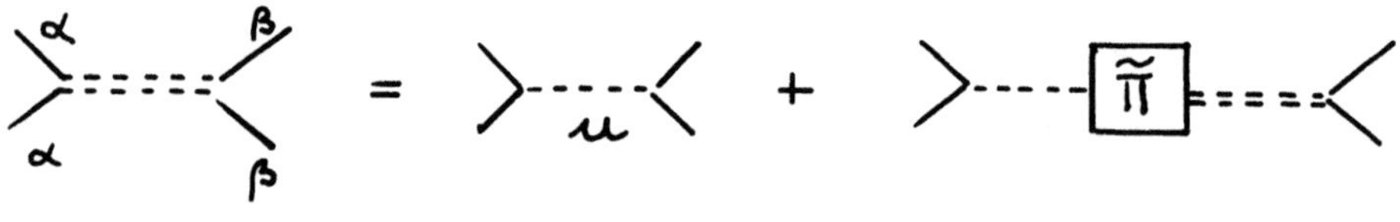

(2.11)

The appearance of $\tilde{\Lambda}$ and $\tilde{\Lambda}'$ vertices in eqs. (2.9) and (2.10), coupling the longitudinal order parameter fluctuations to those of ρ, is a characteristic of the ordered phase.

In deriving eqs. (2.9) and (2.10) the method by Ma and Mazenko[21] to generate perturbation expansion has been applied in ref. 16. Correspondingly, two types of self-energies are present; namely besides the usual ones, self-energy parts terminating in an external field conjugate to the order parameter, instead of a propagator line, also appear. Eqs. (2.9), (2.10) and (2.11) lead to

$$G_{\parallel} = \frac{(1+u\tilde{\pi})(1+\tilde{\Sigma}'_{\parallel}) - u\,\tilde{\Lambda}'\,\tilde{\Sigma}''_{\parallel}}{(G_o^{-1}-\tilde{\Sigma}_{\parallel})(1+u\tilde{\pi}) + u\,\tilde{\Lambda}\tilde{\Lambda}'} \;, \qquad (2.12)$$

$$G_{\rho} = \frac{\tilde{\pi}(G_o^{-1}-\tilde{\Sigma}_{\parallel}) + \tilde{\Lambda}\tilde{\Lambda}'}{(G_o^{-1}-\tilde{\Sigma}_{\parallel})(1+u\tilde{\pi}) + u\,\tilde{\Lambda}\tilde{\Lambda}'} \;. \qquad (2.13)$$

Similarly we can analyse the structure of the order parameter correlation function which can be written, in the spirit of the work by Martin et al.[22], as

$$C_{\parallel}(k,\omega) = \left[\frac{2}{\Gamma^{(o)}} + \Sigma^{(2)}_{\parallel}(k,\omega)\right] |G'_{\parallel}(k,\omega)|^2 \qquad (2.14)$$

where

$$G'^{-1}_{\parallel} = G_o^{-1}-\Sigma_{\parallel}\,, \qquad G_o^{-1} = -\frac{i\omega}{\Gamma^{(o)}} + k^2 \;. \qquad (2.15)$$

Here the self-energy $\Sigma_{\parallel}$ is as above while $\Sigma^{(2)}_{\parallel}$ can graphically be represented as

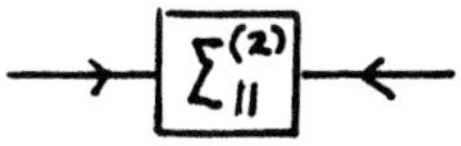

An analysis of the structure of graphs contributing to $\Sigma^{(2)}_{\parallel}$ can be carried out similarly to that of $\Sigma_{\parallel}$ (eq. (2.9)) yielding

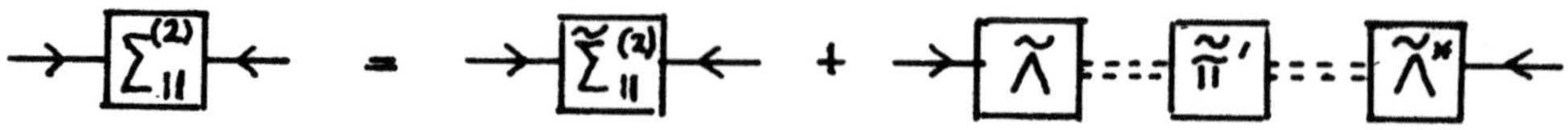

(2.16)

Note that in the case of purely relaxational models Σ' and consequently $\tilde{\Sigma}''$ are zero, so the expression for $G_{\parallel}$ simplifies considerably. Though the correlation function is of the same form in both cases, an essential difference is that, for the purely relaxational model, G' is the true response function and the fluctuation-dissipation theorem[21] leads to the relation

$$\Sigma^{(2)} = \frac{2}{\omega}\,\mathrm{Im}\Sigma \quad . \tag{2.17}$$

After outlining all these rather formal properties of the perturbation series let us see now how the semimacroscopic theory can account for the behaviour of the parallel correlation function discussed on a phenomenological basis in the first section. The form, given in eq. (1.8), for $C_{\parallel}$ means in the language of the diagrams of this section that

$$C_{\parallel} \rightarrow \frac{(n-1)\,\pi^{B}}{2P^{2}} \quad \text{if} \quad k,\omega\rightarrow 0 \tag{2.18}$$

where π^{B} is the contribution of the bubble diagram

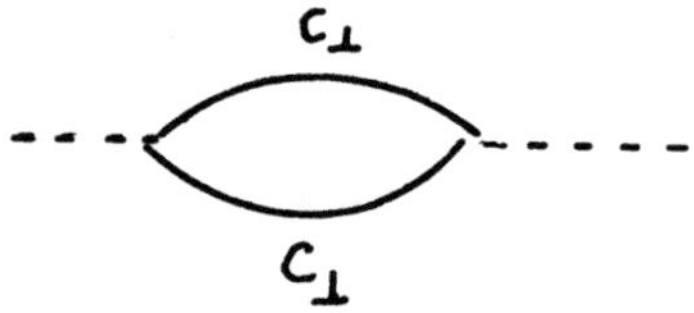

Here $C_{\perp}$ is the fully dressed transverse correlation function. Eq. (2.18) requires that $\frac{2}{\Gamma^{(0)}}$, $\tilde{\Sigma}_{\parallel}$ and $\tilde{\Sigma}_{\parallel}^{(2)}$ be negligible, $\tilde{\Lambda}'\rightarrow 2P$, $\tilde{\Lambda}\rightarrow P$ and

$\tilde{\pi}' \to (n-1)2\tilde{\pi}^B$ in the $k,\omega\to 0$ limit. The proof can be given when some simplifying situation occurs which is the case in the limit $n\to\infty$ or when $d \lesssim 4$. Consider first the large n limit which is well suited to exhibit the desired behaviour since the fluctuations in the magnitude of the order parameter tend to zero in this limit. In case of the relaxational model this limit below T_c has been investigated by Tél[23] (for $2<d<4$), Mazenko[24] (for $2<d<4$, $k=0$) and Mazenko et al.[20] (for $d=3$). In terms of the general structure of diagrams sketched above the $n\to\infty$ limit means enormous simplifications namely $\tilde{\Sigma}_{\parallel}$ and $\tilde{\Sigma}_{\parallel}^{(2)}$ become zero (recall that $\Sigma'_{\parallel}$ and $\tilde{\Sigma}''_{\parallel}$ are anyhow zero in the relaxational model), $\tilde{\Lambda}=P$ and $\tilde{\Lambda}'=2P$. Furthermore $\tilde{\pi}'$ is given exactly by the bubble of two transverse correlation function lines the contribution of which has been calculated by Tél[23] as

$$\tilde{\pi}' = \frac{2}{\omega}\,\mathrm{Im}\,\tilde{\pi}$$

$$\tilde{\pi}(k,\omega) = B_d(1-i\Omega)^{-\varepsilon/2}F\left(\frac{\varepsilon}{2},\ 1-\frac{\varepsilon}{2},\ 2-\frac{\varepsilon}{2};\ \frac{1}{2(1-i\Omega)}\right)k^{-\varepsilon} \qquad (2.19)$$

where $\Omega=\omega/(\Gamma^{(o)}k^2)$. B_d is a numerical constant depending on d and F is the hypergeometric function. Eq. (2.19) shows that $\tilde{\pi}$ is singular in the $k,\omega\to 0$ limit: if $k=0$ $\tilde{\pi}\sim\omega^{-\varepsilon/2}$ and if $\omega=0$ $\tilde{\pi}\to k^{-\varepsilon}$ ($\varepsilon=4-d$). The unperturbed propagator in $G'_{\parallel}$ (eq. (2.15)) and $\frac{2}{\Gamma^{(o)}}$ in the numerator of $C_{\parallel}$ (eq. (2.14)) become negligible in this limit. From all these the behaviour we wanted to verify follows immediately.

The relaxational model has also been investigated near $d=4$ to one loop order[24] and also to two-loop order[25]. The results can be interpreted again as supporting the validity of eq. (1.8) for $C_{\parallel}$. Schäfer[25] used field theoretic renormalised perturbation expansion and introduced in that framework a similar structure of the parallel response function of the order parameter as discussed here. He has pointed out that Ward identities are essential in analysing e.g. the quantity corresponding to our $\tilde{\Sigma}_{\parallel}$. Schäfer's theory clearly demonstrates the importance of the new blocks replacing the self-energies, since in a consistent perturbation expansion these are the quantities to be calculated to the desired loop order

Including mode-coupling terms makes an interesting change regarding the leading term in $G_{\parallel}$ (eq. (2.12)). Namely, $\tilde{\pi}$ remains finite in the small k,ω limit in this case, and the leading contribution comes from $\Sigma''_{\parallel}(k,\omega)$ (Sasvári and Szépfalusy, to be published). The conclusion for $C_{\parallel}$ remains, of course, the same.

Finally, we note that the structures studied for $G_{\parallel}$ and G_{ρ} are not restricted to isotropic multicomponent systems. The eqs. (2.12) and (2.13) (with $\Sigma'_{\parallel} = \Sigma''_{\parallel} = 0$, corresponding to the relaxational model) maintain the same form for n=1.

More generally, when a discrete symmetry is present in case $n \geq 2$, $G_{\parallel}$ can again be analysed in a similar manner[26]. It applies even for models suitable to describe the dynamics of displacive structural phase transitions in which case an irrelevant variable, the momentum conjugate to the displacement, is included since it plays an essential role. On this basis important features of the order parameter dynamics below T_c at the antiferrodistortive transition of $SrTiO_3$ [27] can be understood[26].

3. Longitudinal magnetisation correlation function in the antiferromagnet

Besides the order parameter, another interesting quantity in an antiferromagnet is the total magnetisation. We have seen that the transverse fluctuations of the magnetisation are linearly coupled to the transverse order parameter fluctuations and, as a result, they share the same excitation spectra, namely, the spin waves. For symmetry reasons the longitudinal magnetisation, or more generally in the n-component model, the $L_{\alpha\beta}$ fields with α and β both larger than unity, obey equations of the type:

$$\frac{d}{dt} L_{\alpha\beta k} = - \lambda_{||}(k) k^2 L_{\alpha\beta k} \tag{3.1}$$

Phenomenologically one would expect hydrodynamic wave vector dependence, i.e. normal diffusive behaviour with a $\lambda_{||}$ which is constant in the $k \to 0$ limit. However, it has been found in spin wave theory[28] and in mode coupling calculations[29,12] that the transport coefficient $\lambda_{||}$ is not regular in the limit $k \to 0$:

$$\lambda_{||}(k) \approx k^{-\varepsilon/2} \qquad \varepsilon = 4-d \quad . \tag{3.2}$$

The process responsible for this behaviour is the decay of the longitudinal magnetisation into two propagating transverse order parameter fluctuations:

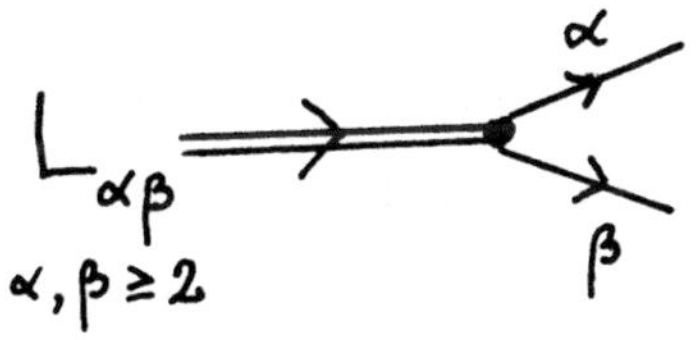

which induces an effective long-range interaction between the longitudinal fluctuations of the magnetisation. As contrasted to the singular behaviour in the longitudinal order parameter correlation function, this effect is purely dynamical since such a coupling leading to the singular behaviour of $\lambda_{||}(k)$ is brought in through the reversible mode-coupling terms.

The correlation function of the $L_{\alpha\beta}$ field is obtained as[17]

$$C_{||}(L_{\alpha\beta};k,\omega) = \frac{\frac{2}{\lambda^{(o)}k^2} + M_2^{(2)}(k,\omega)}{\left|1 - \frac{i\omega}{\lambda^{(o)}k^2} - M_2(k,\omega)\right|^2} \tag{3.3}$$

where $\alpha,\beta \geqslant 2$, $\lambda^{(o)}$ is the bare transport coefficient and the self-energies M_2 and $M_2^{(2)}$ (the notation $M_{\alpha\beta} = M_2$ for $\alpha,\beta \geqslant 2$ is used) are given diagrammatically as

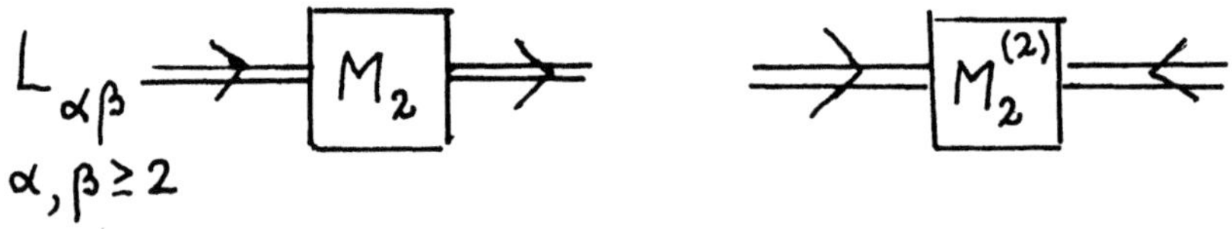

Since the free energy is quadratic in the L field it follows from the fluctuation-dissipation theorem of Deker and Haake[30] that

$$M_2^{(2)}(k,\omega) = -\frac{2}{\lambda^{(o)}k^2}\,\mathrm{Re}\,M_2(k,\omega) \quad . \tag{3.4}$$

The physical transport coefficient $\lambda_{||}(k)$ can be expressed in terms of the self-energy $M_2^{(2)}$ as follows

$$\frac{\lambda_{||}(k)}{\lambda^{(o)}} = 1 + \frac{\lambda^{(o)}k^2}{2} M_2^{(2)}(k,-i\lambda_{||}(k)k^2) \quad . \tag{3.5}$$

Recently there have been renormalisation group investigations of the problem, too[17,20]. Carrying out an $O(\varepsilon)$ calculation the non-hydrodynamic behaviour has been manifested in the sense that, in the limit $k\xi \ll 1$, terms proportional to $\ln k$ as well as to $\ln \xi$ appeared in the expression for $\lambda_{||}$. Though it has provided evidence for a singular behaviour in k, the form to which these logarithms should be exponentiated has not followed from the calculation. A plausible way to resolve the ambiguity of the exponentiation is to calculate the contribution of the same type of diagram which gives the leading term in ε in d dimensions. This implies that the bare coefficients, contained in the functions describing the transverse order parameter fluctuations in the intermediate states, are to be replaced by the corresponding renormalised ones. Such a procedure has been adopted in ref. 17 and the calculation has been carried out near four dimensions. Recently Sasvári and the present author[31] extended the calculation to dimensions $2<d<4$ whose results I am going to present here. The diagram to be evaluated is:

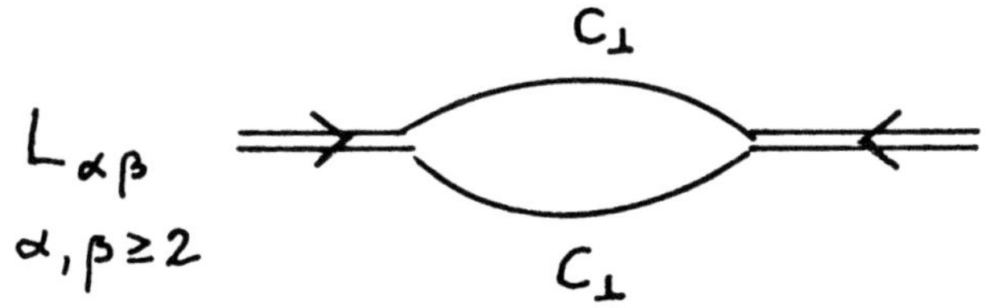

$$M_2^{(2)}(k,\omega) = \frac{g_R^2}{(\lambda^{(o)}k^2)^2}\int\frac{d^d}{(2\pi)^d}\int\frac{d\nu}{2\pi}\left[(k+q)^2-q^2\right]^2 \cdot$$

$$\cdot\; C_\perp(q,\nu)\,C_\perp(k+q,\omega+\nu) \tag{3.6}$$

where, to get the leading term for small k,ω, $C_\perp$ can be substituted by the hydrodynamic expression

$$C_\perp(k,\omega) = \frac{2}{\omega}\,\mathrm{Im}\,G_\perp(k,\omega) \;,$$

$$G_\perp(k,\omega) = \frac{P^2}{\rho_s}\,\frac{\Gamma(\lambda k^2-i\omega)+g^2\rho_s}{g^2\rho_s k^2+\Gamma\lambda k^4-i\omega(\Gamma+\lambda)k^2-\omega^2} \;. \tag{3.7}$$

In eq. (3.6) g_R denotes the renormalised mode-coupling constant. Its renormalisation has to be done in the spirit of mode-coupling theory, i.e., by substituting the true static transverse susceptibility when calculating the mode-coupling vertex[29,12*].

---

*The author is indebted to Professor K. Kawasaki for a discussion on this point.

It yields:

$$g_R = \left(\frac{p^2}{\rho_s}\right)^{-1} g \tag{3.8}$$

where, as before, g is the bare mode-coupling constant which can be used to fix the time scale. We choose the units so that $\chi_L=1$ and correspondingly $c^2=g^2\rho_s$, according to eq. (1.26).

From eqs. (3.5), (3.6), (3.7) and (3.8) we get for $k\xi<<1$

$$\lambda_{\parallel}(k) = \lambda_{\parallel}^{sing}(k) + \lambda_{\parallel}^{reg}(k) \tag{3.9}$$

where

$$\lambda_{\parallel}^{sing}(k) = S_d \frac{2g^2}{D} \left(\frac{2ck}{D}\right)^{-\varepsilon/2} , \tag{3.10}$$

$$S_d = 2^{-\frac{d+2}{2}} K_{d-1} \frac{B\left(\frac{d-1}{2}, \frac{d+2}{4}\right)}{\sin\left(\frac{\pi}{4}d\right)} . \tag{3.11}$$

The result for $\lambda_{\parallel}^{sing}(k)$ can appropriately be discussed by introducing the universal amplitude ratio

$$R \equiv \frac{\lambda_{\parallel}^{sing}(k)\,\xi_{\phi}^{-2}}{c\;\xi_{\phi}^{-1}} (k\xi_{\phi})^{\varepsilon/2} = S_d R_2^{-\frac{d-2}{2}} \tag{3.12}$$

where ξ_{ϕ} is the coherence length, eq. (1.11), and R_2 is defined by eq. (2.5). $\lambda_{\parallel}^{reg}$, the non-singular part of $\lambda_{\parallel}$, has been calculated only to $O(\varepsilon)$[17]

$$\frac{\lambda_{\parallel}^{reg}}{\lambda_{\parallel}^{sing}} = \frac{2n-1}{2n-3} (k\xi)^{\varepsilon/2} . \tag{3.13}$$

The power $k^{-\varepsilon/2}$ in $\lambda_{\parallel}^{sing}$ agrees, of course, with the result of the mode-coupling calculations. Taking the limit $\varepsilon\to 0$ the expression for $\lambda_{\parallel}$, eq. (3.10), matches the result of the ε-expansion[17,20]. It can also be used directly in three dimensions to obtain the amplitude ratio in the physically interesting case. Eq. (3.12) yields

$$R = \frac{1}{10\pi} R_2^{-1/2} \; ; \qquad d=3 \tag{3.14}$$

The magnetisation correlation function has been measured recently by neutron scattering in $RbMnF_3$ below the Néel point[32]. Since $RbMnF_3$ has an extremely small exchange anisotropy it is a very good example for an isotropic antiferromagnet. The experimenters have found a central peak with a width proportional to k^x, $x=1,45 \pm 0,12$ which, within experimental error, is in agreement with the theoretical prediction $\lambda_{\parallel}(k)k^2 \sim k^{3/2}$ (for d=3). A universal amplitude ratio A_o, analogous to R, eq. (3.12), but with ξ_- replacing ξ_ϕ (ξ_- is the coherence length for the fluctuations in the magnitude of the order parameter below T_c), has also been extracted from the measurement:

$$A_o = \begin{cases} 0,17 \text{ for } \xi_- = 3,2\,\mathring{A} \\ 0,22 \text{ for } \xi_- = 1,9\,\mathring{A} \end{cases} \tag{3.15}$$

which gives an experimental value for R as follows

$$R_{exp} = A_o \left(\frac{\xi_-}{\xi_\phi}\right)^{1/2} = 0,11 \tag{3.16}$$

where $\xi_\phi = 7,42\,\mathring{A}$ has been taken for $RbMnF_3$, using the expression for ξ_ϕ given by Hohenberg et al.[13], at the temperature corresponding to the experimental situation.

To achieve a comparison with the theoretical value, eq. (3.14), one needs the amplitude ratio R_2. It has been calculated in a self-consistent way in three dimensions for n=2 by Hohenberg et al.[19]. Their result is $R_2 = 0.09$. Looking at the expression for R_2 (eq.(2.6)) in leading order in ε one sees that its n-dependence is weak. If this property is expected to hold even in three dimensions one can take approximately the above value for R_2. It is also possible to extract an approximate value for R_2 from the measurement of Horn et al.[32]. Namely, from the measured spin-wave peaks in Fig. 1 of ref. 32 one can read off a value for R_2 which turns out to be practically in agreement with the above theoretical result. The experimental value of R_2 has to be taken with some care, however, because - as emphasized by the above experimentalists - the resolution broadening influences the sharp spin-wave peaks considerably. If, nevertheless, we substitue the value $R_2 \approx 0{,}09$ into eq. (3.14), the obtained theoretical value for R practically agrees with the measured one (eq. (3.16)). To understand this striking agreement a number of problems should be clarified. Firstly, one can put forward some convincing arguments that the corrections to the expression of $\lambda_{\parallel}$, as given by eq. (3.9), are negligible in the $k \to 0$ limit[31]. Even so, however, there still remains to be explained why a correction term besides the leading singular one has not been seen in the experiment, although, according to the ε-expansion result (see eq. (3.13)), one would expect a considerable regular term in the region where the measurement has been done.

Moreover, a problem in dynamics like the present one cannot be settled before resolving such contradictions in statics as, e.g., the difference in the critical temperature-dependences of the coherence lengths above and below T_c[13].

4. Longitudinal magnetisation correlation function of isotropic ferromagnets

The transverse and longitudinal fluctuations of the order parameter in the Heisenberg ferromagnet below the Curie point has been studied, by applying renormalisation group techniques, first by Ma and Mazenko[21]. The transverse components are fluctuating as spin wave-modes while the longitudinal one is relaxing to its equilibrium value. In this lecture I want to discuss this relaxational process which exhibits interesting singular behavior.

Let us write the relaxional rate in the usual form

$$\omega_k = \frac{\Gamma_{\parallel}(k)}{\chi_{\parallel}(k)} k^2 \tag{4.1}$$

where $\chi_{\parallel}$ is the longitudinal static susceptibility, and $\Gamma_{\parallel}$ is the corresponding transport coefficient.

A peculiarity of the Heisenberg ferromagnet is that only its static properties follow the conventional theory for d>4, while for its dynamical ones the critical dimensionality $d_c = 6$ [33]. For dimensions d>4 two coherence lengths play a role, namely the transverse one (agreeing with ξ_ϕ as defined by eq. (1.11), now with $\rho_s = P^2$) and the longitudinal one given as

$$\xi_{\parallel} = (u\, P^2)^{1/2} \tag{4.2}$$

where u is the quartic interaction. The latter one enters $\chi_{\parallel}$, having an Ornstein-Zernicke form

$$\chi_{\parallel}^{-1} = 2\xi_{\parallel}^{-2} + k^2 \tag{4.3}$$

The characteristic frequency is expected to have a generalised scaling form [34]

$$\omega_k = k^z\, \Omega(k\xi_{\perp}, k\xi_{\parallel}) \tag{4.4}$$

with dynamical critical exponent $z = 1 + \frac{d}{2}$. [10]

The main contribution to the transport coefficient $\Gamma_{\parallel}$ is coming from the interaction of the longitudinal magnetization with spin-wave modes. Ma and Mazenko [21] have carried out an $O(\varepsilon)$ calculation of $\Gamma_{\parallel}$ ($\varepsilon = 6-d$) and found a singular wave-number dependence of $\Gamma_{\parallel}$ in the hydrodynamic region. They have exponentiated the singular part of $\Gamma_{\parallel}(k)$ into the following form

$$\Gamma_{\parallel}(k) = \Gamma^{(o)} (\Lambda\xi_{\perp})^{\varepsilon/2}(k\xi_{\perp})^{-\varepsilon/6} \; ; \qquad k\xi_{\perp} << 1 \quad . \tag{4.5}$$

Here Λ is the cut-off wave-number, and the formula has been written in the notation of the present paper.

How to exponentiate the ε-expansion result is, however, not unique in this case, similarly to that discussed in the previous section. Sasvári [35] reconsidered the problem by adopting the procedure used in the O(n) symmetric model [17]. In this way he has found a result different from that of Ma and Mazenko, namely

$$\Gamma_{\parallel}(k) = \frac{\Gamma^{(o)}}{2} (\xi_{\perp}\Lambda)^{\varepsilon/2}\left[(k\xi_{\perp})^{-\varepsilon/3} + 1\right], \; k\xi_{\perp} << 1 \quad . \tag{4.6}$$

After substituting eqs.(4.1), (4.3) and (4.6) into eq. (4.4) the scaling function of the relaxational rate of the longitudinal magnetisation to $O(\varepsilon)$ is obtained [35] as

$$\Omega(k\xi_{\perp}, k\xi_{\parallel}) = \frac{\Gamma^{(o)}}{2}\Lambda^{3/2} \frac{(\xi_{\perp}k)^{\varepsilon/2}}{(\xi_{\parallel}k)^{2}} \left[(k\xi_{\perp})^{-\varepsilon/3} + 1\right] \quad . \tag{4.7}$$

Finally, it is worth mentioning that $\Gamma_{\parallel}(k)$, as given by eq. (4.6), predicts in three dimensions a singular k-dependence, k^{-1}, which agrees with the mode-coupling result in d = 3 by Schwabl and Michel [36] and Schwabl [37]. This is not surprising since the diagram which had to be evaluated near 6 dimensions is at the same time the leading mode-coupling diagram.

Acknowledgments

This work was supported in part by Sonderforschungsbereich 130 "Ferroelektrika".
Discussions with G. Meissner, N. Menyhárd and L. Sasvári are gratefully acknowledged. The author is indebted to N. Menyhárd also for critically reading the manuscript.

References

1. B.I. Halperin and P.C. Hohenberg, Phys. Rev. 188, 898 (1969)
2. A.Z. Patashinskii and V.L. Pokrovskii, Zsetf 64, 1445 (1973) (JETP 37, 733, 1973)
3. K. Kawasaki and H. Mori, Prog. Theor. Phys. 25, 1043 (1961)
4. E. Brézin, D.J. Wallace, and K.G. Wilson, Phys. Rev. B7, 232 (1973)
5. E. Brézin and D.J. Wallace, Phys. Rev. B7, 1967 (1973)
6. D.J. Wallace and R.K.P. Zia, Phys. Rev. B12, 5340 (1975)
7. D.R. Nelson, Phys. Rev. B13, 2222 (1976)
8. L. Schäfer and H. Horner, Z. Physik B29, 251 (1978)
9. R.A. Ferrell, N. Menyhárd, H. Schmidt, F. Schwabl and P. Szépfalusy, Annals of Phys. 47, 565 (1968)
10. B.I. Halperin and P.C. Hohenberg, Phys. Rev. 177, 952 (1969)
11. P.C. Hohenberg and B.I. Halperin, Rev. Mod. Phys. 49, 435 (1977)
12. L. Sasvári, F. Schwabl and P. Szépfalusy, Physica 81A, 108 (1975)
13. P.C. Hohenberg, A. Aharony, B.I. Halperin and E.D. Siggia, Phys. Rev. B13, 2986 (1976)
14. K. Kawasaki, in Phase Transitions and Critical Phenomena, ed. C. Domb and M.S. Green (Academic, New York, 1976), Vol. 5a
15. K. Kawasaki and J.D. Gunton, Critical Dynamics, Progress in Liquid Physics, ed. C.A. Croxton (Wiley, New York, 1976) Chap. 5
16. L. Sasvári and P. Szépfalusy, Physica 87A, 1 (1977)
17. L. Sasvári and P. Szépfalusy, Physica 90A, 626 (1978)
18. E.D. Siggia, Phys. Rev. B13, 3218 (1976)
19. P.C. Hohenberg, E.D. Siggia, B.I. Halperin, Phys. Rev. B14, 2865 (1976)
20. G.F. Mazenko, M.J. Nolan, R. Freedman, Phys. Rev. B18, 2281 (1978)
21. S.Ma and G.F. Mazenko, Phys. Rev. B11, 4077 (1975)
22. P.C. Martin, E.D. Siggia and H.A. Rose, Phys. Rev. A8, 423 (1973)
23. T. Tél, Diploma work, 1975 (Roland Eötvös, University, Budapest)
24. G.F. Mazenko, Phys. Rev. B14, 3933 (1976)
25. L. Schäfer, Z. Physik B31, 289 (1978)
26. G. Meissner and P. Szépfalusy, to be published
27. E.F. Steigmeier and H. Auderset, Solid State Commun. 12, 565 (1973)
28. J. Villain, Solid State Commun. 8, 31 (1970)
29. K.H. Michel and F. Schwabl, Z. Phys. 240, 354 (1970)
30. U. Deker and F. Haake, Phys. Rev. A11, 2043 (1975)
31. L. Sasvári and P. Szépfalusy, to be published
32. P.M. Horn, J.M. Hastings and L.M. Corliss, Phys. Rev. Lett. 40, 126 (1978)
33. K. Kawasaki, Ann. Phys. 61, 1 (1970)
34. P.C. Hohenberg, M. DeLeener and R. Resibois, Physica 65, 505 (1973)
35. L. Sasvári, J. Phys. C10, L633, (1977)
36. F. Schwabl and K.H. Michel, Phys. Rev. B2, 189 (1970)
37. F. Schwabl, Z. Physik, 246, 13 (1971)

THE MIGDAL APPROXIMATION AND OTHER NEW METHODS IN THE REAL-SPACE RENORMALIZATION GROUP APPROACH TO CRITICAL DYNAMICS

M. Suzuki

Department of Physics, University of Tokyo
Hongo, Bunkyo-ku, Tokyo, Japan

The Migdal Approximation and Other New Methods in the Real-Space Renormalization Group Approach to Critical Dynamics

Masuo SUZUKI
Department of Physics
University of Tokyo
Hongo, Bunkyo-ku
Tokyo, Japan

Abstract

A historical review of linear and nonlinear critical slowing down is briefly given for the kinetic Ising model. Several recent papers on the real-space RG approach to critical dynamics are also reviewed briefly. The main part of this paper is to present the fundamental idea of the real-space RG approach and to explain explicitly the Migdal approach to critical dynamics.

1. Linear and nonlinear critical slowing down

In this section we review briefly some general aspects of critical slowing down. The instability of the relevant system near the critical point causes an anomalously slow relaxation of the order parameter, as was discussed phenomenologically by van Hove. That is, the relaxation time τ is proportional to the susceptibility $\chi \sim (T - T_c)^{\gamma}$ in his theory. It is well-known that this is not necessarily true even in the simple kinetic Ising model.[1),2)] Our problem is to study how the true critical singularity of slowing down deviates from van Hove's thermodynamic singularity.

For this purpose we define[3)] the linear relaxation time $\tau_A^{(\ell)}$ for an arbitrary hermitian physical quantity A by

$$\tau_A^{(\ell)} = \int_0^{\beta} \Phi_{AA}(t)/\Phi_{AA}(0)\ dt \tag{1.1}$$

where $\Phi_{AA}(t)$ is the relaxation function defined by[4)]

$$\Phi_{AA}(t) = \beta\{(A, A(t)) - \lim_{t\to\infty}(A, A(t)\} \tag{1.2}$$

and (B, A) denotes the following canonical correlation[4)]

$$(B,\ A) = \frac{1}{\beta}\int_0^{\beta} \langle e^{\lambda\mathcal{H}}\, B\, e^{-\lambda\mathcal{H}} A\rangle\, d\lambda \qquad (1.3)$$

for a quantum Hamiltonian $\mathcal{H}$ and (B, A) = <BA> for a classical system. It is easily shown[3)] that $\tau_A^{(\ell)}$ is positive definite for an hermitian operator A. Now the critical exponent of linear slowing down, $\Delta_A^{(\ell)}$, is defined by[3),5),6)]

$$\tau_A^{(\ell)} \sim (T - T_c)^{-\Delta_A^{(\ell)}} \quad . \qquad (1.4)$$

Sometimes $\Delta_A^{(\ell)}$ is denoted simply by Δ. Similarly we can discuss the nonlinear relaxation time $\tau_A^{(n\ell)}$ defined by[7),8)]

$$\tau_A^{(n\ell)} = \int_0^{\infty} (A(t) - \langle A\rangle)/(A(0) - \langle A\rangle)\ dt. \qquad (1.5)$$

Correspondingly the nonlinear critical slowing down exponent $\Delta_A^{(n\ell)}$ is defined[7)] by the following asymptotic form near T_c

$$\tau_A^{(n\ell)} \sim (T - T_c)^{-\Delta_A^{(n\ell)}} \quad . \qquad (1.6)$$

It was first found[7),8)] by the present author that there exists an example in which $\tau^{(n\ell)} \neq \tau^{(\ell)}$. In fact, in the linear anisotropic XY-model, which has been solved exactly, the relaxation time in the nonlinear response takes[7)] a singularity of the form $\tau^{(n\ell)} \sim |H - H_c|^{-1/2}$; $\Delta^{(n\ell)} = \frac{1}{2}$ near the critical field H_c at zero temperature, while we have $\tau^{(\ell)} \sim |H - H_c|^{-1}$; $\Delta^{(\ell)} = 1$. Later, Rácz[9)] found similar results in the Weiss kinetic Ising model using the molecular field expression of the magnetization obtained by Suzuki and Kubo.[2)] Being stimulated by these facts Rácz[10)] and Fisher[11)] found the following general scaling relation

$$\Delta_A^{(n\ell)} = \Delta_A^{(\ell)} - \beta_A \qquad (1.7)$$

where β_A is the exponent characterizing the scaling of A with respect to temperature. This scaling relation (1.7) has been derived[10),11),12)] under the assumptions that dynamic scaling can be extended to nonlinear phenomena and that the extent of the linear regime for a physical quantity A is scaled with the static exponent β_A, as was emphasized by Csépes and Rácz.[13)] The first assumption has been proved by Bausch and Janssen[14)] and by Suzuki[15)]. The above scaling relation (1.7) has also been confirmed by several authors[13),16)∿19)] using the high temperature expansion method for nonlinear critical slowing down formulated first by the present author.[7] The scaling law (1.7) has been proved for some other exactly soluble models[20),21] than those mentioned above.[7),9]

According to van Hove's theory we have $\Delta_M^{(\ell)} = \gamma$ for the relaxation of magnetization M in the kinetic Ising model. We study the difference $(\Delta - \gamma)$ in the succeeding sections.

2. Derivation of the dynamical scaling law by means of Kadanoff's cell analysis and coarse graining of time

The essential idea of the dynamic real-space RGA comes from Kadanoff's cell analysis[22] and from coarse graining of time proposed by Suzuki[23] to derive the dynamic scaling law. It will be instructive to summarize here the main derivation[23] of the dynamic scaling law by means of Kadanoff's cell analysis and coarse graining of time.

For convenience we study the following generalized Ginzburg-Landau model:

$$\frac{\partial s(\mathbf{r})}{\partial t} = -\left(\frac{\Gamma}{k_B T}\right)\frac{\delta \mathcal{H}}{\delta s(\mathbf{r})} + \eta(\mathbf{r}, t) \qquad (2.1)$$

where $\mathcal{H}$ denotes the Hamiltonian of the system and $\eta(\mathbf{r}, t)$ is a Gaussian Markoffian random force with the correlation function

$$\langle \eta(\mathbf{r}, t)\eta(\mathbf{r}', t')\rangle = 2\Gamma\delta(\mathbf{r} - \mathbf{r}')\delta(t - t') \quad . \qquad (2.2)$$

Following Kadanoff [22] we divide the system into cells of length L which is microscopically large but much smaller than the correlation length $\xi(T)$. Then, the scale of length R and reduced variables ε and h are changed as

$$R \rightarrow R' = R/L \;,\; \varepsilon \rightarrow \varepsilon' = \varepsilon L^y \;,\quad h \rightarrow h' = hL^x \qquad (2.3)$$

where $\varepsilon = (T - T_c)/T_c$, $\mu_B H/k_B T = h$ and $R = |\mathbf{r} - \mathbf{r}'|$. Furthermore, according to Kadanoff [22] the spin variable $s(\mathbf{r})$ is scaled as

$$s(\mathbf{r}) \rightarrow s'(\mathbf{r}) = L^{d-x}s(\mathbf{r}); \quad x = d - \beta/\nu \qquad (2.4)$$

with critical exponents β and ν. One of the key-points in <u>dynamic</u> RGA is to divide time region into cells of time length L^z in an appropriate time unit, namely to perform coarse graining of time, as shown in Fig. 1. Here the dynamic critical exponent z has to be determined so that the newly transformed equation of motion corresponding to (2.1) may become again Markoffian and of the same form as (2.1). Otherwise it

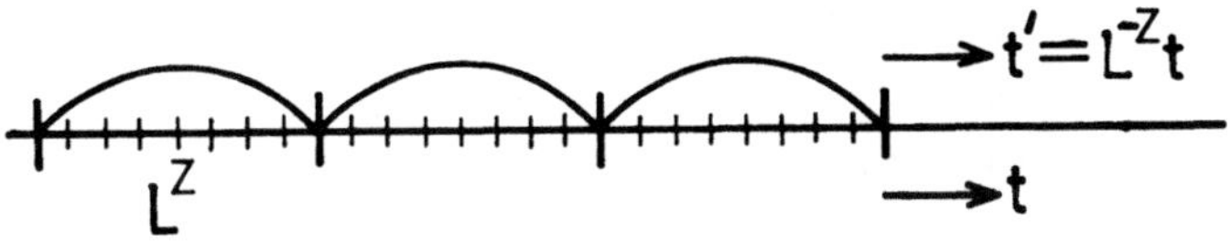

Fig. 1. Coarse graining of time; the time t is divided into cells of a time length L^z according to $t' = L^{-z}t$.

will become non-Markoffian. That is, the time t is changed as

$$t \to t' = L^{-z}t \quad \text{or} \quad t = L^z t' \quad . \tag{2.5}$$

Correspondingly the effective strength of heat-bath interaction Γ and the random force $\eta(\mathbf{r}, t)$ are changed as

$$\Gamma \to \Gamma' = L^{\varphi}\Gamma \quad \text{and} \quad \eta \to \eta' = L^{x+\varphi}\eta \quad . \tag{2.6}$$

Here the critical exponent φ can be also determined from the requirement that the stochastic equation of motion (2.1) should be invariant after the scale transformations (2.4) $\sim$ (2.6). This gives the following relation

$$z = 2x - d + \varphi = \gamma/\nu + \varphi = 2 - \eta + \varphi \quad . \tag{2.7}$$

The above invariance of the equation of motion under the scale transformations yields the following scaling property of the Fourier component of the time correlation $\langle s(0, 0)\, s(\mathbf{r}, t)\rangle \equiv S(R, \varepsilon, h; t)$:

$$S(R, \varepsilon, h; \omega) = L^{2(x-d)+z} S(R/L, L^y\varepsilon, L^x h, L^z\omega); \tag{2.8}$$

The solution of this equation (2.8) takes the following scaling form

$$S(R, \varepsilon, h; \omega) = R^{2(x-d)+z} S(R|\varepsilon|^{1/y}, \varepsilon/|h|^{y/x}, \omega R^z). \tag{2.9}$$

The above general argument suggests an explicit evaluation method of the critical exponent z or φ, by making a cell analysis and coarse graining of time explicitly. In the above arguments both, the deterministic part and fluctuating part in (2.1) are assumed to have the same scaling property, as it should be for critical phenomena. Recently Mori[24] extended the scaling arguments on equations of motion to a more general situation in non-equilibrium systems by admitting the possibility that the deterministic part and fluctuating part are not necessarily scaled by the same scale factor. It should be remarked that quite recently the present author[25] proposed a more generalized nonlinear scaling theory of transient phenomena near the instability point in order to clarify the essential feature of nonlinear fluctuation and onset of macroscopic structure which are very sensitive to initial conditions near the instability point.

The relation $\Delta = \nu z$ is easily shown in the succeeding section where ν is the exponent of the correlation length ξ (i.e., $\xi \sim (T-T_c)^{-\nu}$). Thus, the evaluation of z leads to the determination of the critical exponent Δ.

3. Fundamental formulation of the real-space renormalization group approach to critical dynamics

We start with the following general master equation

$$\frac{\partial}{\partial t}P(\{s(\mathbf{r})\}, t) = \mathcal{L} P(\{s(\mathbf{r})\}, t) \tag{3.1}$$

where P denotes the probability distribution function at time t and $\mathcal{L}$ the temporal evolution operator of the system. If $\mathcal{R}$ denotes an operation to eliminate degrees of freedom in each cell and to eliminate rapid motions corresponding to coarse graining of time, then we have

$$\mathcal{R} P(\{s(\mathbf{r})\}, t) = P'(\{s'(\mathbf{r}')\}, t'), \tag{3.2}$$

or more explicitly we may write

$$\mathcal{R} P(t) = \mathcal{R} (e^{t\mathcal{L}}P(0)) = e^{t'\mathcal{L}'}P'(0). \tag{3.3}$$

The scale transformation of P(0) into P'(0) gives the static renormalization group approach (RGA), and that of $t\mathcal{L}$ into $t'\mathcal{L}'$ yields the dynamic RGA. If K denotes a typical strength of interaction in the system, then the above RGA gives the following general recursion formulae

$$K' = f(K) \text{ and } t' = g(K, t) \simeq t\, g(K) . \tag{3.4}$$

The first equation in (3.4) gives the fixed point $K^*(=K_c)$ determined by $K^* = f(K^*)$. The dynamic critical exponent z is determined by

$$t'/t = g(K^*) = L^{-z} \tag{3.5}$$

for the scale factor L, as was discussed in (2.5). That is,

$$z = - \ell n\, g(K^*)/\ell n\, L . \tag{3.6}$$

The characteristic relaxation time[8)] of the system should satisfy the following relation

$$\tau(K') = L^{-z}\tau(K) \tag{3.7}$$

by definition, irrespective of relevant physical quantities. As before we assume that the relaxation time shows the following power law singularity

$$\tau(K) \simeq C(K - K_c)^{-\Delta} \tag{3.8}$$

with a critical exponent Δ. Equations (3.7) and (3.8) lead to[26)]

$$C(K' - K_c)^{-\Delta} \simeq L^{-z} C(K - K_c)^{-\Delta} \ . \tag{3.9}$$

Noting that

$$K' = f(K) = f(K^* + (K - K^*)) = K_c + \Lambda \cdot (K - K_c) + \ldots \ ; \quad \Lambda = (df(K)/dK)_{K=K^*} \tag{3.10}$$

we obtain

$$\Delta = z \,\ell n\, L/\ell n\, \Lambda = \nu z \ ; \ \nu = \ell n\, L/\ell n\, \Lambda . \tag{3.11}$$

The above formula for the correlation length exponent ν is due to Wilson.[27),28)] Thus, it is sufficient to evaluate the value of z in order to study the critical slowing down exponent Δ.

One of the most difficult problems in the real-space RGA is to take a partial elimination of degrees of freedom in (3.2) or (3.3), because the temporal evolution operator $\mathcal{L}$ is a set of noncommutable local operators.

Recently many authors[29)∼31)] have tried to formulate the real-space RG approach. The main idea of these papers is to renormalize the master equation (3.1) instead of (3.3) or to renormalize[30)] approximately the master operator $\mathcal{L}$ itself. That is, we apply $\mathcal{R}$ to (3.1):

$$\mathcal{R}\frac{d}{dt}P(t) = \mathcal{R}\mathcal{L}P(t). \tag{3.12}$$

If we separate the equilibrium part of $P(t)$ as[5),6),39)] $P(t) = P_{eq}\,\varphi(t)$ then we obtain

$$\mathcal{R}\{P_{eq}\frac{d}{dt}\varphi(t)\} = \mathcal{R}\{P_{eq}\hat{\mathcal{L}}\,\varphi(t)\}, \tag{3.13}$$

where

$$\hat{\mathcal{L}} = -\sum_i W_i(\sigma_i)(1 - P_i) \tag{3.14}$$

and

$$P_i f(\ldots\ldots, \sigma_i, \ldots\ldots) = f(\ldots\ldots, -\sigma_i, \ldots\ldots), \tag{3.15}$$

with the transition probability $W_i(\sigma_i)$ defined by[2)]

$$W_i(\sigma_i) = \frac{1}{2}\alpha(1 - \sigma_i \tanh\beta \sum_j J_{ij}\sigma_j) \tag{3.16}$$

or[32),33)]

$$W_i(\sigma_i) = \exp(-\beta\sigma_i \sum_j J_{ij}\sigma_j); \quad \sigma_j = \pm 1. \tag{3.17}$$

The choice of this form of the transition probability has been believed[2] to be irrelevant to the critical behavior, and it is simply a matter of convenience.

Essentially, the renormalization procedure of (3.13) should be based on that of (3.2) or (3.3). However, the formulation of (3.13) seems to be more practical. The latter formulation is apt to give rather intuitive arguments.

In fact, the formulation (3.12) can be easily transformed[35] to the following method of equations of motion: The distribution function $P(\{\sigma_j\}, t)$ can be expanded in an orthogonal complete set as

$$P(\{\sigma_j\}, t) = c(t) + a_1(t)\sigma_1 + a_2(t)\sigma_2 + \ldots\ldots \\ + a_{12}(t)\sigma_1\sigma_2 + a_{13}(t)\sigma_1\sigma_3 + \ldots + a_{123}(t)\sigma_1\sigma_2\sigma_3 + \ldots \quad . \tag{3.18}$$

Clearly, the expansion coefficients $a_{ij\ldots k}(t)$ are nothing but correlation functions $2^{-N}\langle\sigma_i\sigma_j \ldots \sigma_k\rangle_t$ for the N-spin system. Thus, the master equation (3.1) for $P(\{\sigma_j\}, t)$ is easily transformed into the following equations of motion for correlation functions:

$$\frac{d}{dt}A(t) = \tilde{\mathcal{L}}\, A(t) \tag{3.19}$$

where $\tilde{\mathcal{L}}$ is a linear operator corresponding to $\mathcal{L}$, and $A(t)$ denotes a 2^N-dimensional vector whose components are given by $(c(t), a_1(t), a_2(t), \ldots)$:

$$A(t) = (c(t),\ a_1(t),\ a_2(t),\ \ldots.)^{\dagger} \ . \tag{3.20}$$

In order to renormalize irrelevant degrees of freedom, we divide the components of the vector $A(t)$ into two parts by rearranging as

$$A(t) = (A_1(t),\ A_2(t))^{\dagger} \ . \tag{3.21}$$

Here $A_2(t)$ is a subvector whose components are irrelevant correlation functions containing spin variables inside each cell to be eliminated and $A_1(t)$ is the remaining subvector. The real-space RGA requires to eliminate the part $A_2(t)$ in (3.19). This procedure yields the following type of equations of motion

$$\frac{d}{dt}A_1(t) = \tilde{\mathcal{L}}_1 A_1(t) \quad \text{or} \quad \frac{d}{dt'}A'(t') = \tilde{\mathcal{L}}' A'(t'), \tag{3.22}$$

after appropriate coarse graining of time or Markoffian approximation. The requirement that (3.22) should take the same form as the original one (3.19) gives again the desired scale transformation of time

$$t'/t = L^{-z} \ . \tag{3.23}$$

There are several methods to eliminate $A_2(t)$ explicitly, as will be shown later.

If we are satisfied with a very intuitive RG scheme, the formulation (3.12) may also lead to the idea that the renormalization of time could be obtained approximately by a certain transformation of $\mathcal{L}$ into that of block spins, namely by the evaluation of the lowest eigenvalue of the transformed $\mathcal{L}'$ for block spins, as was discussed by Kinzel[30] and by Achiam and Kosterlitz.[33]

4. Dynamic real-space renormalization for the one-dimensional kinetic Ising model.

In order to explain the main idea presented in the preceding section we discuss here a simple example, namely the exactly soluble linear kinetic Ising model[1),2),40)∿42)] with nearest neighbor interaction. Achiam,[31] Deker,[37] Suzuki et al.[35] have studied independently this one-dimensional case. The Hamiltonian of this Ising chain is given by

$$\mathcal{H} = - J\sum_i \sigma_i \sigma_{i+1} \ . \tag{4.1}$$

The transition probability for the corresponding kinetic Ising chain takes the form[1),2)]

$$W_j(\sigma_j) = \frac{1}{2}\alpha\{1 - \gamma\sigma_j(\sigma_{j-1} + \sigma_{j+1})\} \tag{4.2}$$

with

$$\gamma = \frac{1}{2}\tanh(2K) \quad \text{and} \quad K = J/k_B T \ . \tag{4.3}$$

The equations of motion corresponding to (3.19) are given by

$$\frac{d}{dt}m_j(t) = -m_j(t) + \gamma(m_{j-1}(t) + m_{j+1}(t)) \tag{4.4}$$

with $m_j(t) = \langle\sigma_j(t)\rangle$, where we have put $\alpha = 1$ for simplicity. First note that we have

$$m_j = \gamma(m_{j-1} + m_{j+1}) \tag{4.5}$$

in equilibrium. This is a special example of the following well-known relation among correlation functions:

$$\langle\sigma_j\rangle_{eq} = \langle\tanh(K\sum_{\langle i,j\rangle}\sigma_i)\rangle_{eq} \tag{4.6}$$

or more generally[43]

$$\langle\{f\}\sigma_j\rangle_{eq} = \langle\{f\}\tanh(K\sum_{\langle ij\rangle}\sigma_j)\rangle_{eq} \tag{4.7}$$

where $\{f\}$ denotes an arbitrary function of σ_k's at sites other than the j-th. Then one might consider the following adiabatic approximation

$$m_{j+1}(t) = \gamma(m_j(t) + m_{j+2}(t)) \tag{4.8}$$

to be useful to perform the real-space renormalization group calculation. However, this approximation is too crude to give the correct critical value z = 2 in one dimension. In fact, the above rough approximation (4.8) leads to the following renormalized equation of motion

$$\frac{d}{dt}m_j(t) = (1 - 2\gamma^2)\left\{-m_j(t) + \gamma'(m_{j-2}(t) + m_{j+2}(t))\right\}, \tag{4.9}$$

for the scaling factor $\lambda = 2$, where

$$\gamma' = \gamma^2/(1 - 2\gamma^2) . \tag{4.10}$$

This renormalization transformation is equivalent to $\tanh K' = \tanh^2 K$. Even this rough dynamical RG transformation yields correctly the static one discussed by Nelson and Fisher,[44)] Migdal[45)] and Kadanoff,[46)] and the fixed point is given by $\gamma^* = 1/2$. The time scale corresponding to the general formulation (3.23) is obtained from (4.9) as

$$t'/t = 1 - 2(\gamma^*)^2 = \frac{1}{2} = \lambda^{-z} = 2^{-z} \tag{4.11}$$

at the fixed point $\gamma^* = 1/2$. This gives an incorrect value z = 1. This indicates that the above adiabatic approximation (4.8) is not good for the present dynamic RG approach.

In order to overcome this inconsistency, we have to take into account correctly the feed-back effect of time change of neighboring spins. For this purpose we eliminate $m_{j+1}(t)$ and $m_{j-1}(t)$ by multiplying[35)] the both sides of (4.4) by the operator (d/dt+1) and by using the equations of motion for $m_{j+1}(t)$ and $m_{j-1}(t)$ similar to (4.4). Thus, we arrive at

$$(1 + \frac{d}{dt})^2 m_j(t) = \gamma^2(2m_j(t) + m_{j-2}(t) + m_{j+2}(t)). \tag{4.12}$$

That is,

$$\frac{d}{dt}m_j(t) = \frac{1}{2}(1 - 2\gamma^2)\{-m_j(t) + \gamma'(m_{j-2}(t) + m_{j+2}(t)\} + \frac{1}{2}\frac{d^2}{dt^2}m_j(t) \tag{4.13}$$

with γ' defined by (4.10). If we neglect the term of higher derivative in (4.13), which corresponds to the coarse-graining of time, then we obtain the equation of motion of the same form as (4.4):

$$\frac{d}{dt'}m_j(t') = -m_j(t') + \gamma'(m_{j-2}(t') + m_{j+2}(t')) \tag{4.14}$$

with

$$t' = \frac{1}{2}(1 - 2\gamma^2)t \ . \tag{4.15}$$

Equivalently, this result can also be obtained[31),35),37)] by the Laplace transform of (4.4) as

$$(1 + s)\ m_j[s] = \gamma(m_{j-1}[s] + m_{j+1}[s]) \ . \tag{4.16}$$

Thus the time scale is given by

$$t'/t = \lambda^{-z} = 2^{-z} = \frac{1}{2}(1 - 2(\gamma^*)^2) = 2^{-2}. \tag{4.17}$$

Consequently we obtain[31),35),37)] the correct value $z = 2$.

This treatment can be easily extended to a general case of an arbitrary decimation factor λ. The Laplace transformation method (4.16) is the most convenient for this general case. Thus, we obtain a set of equations of motion

$$(1 + s)m_k = \gamma(m_{k-1} + m_{k+1}) \tag{4.18}$$

for $k = -\lambda + 1, -\lambda + 2, \ldots, 0, 1, \ldots, \lambda - 1$. By eliminating $m_{-\lambda+1}$, $\ldots$ m_{-1}, m_1, $\ldots$, $m_{\lambda-1}$ algebraicly and neglecting terms of higher order in s as before we obtain[35)]

$$\frac{d}{dt'}m_0(t') = -m_0(t') + \gamma'(m_{-\lambda} + m_\lambda) \tag{4.19}$$

where

$$\gamma' = 2\gamma^\lambda(D_{\lambda-1} - 2\gamma^2 D_{\lambda-2})^{-1} = 2\gamma^\lambda(\beta^\lambda + \alpha^\lambda)^{-1} \tag{4.20}$$

and

$$\frac{t'}{t} = \frac{D_{\lambda-1} - 2\gamma^2 D_{\lambda-2}}{D_{\lambda-1} + D'_{\lambda-1} - 2\gamma^2 D'_{\lambda-2}} = \frac{(\beta^\lambda + \alpha^\lambda)(\beta - \alpha)}{\lambda(\beta^\lambda - \alpha^\lambda)} \ . \tag{4.21}$$

Here α and β are two roots of the equation that $t^2 - t + \gamma^2 = 0$ and D_λ is a determinant of a $\lambda \times \lambda$ matrix corresponding to the simultaneous equation (4.18) and it is given explicitly by

$$D_\lambda = (\beta^{\lambda+1} - \alpha^{\lambda+1})/(\beta - \alpha). \tag{4.22}$$

It is easily shown that the recursion relation (4.20) is equivalent to $\tanh K' = (\tanh K)^\lambda$. This is the well-known static recursion relation[44)~46)] in one dimension for a general decimation factor λ. The fixed point K^* is given by $K^* = \infty$, i.e., $\alpha = \beta = \frac{1}{2}$. Consequently, the time scaling exponent z is calculated as

$$z = -\lim_{K \to K^*} \frac{\ln(t'/t)}{\ln\lambda} = -\lim_{\beta \to \alpha = 1/2} \frac{\ln(t'/t)}{\ln\lambda} = 2 \ . \tag{4.23}$$

That is, we obtain the correct value $z = 2$ for an arbitrary scaling factor λ in one dimension, as it should be.

The calculation in the present section will be useful as a preparation of the extended Migdal-Kadanoff approximation to be discussed in the succeeding section.

5. Dynamic Migdal-Kadanoff approximation

In this section we extend the idea by Migdal and Kadanoff for the static real-space RG to dynamical critical phenomena. This has been tried already by Suzuki et al.,[35] Chui et al.[36] and also quite recently by Achiam.[32] Here we review first the essential scheme by Suzuki et al.[35] because it is very simple.

The main procedures in our dynamic Migdal approximation (DMA) are the following two steps:

(i) *Potential moving*; we assume that all spins in each cell with length λ behave effectively quite similarly to each other even in critical dynamics, if λ is much less than the correlation length $\xi(T)$. Thus, the lattice of our system is transformed into another lattice with λ times lattice-spacing and with the interaction λK, as shown in Fig. 2.

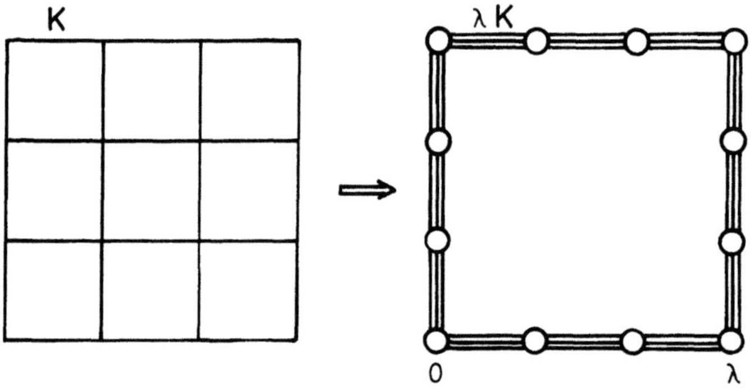

Fig.2a Fig.2b

Fig. 2. Potential moving; (a) the original lattice with the interaction K, (b) the transformed lattice with λ times lattice-spacing and with the interaction λK.

(ii) *Dynamic decimation*; the second step is to eliminate dynamically irrelevant spin variables inside each cell such as $\sigma_1(t)$, $\sigma_2(t)$, ... $\sigma_{\lambda-1}(t)$, as shown in Fig. 3. To perform this procedure we eliminate first spin variables $\sigma_2^1(t)$, ... $\sigma_{\lambda-1}^1(t)$, etc. using equations of motion. From this one-directional dynamic decimation the equation of the average of $\sigma_1^1(t)$ is expressed only by $\sigma_0(t)$ and $\sigma_\lambda^1(t)$. On the other hand, the equation of motion for $\sigma_0(t)$ is coupled with the neighboring four

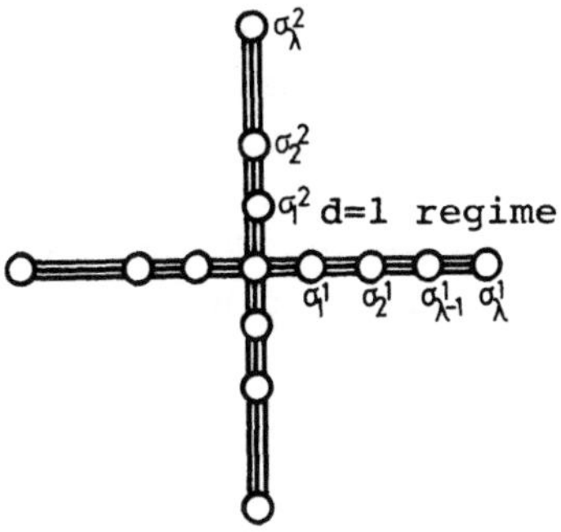

Fig. 3(a)

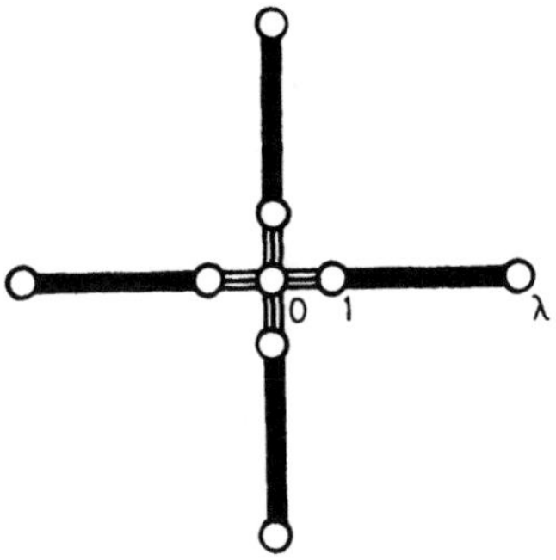

Fig. 3(b)

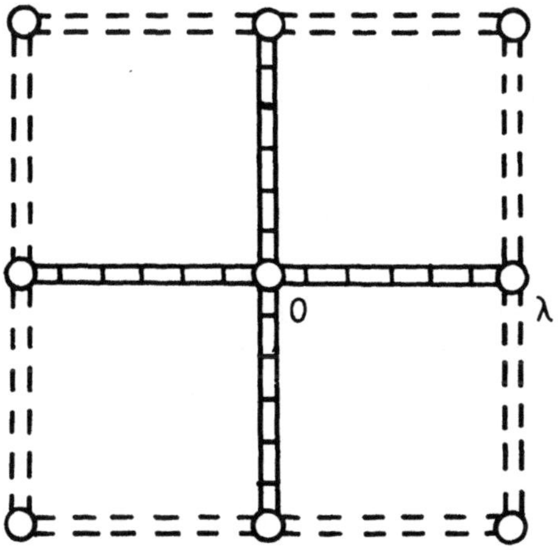

Fig. 3(c)

Fig. 3. Dynamic decimation; (a) a lattice after the potential moving, (b) an intermediate state in which $\sigma_2, \ldots \sigma_{\lambda-1}$ are decimated dynamically, (c) the original moment m_0 is coupled with the new neighboring moments after elimination of the old ones.

spin variables $\sigma_1^1(t)$, $\sigma_1^2(t)$, $\sigma_1^3(t)$ and $\sigma_1^4(t)$. Therefore, by substituting the expressions for these spin variables obtained by the above dynamic decimation into the equation of motion for the central spin $\sigma_0(t)$, we arrive finally at the renormalized kinetic Ising model with a new lattice-spacing ($\propto\lambda$) and with the renormalized coupling constant K' and time vairable t':

$$K' = \mathcal{R}_\lambda(K) \quad \text{and} \quad t' = \mathcal{T}_\lambda(t) \ . \tag{5.1}$$

As discussed generally in §2 the critical temperature $T_c = T^*$, static critical exponent ν and dynamic critical exponent z can be easily calculated from the above recursion formulae (5.1).

Now the kinetic Ising model is described by the master equation (3.1) with the temporal evolution operator $\mathcal{L}$ defined by

$$\mathcal{L} P(\sigma) = -\sum w_j(\sigma_j)P(\sigma) + \sum w_j(-\sigma_j)P(\ldots, -\sigma_j, \ldots) , \tag{5.2}$$

and with the transition probability (3.16). In particular, for the two-dimensional square lattice, the factor $\tanh(K \sum_{<i,j>} \sigma_j)$ in (3.16) takes the form[5),7)]

$$\tanh(K \sum_{j=1}^{4} \sigma_j) = a(K) \sum_{j=1}^{4} \sigma_j + b(K) \sum_{n.n.}^{4 \text{ terms}} \sigma_i\sigma_j\sigma_k \tag{5.3}$$

where

$$a(K) = \frac{1}{8} \tanh(4K) + \frac{1}{4} \tanh(2K)$$

and

$$b(K) = \frac{1}{8} \tanh(4K) - \frac{1}{4} \tanh(2K). \tag{5.4}$$

According to the dynamic Migdal RG procedure proposed above, we can proceed explicitly in the two-dimensional kinetic Ising model as follows.
a) *Decimation in one direction.* From the master equation (3.1) with the one-dimensional transition probability

$$w_j(\sigma_j) = \frac{1}{2}\{1 - \gamma_\lambda\sigma_j(\sigma_{j-1} + \sigma_{j+1})\}; \quad \gamma_\lambda = \frac{1}{2} \tanh(2\lambda K) \tag{5.5}$$

The factor λ in (5.5) has come from the potential moving as explained in Fig. 3. As discussed in §4, the dynamic decimation of spins in one direction can be performed easily by the Laplace transformation method. The equation of motion of the spin σ_1^1 in Fig. 3(a) is thus obtained in the form[35)]

$$D_{\lambda-1}(s)m_1^1 = \gamma_\lambda D_{\lambda-2}(s)m_0 + \gamma_\lambda^{\lambda-1}m_\lambda^1 \tag{5.6}$$

where $m_1^1 = \langle\sigma_1^1\rangle$,

$$D_\lambda(s) = \{\beta(s)^{\lambda+1} - \alpha(s)^{\lambda+1}\}/\{\beta(s) - \alpha(s)\} \tag{5.7}$$

and $\alpha(s)$ and $\beta(s)$ are two roots of the equation that $t^2 - (1+s)t + \gamma_\lambda^2 = 0$. Similar expressions for m_1^i (i=2,3,4) in Fig. 3(a) are obtained.
b) *Renormalized equation of motion*: Now we substitute these expressions for $\{m_1^i\}$ into the equation of motion for the central spin m_0:

$$(1 + s)m_0 = a(\lambda K) \sum_{i=1}^{4} m_1^i + b(\lambda K) \sum_{i=1}^{4} t_1^i \tag{5.8}$$

where $t_1^i \equiv \langle\sigma_1^{i-1}\sigma_1^i\sigma_1^{i+1}\rangle$ with the module $i + 4 \equiv i$. This procedure is shown as the step (b) → (c) in Fig. 3. Thus we arrive at the renormalized equation of motion of the form

$$(1 + s')m_0 = a(K')\sum_{i=1}^{4} m_\lambda^i + b(K')\sum_{i=1}^{4} t_\lambda^i \ , \tag{5.9}$$

up to the order s where

$$s' = \frac{D_{\lambda-1} + D'_{\lambda-1} - 4a_\lambda\gamma_\lambda D'_{\lambda-2}}{D_{\lambda-1} - 4a_\lambda\gamma_\lambda D_{\lambda-2}}\, s \ , \tag{5.10}$$

$$a(K') = a_\lambda\gamma_\lambda^{\lambda-1}/(D_{\lambda-1} - 4a_\lambda\gamma_\lambda D_{\lambda-2}) \ , \tag{5.11}$$

$$b(K') = D_{\lambda-1}b_\lambda/(D_{\lambda-1} - 4a_\lambda\gamma_\lambda D_{\lambda-2}) \tag{5.12}$$

and

$$a_\lambda \equiv a(\lambda K),\ b_\lambda \equiv b(\lambda K),\ D_\lambda = D_\lambda(0),\ D'_\lambda = D'_\lambda(0) \ . \tag{5.13}$$

The recursion equation (5.11) determines the fixed point K* as

$$a(K^*) = \frac{a(\lambda K^*)\{\gamma(2\lambda K^*)\}^{\lambda-1}}{D^*_{\lambda-1} - 4a(\lambda K^*)\gamma(2\lambda K^*)D^*_{\lambda-2}} \tag{5.14}$$

with D^*_λ being the value of D_λ at the fixed point. The thermal exponent $y_T = 1/\nu$ is determined as

$$\lambda^{y_T} = \left[\frac{d}{dK}\mathcal{R}_\lambda(a(K))/\frac{da(K)}{dK}\right]_{K=K^*} \ . \tag{5.15}$$

From the recursion relation (5.10) the dynamic critical exponent z is determined as

$$z = [\ell n(s'/s)]_{K=K^*}/\ell n\lambda . \tag{5.16}$$

Explicit calculations give the following critical values

$$K^* = 0.235,\ z = 1.85,\ \nu^{-1} = 1.31 \text{ for } \lambda = 2,$$

$$K^* = 0.187,\ z = 1.82,\ \nu^{-1} = 1.24 \text{ for } \lambda = 3. \tag{5.17}$$

It is quite interesting to take[35)] here the limit $\lambda \to 1$ of the infinitesimal RG through the analytic continuation of the above results for a general integer λ. For this purpose we study the small deviations of $\mathcal{R}_\lambda(K)$ and $\mathcal{T}_\lambda(t)$ for $\lambda = 1 + \delta\lambda$ as

$$\mathcal{R}_\lambda(a(K)) = a(K) + L(K)\delta\lambda + O[(\delta\lambda)^2] \tag{5.18}$$

and

$$\mathcal{T}_\lambda(t)/t = 1 + M(K)\delta\lambda + O[(\delta\lambda)^2] \ . \tag{5.19}$$

The fixed point in this limit is given by $L(K^*) = 0$ which gives

$K^* = 0.325$. The dynamic critical exponent z for this limit $\lambda \to 1$ is determined as

$$\lambda^{-z} = (1 + \delta\lambda)^{-z} = 1 - z\delta\lambda + \ldots = 1 + M(K^*)\delta\lambda + \ldots \qquad (5.20)$$

and consequently

$$z = -M(K^*) = 1.96. \qquad (5.21)$$

The thermal exponent $y_T = 1/\nu$ is given by

$$y_T = \left(\frac{d}{dK}L(K)/\frac{d}{dK}a(K)\right)_{K=K^*} = 1.51. \qquad (5.21)$$

For the explicit expressions of L(K) and M(K), see ref. 35. The value $z = 1.96$ of the dynamical critical exponent z agrees very well with that obtained in a high temperature expansion method by Yahata and Suzuki,[6] who predicted $z = 2.0 \pm 0.05$ in the present two-dimensional kinetic Ising model. The value of static exponent is not so good, compared with the exact value $\nu = 1$. It should be noted that our results for static physical quantities such as K^* and ν do not agree with those obtained by Migdal and Kadanoff. That is, our method is not a simple straightforward extension of the MK approximation to critical dynamics but it has some new aspects. In fact, we have used only Migdal's fundamental idea i.e., potential moving, and the remaining formulation of our theory is quite dynamical. Therefore, it is not necessary that our static results should agree with those of the MK theory.

If we consider the triplet cluster term Σt^i_λ explicitly in the above arguments on the invariance of equations of motion then we may obtain[35] the value of the magnetic field exponent δ.

Recently, Chui, Forgacs and Frisch[36] and Achiam[32] have studied the same problem in quite similar methods. However, their results are different from ours. Chui et al. have obtained $z = 1.85$ in their first version and $z = 2.064$ in their revised theory.[47] Achiam has obtained $z = 2.2$. It will be interesting to study which is more accurate. However, it is rather difficult at present.

6. Other methods

(i) *Kinzel's method*[30]: He has formulated first a very intuitive ad hoc RG scheme that is, he has assumed that the transition probability

$$w_j(\sigma_j) = \frac{1}{2\tau}(1 - \sigma_j \tanh\beta E_j) \qquad (6.1)$$

with the local energy E_j at the j-th site should be transformed into the same form as (6.1) after renormalization, with a new relaxation time τ'. This τ' was determined as the slowest relaxation time of the temporal evolution operator of block spins. For the triangular lattice Kinzel obtained

$$\tau'/\tau = (1 - \tanh 2K)^{-1} = \frac{1}{2}(e^{4K} + 1). \quad (6.2)$$

Then the dynamic critical exponent z is obtained as

$$z = [\ell n\{(\exp(4K^*) + 1)/2\}]/\ell n\sqrt{3} = 1.776. \quad (6.3)$$

(ii) *Achiam-Kosterlitz method*:[33] In (3.13) the function $\mathcal{G}(\sigma, t)$ may be parametrized in the form

$$\mathcal{G}(\sigma, t) = 1 + h_1(t)\Sigma\sigma_i + h_3(t)\Sigma\sigma_i\sigma_j\sigma_k + \dots \equiv 1 + \vec{O}(\sigma)\cdot\vec{h}(t). \quad (6.4)$$

Equation (3.13) can be written as

$$\frac{d}{dt}\vec{O}(\sigma)\cdot\vec{h}(t) = -\tilde{\mathcal{L}}(\sigma)\vec{O}(\sigma)\cdot\vec{h}(t). \quad (6.5)$$

This is transformed under the RG to the following equation

$$\frac{d}{dt}\vec{O}(\mu)\cdot\Lambda\vec{h}(t) = -\tilde{\mathcal{L}}'(\mu)\vec{O}(\mu)\cdot\Omega\vec{h}(t) \quad (6.6)$$

for renormalized spins μ where Λ and Ω are matrices defined by the RG transformation. Achiam and Kosterlitz have calculated the dynamic critical exponent z through the plausible relation

$$L^z = \lambda_{max}/\omega_{max} \quad (6.7)$$

for the scaling factor L where λ_{max} and ω_{max} are the largest eigenvalues of the matrices Λ and Ω, respectively. In actual calculations they have restricted the Hilbert space of $\vec{O}(\sigma)$, for example, into $\vec{O}(\sigma) = \Sigma\sigma_i$. Thus they have obtained z = 2.19. This theory has double approximations. That is, the relation (6.7) is not justified microscopically and the restriction of $\mathcal{G}(\sigma, t)$ to the above simple form seems to be very severe.

(iii) *Dynamic Niemeijer-van Leeuwen approach*:[35] As a simple example to illustrate our main idea[35] we consider cells shown in Fig. 4. Following Niemeijer and van Leeuwen[48] we assign a new set of spin variables σ'_α, σ'_β, σ'_γ, σ'_δ, etc. to each cell and find approximately equations of motion for these cell variables by eliminating dynamically old spin variables in each cell by the help of the Markoffian approximation. The idea is quite analogous to those by Achiam and Kosterlitz[33] and by Mazenko, Nolan and Valls.[34] The principle of our method is simple but explicit calculations are very tedious. We eliminate spins in a cell by the help of equations of motion as in §3 and §4, in contrast to those by Achiam and Kosterlitz and by Mazenko et al. For details see ref. 35. The results thus obtained are not conclusive at present for the above small cell division.

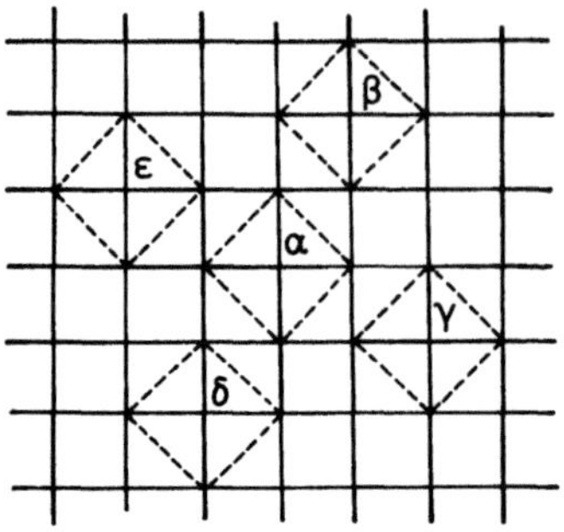

Fig. 4
Division of the two-dimensional square lattice into cells, and renormalized lattice points α, β, γ, δ and ε.

(iv) *Path integral method:*[49),50)] Quite recently Ma[49)] formulated a phenomenological theory of dynamic RG by introducing a discrete time sequence of probability distributions for configurations in the momentum space. He has obtained the ε-expansion of z as $z = 2 + \varepsilon$ up to the first order in ε near one dimension, $\varepsilon = d - 1$. Shiwa[38)] has extended Ma's method to the real-space RG and obtained $z = 2.09 \pm 0.04$ by applying also the variational method to a triangular lattice.

The present author[50)] has also formulated independently a path integral approach to the real space-time RG. The path integral method has been used successfully in many stochastic problems.[51)∿53)]

In order to explain our main idea we discuss first the linear Brownian motion (or Gaussian process) of the following simplest form

$$\frac{d}{dt}x = \gamma x + \eta(t) \quad ; \quad \gamma \gtrless 0 \tag{6.8}$$

where η(t) is a Gaussian white noise satisfying the relation

$$\langle \eta(t)\, \eta(t') \rangle = 2\varepsilon\delta(t - t') \ . \tag{6.9}$$

As in Ma's argument we divide the time into discrete steps : $t = n\Delta t$ $(n=0,1,2,\dots)$ and we specify x at the "time" n by x_n. Then the probability function $P(x_1, x_2 \dots) \equiv P(\{x_n\})$ is expressed by

$$P(\{x_n\}) \propto \exp[-\sum_n (x_{n+1} - x_n - \gamma\Delta t x_n)^2/(2\varepsilon\Delta t)] \ . \tag{6.10}$$

This is zero-dimensional in real space and one-dimensional in time. Although this is an exactly soluble trivial example it is very instructive. We perform here the following real-time RG scheme, namely we integrate $P(\{x_n\})$ over variables $\{x_{2n}\}$ as

$$\int \prod_n dx_{2n}\ P(\{x_n\}) \propto P'(\{x'_n\}) \propto \exp \sum_n [-(x'_{n+1} - (1+\gamma\Delta t)^2 x'_n)^2/(2\varepsilon\Delta t')] \tag{6.11}$$

where

$$x'_n = x_{2n+1} \quad \text{and} \quad \Delta t' = \{1 + (1+\gamma\Delta t)^2\}\Delta t \ . \tag{6.12}$$

For small Δt the renormalized $\Delta t'$ is reduced to $\Delta t' = 2\Delta t = L'\Delta t$ with $L' = 2$, as it should be.

For a general case we may write the nonlinear Langevin equation as

$$\frac{d}{dt}\varphi(\mathbf{r}, t) = \nabla^2\varphi(\mathbf{r}, t) + f(\varphi(\mathbf{r}, t)) + \eta(\mathbf{r}, t) \quad . \tag{6.13}$$

The path integral corresponding to the above stochastic process may be essentially expressed[52)~54)] by

$$P(\varphi(\mathbf{r}), t; \varphi_0(\mathbf{r}), t_0) = \int_{\varphi(\mathbf{r},t_0)=\varphi(\mathbf{r})}^{\varphi(\mathbf{r},t)=\varphi(\mathbf{r})} D\varphi(\mathbf{r}, \tau)\exp[-\frac{1}{2\varepsilon}\int_{t_0}^{t} d\tau \mathcal{L}] \tag{6.14}$$

where

$$\mathcal{L} = (\dot{\varphi}(\mathbf{r}, t) - \nabla^2\varphi(\mathbf{r}, t) - f(\varphi(\mathbf{r}, t)))^2 \quad . \tag{6.15}$$

The discrete version of this path integral will be clear. The scale transformation of the cell sizes $\Delta\mathbf{r}$ and Δt in real space and time will be obtained as

$$\Delta\mathbf{r} \to (\Delta\mathbf{r})' = L(\Delta\mathbf{r}), \ \Delta t \to (\Delta t)' = L'(\Delta t) \ , \tag{6.16}$$

after appropriate real-space RG, and the dyanmic critical exponent z can be calculated through the relation

$$z = \ln L'/\ln L \quad . \tag{6.17}$$

However, actual calculations are very complicated and they are still in progress. The above formulation will be also useful for Monte Carlo simulation.

7. Related problems in future and discussion

(i) *Improved systematic real-space dynamic RG*: As we have shown in this text, all the methods proposed up to now are far from a desired systematic approach. For example, the Migdal approximation extended to critical dynamics is very simple and intuitive, but it is quite difficult to improve this approximation systematically. The situation is almost the same in other methods discussed in this article.

(ii) *Critical slowing down near phase transitions far from equilibrium.*

In section 1, we have formulated the ordinary critical slowing down in the linear response theory[4], so-called linear critical slowing down, and also the nonlinear critical slowing down.[7] The critical slowing down in the first category is generally expected to appear even near phase transitions far from equilibrium, quite analogously to that in equilibrium.

In many cases phenomena far from equilibrium are described by the nonlinear Langevin equation (6.13) or by the Fokker-Planck equation. If we consider the time dependent "order parameter" $x(t)$ describing bifurcation phenomena of the system then the correlation time of the correlation function $\langle x(t' + t)x(t')\rangle$ will be anomalously enhanced near the transition point of the second order. This situation is immediately realized in a linear Langevin equation of the form $\dot{x}(t) = -\gamma x + \eta(t)$. The relaxation time is proportional to γ^{-1} where $\gamma = 0$ is a typical bifurcation point. To study more general nonlinear cases will be a problem of the future.

This kind of critical slowing down has been also found[25),55)] in transient phenomena near the instability point.

References

1. R.J. Glauber, J. math. Phys. 4 (1963) 294.
2. M. Suzuki and R. Kubo, J. Phys. Soc. Japan 24 (1968) 51.
3. M. Suzuki, Prog. Theor. Phys. 43 (1970) 882.
4. R. Kubo, J. Phys. Soc. Japan 12 (1957) 570.
5. M. Suzuki, H. Ikari (Yahata) and R. Kubo, J. Phys. Soc. Japan 26 (1969) supplement, 153.
6. H. Yahata and M. Suzuki, J. Phys. Soc. Japan 27 (1969) 1421.
7. M. Suzuki, Int. J. Magnetism, 1 (1971) 123.
8. M. Suzuki, in *Critical Phenomena in Alloys, Magnets, and Superconductors* (ed. R.E. Mills, E. Ascher and R.I. Jaffee; 1971) Mcgraw-Hill Book Co., New York.
9. Z. Rácz, Phys. Lett. 53A (1975) 433.
10. Z. Rácz, Phys. Rev. B13 (1976) 263.
11. M.E. Fisher and Z. Rácz, Phys. B13 (1976) 5039.
12. Z. Rácz, ref. 10.
13. Z. Csépes and Z. Rácz, J. Phys. A: Math. Gen. Vol. 11 (1978) No.3, 575.
14. R. Bausch and H.K. Janssen, Z. Phys. B25 (1976) 275.
K. Binder, Phys. Rev. B8 (1973) 3423.
15. M. Suzuki, Phys. Lett. 58A (1976) 435.
16. Z. Rácz and M.F. Collinz, Phys. Rev. B13 (1976) 3074.
17. H. Ikeda, Prog. Theor. Phys. 55 (1976) 2027; ibid 57 (1977) 687.
18. N.J. White, J. Phys. C: Solid St. Phys. 9 (1976) L187.
19. M. Suzuki and H. Ikeda, Prog. Theor. Phys. 55 (1976) 2041.
20. R. Kretschmer, K. Binder and D. Stauffer, J. Stat. Phys. 15 (1976) 267.
21. A. Rácz and T. Tél, Phys. Lett. 60A (1977) 3.
22. L.P. Kadanoff, Physics 2 (1966) 263.
23. M. Suzuki, Prog. Theor. Phys. 51 (1974) 1257.
24. H. Mori, Prog. Theor. Phys. 52 (1974) 433; 53 (1975) 1617; 57 (1977) 785.
25. M. Suzuki, Prog. Theor. Phys. 56 (1976) 77; 56 (1976) 477; 57 (1977) 380; J. Stat. Phys. 16 (1977) 11; 16 (1977) 477.
26. M. Suzuki, K. Sogo, I. Mastuba, H. Ikeda, T. Chikama and H. Takano, Prog. Theor. Phys. 61 (1979) No.3.
27. K.G. Wilson, Phys. Rev. B4 (1971), 3174, 3184.
28. K.G. Wilson and J. Kogut, Phys. Rep. C12 (1974) 75; M.E. Fisher, Rev. Mod. Phys. 46 (1974) 597.
29. S. Ma, Phys. Rev. Lett. 37 (1976) 461.
30. W. Kinzel, Z. Phys. B29 (1978) 361.
31. Y. Achiam, J. Phys. A11 (1978) 975.
32. Y. Achiam, J. Phys. A11 (1978) L129; preprint.
33. Y. Achiam and J.M. Kosterlitz, Phys. Rev. Lett. 41 (1978) 128.
34. G.F. Mazenko, M.J. Nolan and O.T. Valls, Phys. Rev. Lett. 41 (1978) 500.
35. M. Suzuki et al. ref. 26.
36. S.T. Chui, G. Forgacs and H.L. Frisch, preprint.
37. U. Deker, Z. Phys. B31 (1978) 283.
38. Y. Shiwa, preprint.
39. R. Abe, Prog. Theor. Phys. 39 (1968) 947.
40. B.U. Felderhof, Reports on math. Phys. 1 (1970) 215.
41. B.U. Felderhof and M. Suzuki, Physics 56 (1971) 43.
42. H.J. Hilhorst, M. Suzuki and B.U. Felderhof, Physica 60 (1972) 199.
43. M. Suzuki, Phys. Lett. 19 (1965) 267.
44. D.R. Nelson and M.E. Fisher, Ann. Phys. (N.Y.) 91 (1975) 266.
R.G. Priest, Phys. Rev. B11 (1975) 3461.
45. A.A. Migdal, Zh. Eksp. Teor. Fiz. 69 (1975) 1457; ibid. 69 (1975) 810.
46. L.P. Kadanoff, Ann. Phys. (N.Y.) 100 (1976) 359.

47. The present author has received a letter to our critical comments to the first version of Chui et al.
48. Th. Niemeijer and J.M.J. van Leeuwen, Phys. Rev. Lett. 31 (1973) 1411; Physica 71A (1974) 17.
49. S. Ma, preprint.
50. M. Suzuki, unpublished.
51. R. Kubo, K. Matsuo and K. Kitahara, J. Stat. Phys. 9 (1973) 51.
52. H.K. Janssen, Z. Phys. B23 (1976) 377.
53. F. Langouche, D. Roekaerts and E. Tirapequi, preprints.
54. C.P. Enz, Lecture Notes in Physics, Vol. 54, p.79, Springer.
55. M. Suzuki, Phys. Lett. 67A (1978) 339; Prog. Theor. Phys. 64 (1978) supplement; in Synergetics: *Far from Equilibrium*, ed. A. Pacault and C. Vidal, Springer-Verlag 1979; Proceedings of the 17th International Solvay Conference, November 1968, in press.

APPLICATION OF REAL-SPACE RENORMALIZATION GROUP APPROACH TO CRITICAL DYNAMICS

G.F. Mazenko*

The James Franck Institute and Department of Physics
The University of Chicago, Chicago, Illinois 60637

* Work supported by the National Science Foundation, NSF DMR77-12637, and an Alfred P. Sloan Foundation Fellowship.

APPLICATION OF REAL-SPACE RENORMALIZATION GROUP APPROACH TO CRITICAL DYNAMICS

Gene F. Mazenko*

The James Franck Institute and Department of Physics
The University of Chicago, Chicago, Illinois 60637

Since the work of Niemeijer and van Leeuwen in 1973 on the real-space renormalization group (RSRG) method, it has seemed natural to apply these ideas to dynamic problems. Progress, however, has been slow. The reason the problem is difficult is associated with the global nature of RSRG transformations and the generation of non-Markovian behavior in each iteration of the transformation. Technically, the problem is to map a time evolution operator for one problem onto a corresponding operator for the associated coarse-grained lattice.

We discuss how to set up this problem of mapping operators onto operators. For an arbitrary transformation the new time evolution operator shows non-Markovian behavior. We show how one can devise restricted RSRG transformations which generate new time evolution operators which are completely Markovian.

We show how these ideas can be applied to the two-dimensional kinetic Ising model.

*Work supported by the National Science Foundation, NSF DMR77-12637, and an Alfred P. Sloan Foundation Fellowship.

I. Introduction

In this paper, we discuss a method for systematically implementing a real space dynamic renormalization group (RSDRG) transformation. This method will be applied to the problem of treating the critical dynamics of kinetic Ising models. We believe, however, that many of the ideas we develop are more general and could have applications in a number of other dynamical problems. The work reported here was carried out in collaboration with Dr. Michael J. Nolan and Professor Oriol T. Valls. A brief report of our work has already appeared[(1)] and more detailed papers will appear elsewhere.[(2)]

The continuum renormalization group method due to Wilson[(3)] and the mode coupling theory due to Fixman,[(4)] Kawasaki[(5)] and others[(6)] have been successfully married into a dynamic renormalization group theory that works quite well to explain dynamic phenomena in a variety of systems. Many of these successes are discussed in the recent review by Hohenberg and Halperin.[(7)] This method is designed to work in the critical scaling region and near the upper critical dimensionality where fluctuations are "weak." In the scaling region one can invoke the universality hypothesis to justify using the most convenient calculational model and ignore all microscopic details. Within these limitations the theory is very powerful.

While the continuum approach has been very successful in certain circumstances the limitations can be severe. Since one ignores local details from the very beginning of the theory it is only useful for determining asymptotic properties. Thus, while it will give you the critical exponents, it will not tell you the value of the transition temperature or how one approaches asymptopia. Similarly, the continuum formulation essentially assumes there is a second order phase transition and only works near such a transition. It is not useful for looking at global properties over the entire thermodynamic plane and can not be used to systematically investigate first order phase transitions.

Practically speaking, we only know how to treat the model field theories studied in the continuum theories within the framework of perturbation theory where fluctuations are treated as small. Thus, we must work "near" 4 or 6 dimensions or use sophisticated[(8)] resummation techniques. Clearly these techniques are not the most appropriate for treating two-dimensional systems with strong fluctuations.

Suppose we are interested in the dynamics of a lattice model in two dimension. We may be interested in nonuniversal quantities, like the transition temperature, T_c, and in global thermodynamic questions.[(9)] In this case, the continuum RG is not well suited to the problem. The real space renormalization group (RSRG) method is a better candidate for treating this situation.

The RSRG method, which takes the original block spin picture of Kadanoff[(10)] very seriously, was first successfully implemented by Niejmeier and van Leeuwen[11](NvL) in case of the two-dimensional Ising model. This approach allows for a determination of non-universal quantities (like the transition temperature), works in lower dimensions and seems generally complementary to the continuum RG. It has seemed highly desirable since the work of NvL to extend their ideas to dynamical problems. Progress has been very slow. In this paper, we want to point out the difficulties hindering progress and, we believe, the resolution of these difficulties.

In the next section we introduce the general properties of kinetic Ising models we will need in our discussion. In the third section we sketch the basic ideas in the static RSRG necessary for our discussion of dynamics. In section IV, we introduce our RSDRG transformation with special attention to the role of non-Markovian effects in the development. The implementation of these formal ideas to the kinetic Ising model on a two-dimensional triangular lattice will be discussed in section V. We conclude the paper in section VI with a few comments.

II. Kinetic Ising Models

Kinetic Ising models[12,13] are intermediate between completely microscopic and macroscopic dynamical models. On the one hand the static or equilibrium behavior is controlled by a microscopic Ising Hamiltonian. On the other hand, one can not define a Poisson bracket for a system containing only Ising spins, and therefore one does not have a microscopic dynamics generated by the Hamiltonian. Alternately, we must introduce a stochastic time evolution operator. It is conventional in motivating the dynamic evolution of a set of Ising spins to picture the system as being in contact with a heat reservoir. The interactions with the heat reservoir provide the dynamics. Specifically, let us consider a system of N Ising spins (in any dimension) interacting with a heat bath, and let i be an index which numbers the spins and let $\{\sigma\}$ stand for a given spin configuration $\{\sigma\} \equiv \{\sigma_1 \ldots \sigma_i \ldots \sigma_N\}$. The equilibrium probability distribution is given by

$$P[\sigma] = e^{H[\sigma]}/Z \tag{2.1}$$

and $H[\sigma]$ is the Ising Hamiltonian (multiplied by $-\beta$), and Z is the partition function. We write

$$H[\sigma] = \frac{K}{2} \sum_{i=1}^{N} \sum_{\delta} \sigma_i \sigma_{i+\delta} \tag{2.2}$$

where K is the coupling constant ($K = +\beta J$ where J is the exchange interaction), and the δ sum is over basis vectors connecting a spin at site i with its nearest neighbors. The equilibrium spin-spin time correlation function can be defined in this case as

$$C_{ij}(t) = \sum_{\{\sigma\}} P[\sigma]\, \sigma_j\, e^{\tilde{D}_\sigma t} \sigma_i \tag{2.3}$$

where $\tilde{D}_\sigma$ is an operator similar to a Liouville operator in a fully microscopic theory. In the case of kinetic Ising models $\tilde{D}_\sigma$ generates a stochastic dynamics (includes damping effects) and is a pseudo-Liouville operator. In the context of models we study here, we will refer to $\tilde{D}_\sigma$ as a spin-flip operator (SFO). An operator in this case is a matrix which rotates one spin configuration $\{\sigma'\}$ into another spin configuration $\{\sigma\}$:

$$\tilde{D}_\sigma A[\sigma] \equiv \sum_{\{\sigma'\}} \tilde{D}[\sigma|\sigma'] A[\sigma'] \quad . \tag{2.4}$$

The adjoint operator D_σ is defined by

$$D[\sigma|\sigma'] \equiv \tilde{D}[\sigma'|\sigma] \quad . \tag{2.5}$$

We will want to demand that $\tilde{D}_\sigma$, or equivalently D_σ , satisfy two basic properties. First we require that the equilibrium probability distribution is stationary under time translations. This requires that

$$e^{D_\sigma t} P[\sigma] = P[\sigma] \tag{2.6}$$

or

$$\sum_{\{\sigma'\}} D[\sigma|\sigma'] P[\sigma'] = 0 \quad . \tag{2.7}$$

We refer to Eq. (2.7) as the detailed balance condition. A second symmetry condition follows from demanding that a time correlation functions of interest satisfy

$$C_{AB}(t) = C_{BA}(t) \quad . \tag{2.8}$$

This condition is satisfied in a fully-microscopic theory if A and B have the same signature under time reversal invariance. We would like our stochastic dynamical model to satisfy this symmetry constraint. It is easy to see that this constraint will be satisfied if the SFO D_σ satisfies the symmetry condition

$$D[\sigma|\sigma'] P[\sigma'] = D[\sigma'|\sigma] P[\sigma] \quad . \tag{2.9}$$

Note that Eqs. (2.7) and (2.9) give us the useful identity for any $f(\sigma)$:

$$\sum_{\{\sigma'\}} D[\sigma|\sigma'] f[\sigma'] P[\sigma'] = P[\sigma] \sum_{\{\sigma'\}} \tilde{D}[\sigma|\sigma'] f[\sigma'] \quad . \tag{2.10}$$

Together the detailed balance condition (2.7) and the symmetry relation (2.9) constrain the possible operators D_σ significantly. There is, however, still a great deal of latitude in choosing the appropriate operators and operationally one is guided by argreements of locality and simplicity. In this paper, we will focus on a particular class of operators satisfying (2.16) and (2.19). We call this class of operators "spin-flip operators " because they have the property that they are a sum of terms of the form:

$$\sum_{i_1}\sum_{i_2}\cdots\sum_{i_n} \Lambda_{\sigma,\sigma'}^{[i_1,i_2,\dots,i_n]} F[\sigma|\sigma']$$

where

$$\Lambda_{\sigma,\sigma'}^{[i_1,i_2,\dots,i_n]} = \prod_{\ell \neq (i_1,i_2,\dots,i_n)} \delta_{\sigma_\ell,\sigma'_\ell} \tag{2.11}$$

is the operator that sets $\sigma = \sigma'$ on all lattice sites except $i_1, i_2, \dots, i_n$. Thus, D_σ will "operate" on only a restricted set of n spins. D_σ is diagonal as far as the other $N - n$ spins are concerned. There are other models one could consider which do not have this property.

The simplest SFO involves operating on one spin and is of the form

$$\tilde{D}[\sigma|\sigma'] = -\frac{\alpha}{2}\sum_i \Lambda_{\sigma,\sigma'}^{[i]} W_i[\sigma]\,\sigma_i\,\sigma_i' \tag{2.12}$$

where α^{-1} is some characteristic spin-flip time, and $\alpha W_i[\sigma]$ has the physical interpretation of the probability per unit time that the ith spin will flip from σ_i to $-\sigma_i$. We have yet to fully specify $W_i[\sigma]$. From detailed balance (2.7) and the symmetry condition (2.9) we have

$$\sum_{\{\sigma'\}}\sum_i \Lambda_{\sigma,\sigma'}^{[i]} W_i[\sigma']\,\sigma_i\,\sigma_i'\,P[\sigma'] = 0 \tag{2.13}$$

and

$$\sum_i \Lambda_{\sigma,\sigma'}^{[i]} W_i[\sigma']\,\sigma_i\,\sigma_i'\,P[\sigma'] \tag{2.14}$$

$$= \sum_i \Lambda_{\sigma,\sigma'}^{[i]} W_i[\sigma]\,\sigma_i\,\sigma_i'\,P[\sigma] .$$

Both equations are satified as long as $W_i[\sigma]\,P[\sigma]$ is independent of the spin at site i. The simplest choice for $W_i[\sigma]$ is

$$W_i[\sigma] = e^{-\sigma_i E_i[\sigma]} \tag{2.15}$$

where

$$E_i[\sigma] = K\sum_\delta \sigma_{i+\delta} \tag{2.16}$$

gives the interaction energy of the spin at site i with its nearest neighbors.

We will be interested in expressions for $D[\sigma\,|\,\sigma']$ more general than (2.12). The reason is that we expect that in a RG transformation starting with a given operator $\tilde{D}_\sigma$ we may, at a given order in perturbation theory, generate new terms. If we want the procedure to "close" we must start with a general enough operator D_σ. A general SFO can be written in the form

$$D[\sigma|\sigma'] = \frac{1}{2}\sum_i \Lambda^{[i]}_{\sigma,\sigma'}\, \sigma_i \sigma_i'\, W_i[\sigma']\, V_1^{[i]}[\sigma']$$

$$+\frac{1}{4}\sum_{i,j} \Lambda^{[i,j]}_{\sigma,\sigma'} W_{ij}[\sigma']\left(\sigma_i \sigma_j' V_2^{[ij]}[\sigma'] + \sigma_i \sigma_j \sigma_i' \sigma_j' V_3^{[ij]}[\sigma']\right) \tag{2.17}$$

$$+\frac{1}{8}\sum_{ijk} \Lambda^{[ijk]}_{\sigma,\sigma'} W_{ijk}[\sigma']\left(\sigma_i \sigma_i' \sigma_j \sigma_k' V_4^{[ijk]}[\sigma'] + \sigma_i \sigma_i' \sigma_j \sigma_j' \sigma_k \sigma_k' V_5^{[ijk]}[\sigma']\right)$$

where we have included up to three spin terms. The $W_{ij\ldots}[\sigma]$ in Eq. (2.17) are defined as follows: let $H^{ijk\ldots}$ be the sum of all terms in the Hamiltonian which involve $\sigma_i \sigma_j \sigma_k \ldots$ for example:

$$H^i = \sigma_i E_i[\sigma] \tag{2.18}$$

or, if $\sigma_i \sigma_j$ are neighboring spins

$$H^{ij} = \sigma_i E_i[\sigma] + \sigma_j E_j[\sigma] - K \sigma_i \sigma_j \tag{2.19}$$

(the last term avoids double counting). Then we define:

$$W_{ij\ldots}[\sigma] = e^{-H^{ij\ldots}}\,. \tag{2.20}$$

Finally, the functions $V^{[ij\ldots]}(\sigma')$ are arbitrary functions of σ', except that they must be independent of the set of spins $\sigma_i' \sigma_j' \ldots$. The V's have dimensions of an inverse time, whereas we take the W's to be dimensionless.

It is easy to show that (2.17) satisfies the detail balance condition (2.7) and the symmetry condition (2.9). The key point is that $V_m^{[ij\ldots]} W_{ij\ldots}[\sigma]\, P[\sigma]$ does not depend on the spins $\sigma_i \sigma_j \ldots$. Direct summation over σ' of $D[\sigma|\sigma']\, P[\sigma']$ gives zero since there will always be a term linear in σ_i' . Since $V_m^{[ij\ldots]} \Lambda^{[ij\ldots]}_{\sigma\sigma'} W_{ij\ldots}(\sigma\ \sigma \ldots)\, P(\sigma)$ is symmetric with respect to the interchange $\sigma \rightleftarrows \sigma'$, it follows that the symmetry relation (2.9) is satisfied by (2.17). For a given set of functions $\{W_i(\sigma)\ \ldots,\ W_{ij}(\sigma), \ldots\}$ one can develop a systematic method for extracting the coefficients $V_1^{[i]}[\sigma]$, $V_2^{[i,j]}[\sigma]$, etc. from a given operator $D[\sigma \mid \sigma']$. The details of this procedure will be described elsewhere.

III. The Real Space Renormalization Group

In this section, we will review some well known[11] properties of the RSRG that will be needed in our dynamical treatment. Let us consider a set of Ising spin variables $\{\sigma\}$ on a lattice. The RG transformation function $T[\mu|\sigma]$ maps the set $\{\sigma\}$ onto a new set of spins $\{\mu\}$ located on a lattice geometrically similar to the initial one, but having a larger lattice constant. If the initial Hamiltonian is designated by $H[\sigma]$, one defines the Hamiltonian $H[\mu]$ on the new μ-lattice by the relation

$$e^{H[\mu]} = \sum_{\{\sigma\}} e^{H[\sigma]} T[\mu|\sigma] . \tag{3.1}$$

The demand that the partition function be invariant under this transformation

$$\begin{aligned} Z &= \sum_{\{\sigma\}} e^{H[\sigma]} \\ &= \sum_{\{\mu\}} e^{H[\mu]} \end{aligned} \tag{3.2}$$

requires that the transformation function satisfy the normalization condition

$$\sum_{\{\mu\}} T[\mu|\sigma] = 1 . \tag{3.3}$$

With the exception of this normalization condition there is no restriction on the choice of $T[\mu|\sigma]$. Let us consider a system of Ising spins on a two dimensional triangular lattice. This system is discussed extensively in Ref. 11 and is the system we concentrate on in this paper. Let us divide the lattice into triangular cells, drawn so as to preserve the symmetry of the lattice.

Figure 1. A two-dimensional triangular lattice divided into three-spin cells labeled by a cell index i and a set of basis vectors $\vec{a}_1, \vec{a}_2, \vec{a}_3$.

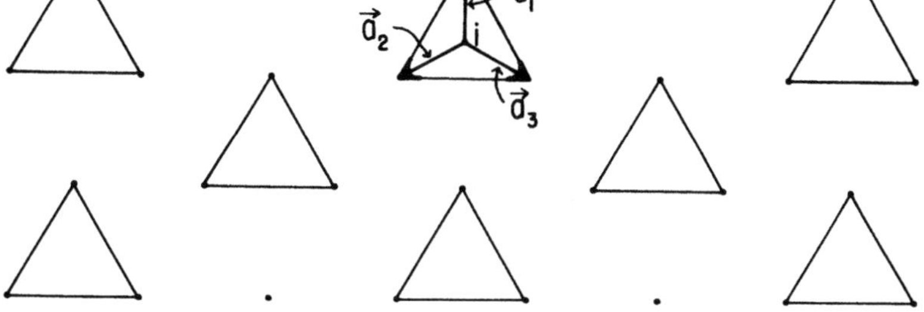

Each cell will correspond to a new spin, μ. We will use an index i to designate each cell (and hence each μ). The three σ spins in cell i are designated $\sigma_{i,\vec{a}_1}$, $\sigma_{i,\vec{a}_2}$ and $\sigma_{i,\vec{a}_3}$, where the vectors a_1, a_2, and a_3 are defined in Fig. 1. Following NvL[11] we will consider all intracell interactions to be of zeroth order and all nearest neighbor intercell couplings to be of first order in ϵ (ϵ to be set equal to one at the end of the calculation).

A widely studied class of RG transformation functions corresponds to assuming that T can be written as a product over cells, i.e.,

$$T_o[\mu|\sigma] = \prod_i T^i[\mu|\sigma] . \tag{3.4}$$

The index o signifies that T does not include any intercell couplings. The most general form that we can write down for $T_i[\mu|\sigma]$, depending only on the spin μ_i and the three σ-spins in the ith cell, is

$$T^i[\mu|\sigma] = \frac{1}{2}\left(1 + \mu_i \varphi_i(\sigma)\right) \tag{3.5}$$

which satisfies the nomalization condition

$$\sum_{\mu_i} T^i[\mu|\sigma] = 1 . \tag{3.6}$$

It is desirable, in order to preserve the symmetry $H[-\sigma] = H[\sigma]$, to choose $T^i[\mu|\sigma]$ to be even under the flip of all the μ and σ spins in a cell. Thus we want $\varphi_i(\sigma)$ to be an odd function of the σ spins in the cell i. We will choose the $\varphi_i(\sigma)$ to satisfy the normalization condition.

$$\langle \varphi_i^2(\sigma) \rangle_o = 1 \tag{3.7}$$

where the average is defined by

$$\langle f(\sigma) \rangle_o = \sum_{\{\sigma\}} P_o[\sigma] f(\sigma) \tag{3.8}$$

where $P_o[\sigma]$ is the equilibrium probability destribution corresponding to uncoupled cells. If we further demand that φ_i be symmetric function of the three spins in the cell then it can be written in the form

$$\varphi_i(\sigma) = N(\sigma_i^T - f\,\sigma_i^c) \tag{3.9}$$

where N and f are constants related through the normalization condition (3.7), σ_i^T is the total spin for the ith cell,

$$\sigma_i^T = \sigma_{i,a_1} + \sigma_{i,a_2} + \sigma_{i,a_3} \tag{3.10}$$

and σ_i^c is the product of the three spins in the ith cell,

$$\sigma_i^c = \sigma_{i,a_1} \sigma_{i,a_2} \sigma_{i,a_3} \quad . \tag{3.11}$$

With this choice for T_o, one can calculate $H[\mu]$ as a cumulant expansion in powers of the intercell coupling ϵ. The result to first order in ϵ is that $H[\mu]$ is of the same form as $H[\mu]$ with only nearest neighbor couplings and the new coupling constant is related to the old by

$$K' = 2v^2 K \tag{3.12}$$

where

$$v = \langle \sigma_{i,a} \varphi_i(\sigma) \rangle_o \quad . \tag{3.13}$$

Since we assume $\varphi_i(\sigma)$ is symmetric in the three spins in cell i, ν will not depend on the particular basis vector a. In a second order calculation, one generates couplings between second and third nearest neighbors and the corresponding recursion relations involve three parameters. One appealing choice for $\varphi_i(\sigma)$ is given by the "majority rule" due to NvL where

$$\varphi_i(\sigma) = \mathrm{Sgn}\, \sigma_i^T = \frac{1}{2}\left(\sigma_i^T - \sigma_i^c\right) \tag{3.14}$$

and corresponds to $N = {}^1/2$ and $f = 1$ in Eq. (3.9).

IV. The Real Space Dynamic Renormalization Group

In this section, we will develop our scheme for implementing the RG for dynamics in position space.

A. Basic RSRG Transformation for the Time-Evolution Operator.

In the dynamical problems of interest the system is specified by the Hamiltonian $H[\sigma]$ and SFO $D[\sigma \mid \sigma']$. If we want to apply RG ideas to this problem, we want to design transformations which map the old Hamiltonian $H[\sigma]$ and SFO $D[\sigma \mid \sigma']$ onto their counterparts $H[\mu]$ and $D[\mu \mid \mu']$ defined on the μ-lattice. We "understand" the mapping $H[\sigma] \to H[\mu]$. The problem is with the mapping $D_\sigma \to D_\mu$. The difficulty is that we now have to map an operator onto an operator.

It is useful, in developing the appropriate transformation, to introduce the time evolution matrix

$$G_t[\sigma|\sigma'] = e^{D_\sigma t}\,\delta_{\sigma,\sigma'}\,P[\sigma] \tag{4.1}$$

where

$$\delta_{\sigma,\sigma'} = \prod_i \delta_{\sigma_i,\sigma_i'} \tag{4.2}$$

is the matrix setting $\sigma = \sigma'$ on all lattice sites. From this quantity we can construct any time correlation function of the form $\langle \sigma_i\, \sigma_j \ldots \sigma_k\, e^{\tilde{D}_\sigma t} \sigma_m \ldots \sigma_n \rangle$. We see, for example, that the spin-spin correlation function is given by

$$C_{ij}(t) = \sum_{\{\sigma\}} \sum_{\{\sigma'\}} \sigma_j\, G_t[\sigma|\sigma']\, \sigma_i' = \sum_{\{\sigma\}} \sigma_j\, e^{D_\sigma t} \sigma_i\, P[\sigma] \,. \tag{4.3}$$

Remembering that the RG transformation for the probability distribution is given by

$$P[\mu] = \sum_{\{\sigma\}} P[\sigma]\, T[\mu|\sigma] \tag{4.4}$$

and noticing that $G_t[\sigma|\sigma']$ is a matrix depending on two sets of spins, it is natural to assume that the new time evolution matrix for the μ-lattice is defined by

$$G_t[\mu|\mu'] = \sum_{\{\sigma\}} \sum_{\{\sigma'\}} T[\mu|\sigma]\, G_t[\sigma|\sigma']\, T[\mu'|\sigma'] \,. \tag{4.5}$$

In this case there are two constraints on the mapping functions T. The first constraint is the same as in the static case and given by Eq. (3.3). The second constraint follows from requiring that the zero time form of $G_t[\sigma|\sigma']$,

$$G_0[\sigma|\sigma'] = \delta_{\sigma,\sigma'} P[\sigma] \tag{4.6}$$

be preserved by the RG transformation,

$$G_0[\mu|\mu'] = \sum_{\{\sigma\}} T[\mu|\sigma]\, T[\mu'|\sigma]\, P[\sigma] \equiv \delta_{\mu,\mu'} P[\mu] . \tag{4.7}$$

Notice that together Eqs. (3.3) and (4.7) imply that Eq. (4.4) still holds. This definition of $G_t[\mu|\mu']$ is a physically sensible dynamical generalization of the static RSRG procedure. To see this, consider the spin-spin correlation function generated by $G_t[\mu|\mu']$:

$$\begin{aligned} C'_{ij}(t) &= \sum_{\{\mu\}} \sum_{\{\mu'\}} \mu_i\, G_t[\mu|\mu']\, \mu'_j \\ &= \sum_{\{\mu\}} \sum_{\{\mu'\}} \mu_i \sum_{\{\sigma\}} \sum_{\{\sigma'\}} T[\mu|\sigma]\, G_t[\sigma|\sigma']\, T[\mu'|\sigma']\, \mu'_j . \end{aligned} \tag{4.8}$$

If the mapping function T is of the product form given by Eq. (3.4) then

$$\sum_{\{\mu\}} \mu_i\, T_0[\mu|\sigma] = \varphi_i(\sigma) \tag{4.9}$$

and

$$C'_{ij}(t) = \langle \varphi_j(\sigma)\, e^{\tilde{D}_\sigma t}\, \varphi_i(\sigma) \rangle \tag{4.10}$$

where we have defined

$$\langle A(\sigma) \rangle = \sum_{\{\sigma\}} P[\sigma]\, A(\sigma) . \tag{4.11}$$

$C'_{ij}(t)$ is, as desired, the time correlation for a set of coarse grained spin variables. This means that we must choose φ_i to represent the effective spin for a cell. Physically it makes sense to identify φ_i as the dynamically slowest varying odd spin variable in a cell. We will return to this point later.

Before going further we should comment on the constraint on $T[\mu|\sigma]$ given by (4.7). In the case of the majority rule choice for $\varphi_i(\sigma)$ given by (3.14) one has that

$$T^i[\mu|\sigma]\, T^i[\mu'|\sigma] = \delta_{\mu_i,\mu_i'}\, T^i[\mu|\sigma] \tag{4.12}$$

and (4.7) is satisfied directly. For other choices of φ_i the quantity

$$\tilde{G}[\mu|\mu'] = \langle T[\mu|\sigma]\, T[\mu'|\sigma] \rangle \tag{4.13}$$

is not diagonal. However, since $\tilde{G}[\mu|\mu']$ is symmetric in μ and μ', $\tilde{G}$ can be diagonalized through a rotation in μ-space. One must, however, exercise care that this rotation does not violate the condition (3.3).

At the beginning of this section we indicated that we are interested in finding the SFO $D[\mu|\mu']$ appropriate to the μ-lattice. At present we have only an expression for the time-evolution operator $G_t[\mu|\mu']$. We must now indicate how one can extract D_μ from $G_t[\mu|\mu']$. This indentification of D_μ requires some formal development. Rather than working directly with $G_t[\mu|\mu']$ it is formally more convenient to introduce the Laplace transform

$$G_z[\sigma|\sigma'] = -i \int_0^{+\infty} dt\, e^{+izt}\, G_t[\sigma|\sigma'] . \tag{4.14}$$

Using the definition of G_t given by (4.1), we can easily carry out the time integration to obtain

$$G_z[\sigma|\sigma'] = R_\sigma(z)\, \delta_{\sigma,\sigma'}\, P[\sigma] \tag{4.15}$$

where $R_\sigma(z)$ is the resolvent operator,

$$R_\sigma(z) = \left[z - i D_\sigma\right]^{-1} . \tag{4.16}$$

Using the operator identity

$$z R_\sigma(z) = 1 + i R_\sigma(z) D_\sigma , \tag{4.17}$$

we find that $G_z[\sigma|\sigma']$ satisfies the equation of motion

$$\sum_{\{\bar{\sigma}\}} \left[z\, \delta_{\sigma,\bar{\sigma}} - i\, D[\sigma|\bar{\sigma}] \right] G_z[\bar{\sigma}|\sigma'] = \delta_{\sigma,\sigma'}\, P[\sigma] . \tag{4.18}$$

It is then natural, taking into account (4.7), to define the operator $D_z[\mu|\mu']$ by the relation:

$$\sum_{\{\bar{\mu}\}} \left[z\, \delta_{\mu,\bar{\mu}} - i\, D_z[\mu|\bar{\mu}] \right] G_z[\bar{\mu}|\mu'] = \delta_{\mu,\mu'}\, P[\mu] \tag{4.19}$$

where we have taken into account the fact that $D_z[\mu|\mu']$ may depend on z. We are now left with the task of deriving an expression for $D_z[\mu|\mu']$ in terms of $D[\sigma|\sigma']$ and $T[\mu|\sigma]$. This derivation parallels methods used in the memory function formalism for fluids[14]. One obtains finally that

$$D_z[\mu|\mu'] = D^s[\mu|\mu'] + D_z^c[\mu|\mu'] \tag{4.20}$$

where we have specified that all the z dependence (i.e., the non-Markovian part) of D_μ is contained in D^c. D^s obeys the equation

$$D^s[\mu|\mu']P[\mu'] = \langle T[\mu|\sigma]\,\tilde{D}_\sigma\, T[\mu'|\sigma]\rangle \tag{4.21}$$

while

$$\begin{aligned} D_z^c[\mu|\mu']P[\mu'] &= i\langle(\tilde{D}_\sigma T[\mu|\sigma])\tilde{R}_\sigma(z)(\tilde{D}_\sigma T[\mu'|\sigma])\rangle \\ &- i\langle(\tilde{D}_\sigma T[\mu|\sigma])\tilde{R}_\sigma(z) T[\bar{\mu}|\sigma]\rangle G_z^{-1}[\bar{\mu}|\bar{\mu}']\langle T[\bar{\mu}'|\sigma]\tilde{R}_\sigma(z)(\tilde{D}_\sigma T[\mu'|\sigma])\rangle \end{aligned} \tag{4.22}$$

where in the last line of (4.22) summation over $\bar{\mu}$ and $\bar{\mu}'$ is implied and $\tilde{R}_\sigma$ corresponds to $R_\sigma(z)$ with $D_\sigma \rightarrow \tilde{D}_\sigma$. Equations (4.20), (4.21), and (4.22) give one well defined statistical mechanical expressions for the new SFO on the μ-lattice. We note immediately that $D_z[\mu|\mu']$ satisfies the fundamental constraints imposed originally on $D[\sigma|\sigma']$. This is, if $D[\sigma|\sigma']$ satisfies Eqs. (2.7) and (2.9), then one can easily show that

$$\sum_{\{\mu'\}} D_z[\mu|\mu']\,P[\mu'] = 0 \tag{4.23}$$

and

$$D_z[\mu|\mu']P[\mu'] = D_z[\mu'|\mu]P[\mu]\,. \tag{4.24}$$

B. The RSDRG to Zeroth Order in the Cell Coupling.

We have developed our general formalism for carrying out dynamical RSRG transformations. We have seen how the new generator of correlation function, $G_z[\mu|\mu']$, and the new SFO $D_z[\mu|\mu']$ are defined in terms of $D[\sigma|\sigma']$ and the RG transformation function $T[\mu|\sigma]$. Several important points must be sorted out: The first, and technically bothersome point, concerns the non-Markovian behavior of the new SFO $D_z[\mu|\mu']$. From a RG point of view we must be sensitive to the fact that our original problem defined on the σ-lattice, is Markovian ($D[\sigma|\sigma']$ is time independent) and we desire to map this problem onto as similar a problem as possible on the μ-lattice. Is this non-Markovian behavior an essential problem? Two other points concern the best choice for $T[\mu|\sigma]$ in the dynamical case, and the degree to which we can explicitly calculate the frequency dependent part of the $D_z^c[\mu|\mu']$? We shall see that all three points are interrelated.

We can gain some understanding of our problem by considering the calculation of $G_t[\mu|\mu']$ for the case where we divide our lattice up into cells and treat the cells as uncoupled. Such a calculation gives one the zeroth order contribution for an expansion in the intercell coupling. In this case the SFO $\tilde{D}_\sigma^o$ involves only intracell couplings, so we can write

$$\tilde{D}_\sigma^o = \sum_i \tilde{D}_\sigma^{i,o} \tag{4.25a}$$

or

$$\tilde{D}^o[\sigma|\sigma'] = \sum_i \Lambda_{\sigma,\sigma'}^{[i]} \tilde{D}^{i,o}[\sigma|\sigma'] \tag{4.25b}$$

where $\tilde{D}_\sigma^{i,o}$ operates on the cell i only. The static properties of this system are governed by the Hamiltonian

$$H_o[\sigma] = \sum_i H_o^i[\sigma] \quad . \tag{4.26}$$

The associated probability distribution is a product of contributions from each cell

$$P_o[\sigma] = \prod_i P_o^i[\sigma] \tag{4.27}$$

where

$$P_o^i[\sigma] = e^{H_o^i[\sigma]} \Big/ \sum_{\{\sigma\}}^i e^{H_o^i[\sigma]} \tag{4.28}$$

and Σ^i means we sum over the degrees of freedom of the spins in the cell i . The RSRG transformation for the statics is relatively trivial in this case,

$$\begin{aligned} P_o[\mu] &= \sum_{\{\sigma\}} P_o[\sigma]\, T_o[\mu|\sigma] \\ &= \prod_i \sum_{\{\sigma\}}^i P_o^i[\sigma]\, T^i[\mu|\sigma] = \prod_i \frac{1}{2} = 2^{-N/n} \end{aligned} \tag{4.29}$$

where n is the number of spins per cell and $P_o[\mu]$ is independent of μ. If matters are arranged properly, then $\tilde{D}_\sigma^{i,o}$ and $H_o^i[\sigma]$ are compatible in that the detailed balance condition Eq. (2.7), and the symmetry condition Eq. (2.9) are satisfied cell by cell. For a cell with three spins the operator $\tilde{D}_\sigma^{i,o}$ is a $2^3 \times 2^3$ matrix. In practice it is relatively easy to diagonalize this matrix in order to obtain the eigenfunctions and eigenvalues defined by

$$\tilde{D}_\sigma^{i,o} \Psi_i^n(\sigma) = -\lambda_n \Psi_i^n(\sigma) . \tag{4.30}$$

It is easy to see that the Ψ_i's can be chosen to form a complete and orthonormal set for a given cell with respect to the weight function $P_o^i[\sigma]$:

$$\langle \Psi_i^n(\sigma) \Psi_i^{n'}(\sigma) \rangle_o = \delta_{n,n'} \tag{4.31}$$

$$\sum_n \Psi_i^n(\sigma) \Psi_i^n(\sigma') P_o[\sigma'] = \delta_{\sigma,\sigma'}^i \tag{4.32}$$

where $\delta_{\sigma,\sigma'}^i$ means that $\sigma = \sigma'$ for all spins in the cell i. Using the completeness of the cell eigenfunctions, we can write

$$\begin{aligned} T^i[\mu|\sigma] &= \frac{1}{2}\left(1 + \mu_i \varphi_i(\sigma)\right) \\ &= \frac{1}{2}\left(1 + \mu_i \sum_n C_n \Psi_i^n(\sigma)\right) \end{aligned} \tag{4.33}$$

where

$$C_n = \langle \varphi_i(\sigma) \Psi_i^n(\sigma) \rangle_o \ . \tag{4.34}$$

Using the eigenvalue equation (4.30) and the completeness condition (4.32) (remembering that $\varphi_i(\sigma)$ is odd in σ, so the Ψ_i^n in (4.33) are odd, and therefore $\langle \Psi_i^n \rangle_o = 0$), one can explicitly average over the σ-variables in (4.5) to obtain

$$G_t^o[\mu|\mu'] = \prod_i \frac{1}{2}\left(1 + \mu_i \mu_i' \sum_n e^{-\lambda_n t} C_n^2\right) P_o[\mu'] \tag{4.35}$$

where we have used (4.29) . Note that at $t = 0$, using (4.32), (4.34), and (3.7) that

$$\sum_n C_n^2 = 1 \tag{4.36}$$

and

$$G_o^o[\mu|\mu'] = \delta_{\mu,\mu'} P_o[\mu'] \tag{4.37}$$

as expected.

We can analyze $G_t^o[\mu|\mu']$ using the Laplace transform method discussed in the last section in which case one finds that

$$D_z[\mu|\mu'] = -\sum_i \mu_i \mu_i' P_0[\mu]\Delta_1(z) - \frac{1}{2}\sum_{i,j}\mu_i\mu_j\mu_i'\mu_j' P_0[\mu]\Delta_2(z) + \cdots \tag{4.38}$$

where

$$\Delta_1(z) = \sum_n \frac{c_n^2 \lambda_n}{z + i\lambda_n} \Big/ \sum_n \frac{c_n^2}{z + i\lambda_n} \tag{4.39}$$

and

$$\Delta_2(z) = \sum_{n,n'} \frac{c_n^2 c_{n'}^2 (\lambda_n + \lambda_{n'})}{z + i(\lambda_n + \lambda_{n'})} \Big/ \sum_{n,n'} \frac{c_n^2 c_{n'}^2}{z + i(\lambda_n + \lambda_{n'})} . \tag{4.40}$$

Alternatively, we can work directly in the time representation and write

$$G_t^0[\mu|\mu'] = e^{D_t^0(t)t} P_0[\mu]\delta_{\mu,\mu'} \tag{4.41}$$

where we can identify

$$D_t^0[\mu|\mu'] = -\sum_i \frac{\alpha'(t)}{2}\mu_i\mu_i' \Lambda_{\mu,\mu'}^{[i]} \tag{4.42}$$

and we now have a time-dependent inverse scattering time which can be written as

$$\alpha'(t) = \lambda_1 - \frac{1}{t}\ln(c_1)^2 - \frac{1}{t}\ln\left(1 + \sum_{n>1}(c_n/c_1)^2 e^{-(\lambda_n - \lambda_1)t}\right) . \tag{4.43}$$

Notice that $D_t^0[\mu|\mu']$ is of precisely the same form as $D[\sigma|\sigma']$ given by (2.12) with the interaction between spins set to zero. For long times $\alpha'(t) = \lambda_1$ (the eigenvalue for the slowest mode) and the last term in (4.43) decays exponentially. The worrysome term is the 1/t term which indicates a long-range in time interaction not included in the original problem.

In the case of the Laplace transform representation, one sees that the operator $D_\mu^0(z)$ is not of the same form as $D_\mu^0(t)$. One sees that there are now long range spatial interactions generated. The form of the renormalized operator D_μ looks different if we work in the time or the frequency domain. In the time representation we see that the form of operator for D_μ reduces to the form for D_σ as $K \to 0$. The difficulty is that the inverse flipping time is now itself time dependent. More worrysome is the fact that there is a long range aspect to this time dependence. If we work in the frequency representation, we find that we can compute the matrix elements of the operator $D_\mu(z)$ directly and they are local in time in that the small z limit is well behaved. Even though the individual matrix elements of $D_z[\mu|\mu']$ are well behaved, there are an infinite number of distinct matrix elements which correspond to couplings which build up in time. These new couplings are highly non-local in space and very undesirable from a RG point of view. The time representation and frequency representations are complementary since $D_\mu(t)$ is a local operator in space but non-local in time, while $D_\mu(z)$ is local in time, but non-local in space. Note that we obtain different prescriptions for the renormalized parameters (say α') using the two different methods. Either way we look at the problem we see that we are in trouble without further constraints on our RG procedure.

Summarizing these results, we see that for a given SFO $\tilde{D}^o_\sigma$ and a given mapping function $T_o[\mu|\sigma]$ the RSRG leads to complicated and physically unsatisfying results to zeroth order in the coupling between cells. The problem is that we can not choose the mapping function $T[\mu|\sigma]$ arbitrarily once we have an operator D^o_σ.

Our technical problems in zeroth order are remedied if we simply choose the function $\varphi_i(\sigma)$ in $T^i[\mu|\sigma]$ to be one of the odd eigenfunctions of $\tilde{D}^{i,o}_\sigma$. Physically we want to choose $\varphi_i = \Psi^1_i$ where Ψ^1_i is the odd eigenfunction whose eigenvalue λ_1 is the smallest in magnitude. We then have the mapping function (let $\Psi_i \equiv \Psi^1_i$)

$$T^i[\mu|\sigma] = \frac{1}{2}\left(1 + \mu_i \Psi_i(\sigma)\right) \tag{4.44}$$

We see from Eq. (4.10) that to lowest order $G_t[\mu|\mu']$ generate the time correlation functions between the slowest modes in a cell. Physically, this is precisely what one wants, and is in keeping with our identification of the μ-variables as representing an average spin for a cell. It is convenient, in order to appreciate the utility of Eq. (4.44) to look at the effect of applying $\tilde{D}^o_\sigma$ to $T_o[\mu|\sigma]$:

$$\tilde{D}^o_\sigma T_o[\mu|\sigma] = -\frac{1}{2}\sum_i T_o^{[i]}[\mu|\sigma]\, \lambda_1 \Psi_i(\sigma)\mu_i \tag{4.45}$$

$$T_o^{[i]}[\mu|\sigma] = \prod_{k(\neq i)} T^k[\mu|\sigma] \ . \tag{4.46}$$

The above expression can be rewritten as:

$$\tilde{D}^o_\sigma T_o[\mu|\sigma] = -\frac{1}{2}\sum_i \sum_{\{\bar{\mu}\}} T_o[\bar{\mu}|\sigma]\mu_i\bar{\mu}_i \Delta^{[i]}_{\mu,\bar{\mu}} \tag{4.47}$$

where we have made use of Eq. (4.9). Notice then that (4.47) can be written in the form

$$\tilde{D}^o_\sigma T_o[\mu|\sigma] = D^o_\mu T_o[\mu|\sigma] \tag{4.48}$$

where

$$D^o[\mu|\mu'] = -\sum_i \frac{\lambda_1}{2} \Delta^{[i]}_{\mu,\mu'} \mu_i \mu'_i \ . \tag{4.49}$$

Notice that this result agrees with Eqs. (4.42) and (4.43) with $C_n = \delta_{n,1}$. Equation (4.48) is in the form of an eigenvalue problem in σ-space where the eigenvalues are operators in μ-space. This equation is extremely useful. If we return to Eq. (4.5), in the case of no intercell coupling, we see, on repeated use of (4.48), that

$$\begin{aligned} G_t^o[\mu|\mu'] &= e^{D_\mu^o t} \langle T_o[\mu'|\sigma]\, T_o[\mu|\sigma]\rangle_o \\ &= e^{D_\mu^o t}\, \delta_{\mu,\mu'}\, P_o[\mu] \; . \end{aligned} \tag{4.50}$$

Similarly, if we return to the frequency representation for D_μ, we find using (4.48) in (4.21) that

$$D_o^s[\mu|\mu']\, P_o[\mu'] = D^o[\mu|\mu']\, P_o[\mu'] \tag{4.51}$$

where D_μ^o is given by (4.49). Turning to the calculation of $D_o^c[\mu|\mu']$, we see that the last term on the right in (4.22) can be written, using (4.48), as

$$\begin{aligned} &-i\, D^o[\mu|\bar{\mu}]\langle T_o[\bar{\mu}|\sigma]\, \tilde{R}_\sigma^o(z)\, T_o[\bar{\mu}''|\sigma]\rangle_o\, G_z^{o-1}[\bar{\mu}''|\bar{\mu}']\langle T_o[\bar{\mu}'|\sigma]\tilde{R}_\sigma^o(z)(\tilde{D}_\sigma^o T_o[\mu'|\sigma])\rangle_o \\ &= -i\, D^o[\mu|\bar{\mu}]\, G_z^o[\bar{\mu}|\bar{\mu}'']\, G_z^{o-1}[\bar{\mu}''|\bar{\mu}']\langle T_o[\bar{\mu}'|\sigma]\tilde{R}_\sigma^o(z)(\tilde{D}_\sigma^o T_o[\mu'|\sigma])\rangle_o \\ &= -i\, D^o[\mu|\bar{\mu}]\langle T_o[\bar{\mu}|\sigma]\tilde{R}_\sigma^o(z)(\tilde{D}_\sigma^o T_o[\mu'|\sigma])\rangle_o \; . \end{aligned}$$

This cancels the first term on the right in Eq. (4.22), and therefore D_o^c vanishes. With the choice for T_o given by Eq. (4.44) the renormalized operator $D^o[\mu|\mu']$ is given by (4.49) using both the time and frequency representations.

We have seen, therefore, how in this particular example it is possible to choose a transformation $T_o[\mu|\sigma]$ which is not only clearly very appropriate from a physical point of view, but also, from a calculational point of view. It is extremely convenient, because all non-Markovian terms (regarding D^c) are automatically eliminated. As soon as the choice of $\varphi_i(\sigma)$ in $T[\mu|\sigma]$ becomes tied to the dynamical operator $\tilde{D}_\sigma^o$ we lose the freedom of adjusting the statics independent of the dynamics. Consider the recursion relation for the statics given, for example, by Eq. (3.12). The parameter ν defined by (3.13) depends on $\Psi_i(\sigma)$ and in turn on the dynamical parameters specifying $D^o[\sigma|\sigma']$. This means that the recursion relation for the statics will couple to those for the dynamic parameters.

C. The General Eigenvalue Method.

In the last section, we showed that through a particular choice of $T_o[\mu|\sigma]$ we could eliminate all non-Markovian behavior to zeroth order in our RSRG transformation. We show in this section that one can eliminate non-Markovian behavior to all orders in perturbation theory. Similar ideas were developed elsewhere[15] in the context of RSRG methods for treating quantum spin systems.

Let us assume that we can find a $T[\mu|\sigma]$ such that

$$\sum_{\{\bar{\sigma}\}} \tilde{D}[\sigma|\bar{\sigma}]\, T[\mu|\bar{\sigma}] = \sum_{\{\bar{\mu}\}} \Sigma[\mu|\bar{\mu}]\, T[\bar{\mu}|\sigma] \tag{4.52}$$

where $\tilde{D}[\sigma|\sigma']$ is a given SFO and $\Sigma[\mu|\mu']$ is an operator in μ-space. Equation (4.52) is the fully interacting generalization of Eq. (4.48). It is easy to show, using Eqs. (4.20), (4.21), and (4.22), (4.7), and (4.52) that

$$D[\mu|\mu']P[\mu'] = D^S[\mu|\mu']P[\mu'] \tag{4.53}$$

$$= \Sigma[\mu|\mu']P[\mu'] , \tag{4.54}$$

$$D_z^c[\mu|\mu']P[\mu'] = 0 \tag{4.55}$$

and

$$G_t[\mu|\mu'] = e^{D_\mu t}\, \delta_{\mu,\mu'}\, P[\mu] . \tag{4.56}$$

The question now is whether we can practically find a $T[\mu|\sigma]$ satisfying (4.52) and (4.7). We discuss here the construction of T using perturbation theory.

Let us assume that $\tilde{D}_\sigma$ can be written as a power series in ε, where ε is a "small" parameter. For definiteness, we will assume that our system is coarse grained into cells, and ε is related to the intercell coupling:

$$\tilde{D}[\sigma|\sigma'] = \sum_{n=0}^{\infty} \varepsilon^n\, \tilde{D}^{(n)}[\sigma|\sigma'] . \tag{4.57}$$

We then assume expansions for T and D_μ :

$$T[\mu|\sigma] = \sum_{n=0}^{\infty} \varepsilon^n\, T^{(n)}[\mu|\sigma] \tag{4.58}$$

$$D[\mu|\mu'] = \sum_{n=0}^{\infty} \varepsilon^n\, D^{(n)}[\mu|\mu'] . \tag{4.59}$$

Inserting all the expansions into (4.52) and equating powers of ε, we obtain:

$$\tilde{D}^0[\sigma|\bar{\sigma}]\, T_0[\mu|\bar{\sigma}] = D^0[\mu|\bar{\mu}]\, T_0[\bar{\mu}|\sigma] \tag{4.60}$$

$$\tilde{D}^{o}[\sigma|\bar{\sigma}]\, T^{(1)}[\mu|\bar{\sigma}] - D^{o}[\mu|\bar{\mu}]\, T^{(1)}[\bar{\mu}|\sigma] = D^{(1)}[\mu|\bar{\mu}]\, T_o[\bar{\mu}|\sigma] - \tilde{D}^{(1)}[\sigma|\bar{\sigma}]\, T_o[\mu|\bar{\sigma}] \tag{4.61}$$

$$\tilde{D}^{(2)}[\sigma|\bar{\sigma}]\, T_o[\mu|\bar{\sigma}] + \tilde{D}^{(1)}[\sigma|\bar{\sigma}]\, T^{(1)}[\mu|\bar{\sigma}] + \tilde{D}^{o}[\sigma|\bar{\sigma}]\, T^{(2)}[\mu|\bar{\sigma}]$$

$$= D^{(2)}[\mu|\bar{\mu}]\, T_o[\bar{\mu}|\sigma] + D^{(1)}[\mu|\bar{\mu}]\, T^{(1)}[\bar{\mu}|\sigma] + D^{o}[\mu|\bar{\mu}]\, T^{(2)}[\bar{\mu}|\sigma] \tag{4.62}$$

etc.

where summation over barred spin indices is implied. We assume that (4.60) is equivalent to the zeroth order result (4.48) and we assume, therefore, that $T_o[\mu|\sigma]$ and $D^o[\mu|\mu']$ are known. The problem is then completely analogous to standard Schrödinger perturbation theory, D_μ corresponding to the eigenvalue, and T to the eigenfunction. In the ordinary perturbation case, one can always choose the higher order corrections to a given eigenfunction to be orthogonal to the zeroth order part. This is also the case here. Starting from any zero order transformation, T_o, we can build an operator $\bar{T}[\mu|\sigma]$ which satisfies Eqs. (4.60)-(4.62) with $D_\mu^{(n)} \to \bar{D}_\mu^{(n)}$ and the normalization conditions

$$\langle \bar{T}^{(n)}[\mu|\sigma]\, T_o[\mu'|\sigma] \rangle_o = 0 \qquad n > 0 \tag{4.63}$$

$$\langle T_o[\mu|\sigma]\, T_o[\mu'|\sigma] \rangle_o = \delta_{\mu,\mu'}\, P_o[\mu] \tag{4.64}$$

and where $\bar{T}_o = T_o$. If we then multiply (4.61) by $T_o[\mu'|\sigma]$ and average over $P_o[\sigma]$ we obtain

$$\langle T_o[\mu'|\sigma]\, \tilde{D}^o_\sigma\, \bar{T}^{(1)}[\mu|\sigma] \rangle_o - D^o[\mu|\bar{\mu}] \langle T_o[\mu'|\sigma]\, \bar{T}^{(1)}[\bar{\mu}|\sigma] \rangle_o$$

$$= \bar{D}^{(1)}[\mu|\bar{\mu}] \langle T_o[\mu'|\sigma]\, T_o[\bar{\mu}|\sigma] \rangle_o - \langle T_o[\mu'|\sigma]\, \tilde{D}^{(1)}_\sigma\, T_o[\mu|\sigma] \rangle_o \tag{4.65}$$

By using the adjoint property of $\tilde{D}^o$, Eqs. (4.60) and (4.63), it is found that the first term on the left of (4.65) vanishes. The second term vanishes also because of (4.63), and we are left with:

$$\bar{D}^{(1)}[\mu|\mu']\, P_o[\mu'] = \langle T_o[\mu'|\sigma]\, \tilde{D}^{(1)}_\sigma\, T_o[\mu|\sigma] \rangle_o \quad . \tag{4.66}$$

Knowing $\bar{D}_\mu^{(1)}$, we can go back to (4.61) and determine $\bar{T}^{(1)}$. This requires inverting the matrix $\tilde{D}^o(\sigma|\sigma')\,\delta_{\mu,\mu'} - D^o(\mu|\mu')\,\delta_{\sigma,\sigma'}$ all of which can be done by using straightforward perturbation theory techniques. The calculations are rather long, and they are carried out in Ref. 2 where the general expression for $\bar{T}^{(1)}$ is obtained. Of course, to calculate D_μ

to first order, one does not need $\bar{T}^{(1)}$, just as one does not need the first order eigenfunctions to calculate the energy levels to first order. $\bar{T}^{(1)}$ is needed to obtain $D[\mu|\mu']$ to second order in ϵ. Once $\bar{T}^{(1)}$ is found, an analysis of Eq. (4.62) identical to that carried out above for Eq. (4.61) yields:

$$\bar{D}^{(2)}[\mu|\mu']P_0[\mu'] = \langle T_0[\mu'|\sigma]\tilde{D}_\sigma^{(2)}T_0[\mu|\sigma]\rangle_0 \\ + \langle T_0[\mu'|\sigma]\tilde{D}_\sigma^{(1)}\bar{T}^{(1)}[\mu|\sigma]\rangle_0 . \tag{4.67}$$

It is clear that this procedure can be carried on to any order desired. There is, however, an additional problem left and that is the mapping function $\bar{T}$ we have constructed, satisfying, the normalization conditions (4.63) and (4.64), will not in general satisfy Eq. (4.7). This complication can be remedied by rotating $\bar{T}[\mu|\sigma]$ in μ-space into $T[\mu|\sigma]$:

$$T[\mu|\sigma] = \sum_{\{\bar{\mu}\}} S[\mu|\bar{\mu}]\,\bar{T}[\bar{\mu}|\sigma] \tag{4.68}$$

where S is a symmetric operator, to be determined from the diagonalization condition (4.7). In (4.68) and below summation over repeated barred indices is implied; of course, we must carefully preserve Eq. (3.3) as well. We note then that

$$\tilde{D}_\sigma T[\mu|\sigma] = S[\mu|\bar{\mu}]\,\tilde{D}_\sigma \bar{T}[\bar{\mu}|\sigma] = D[\mu|\bar{\mu}]\,T[\bar{\mu}|\sigma] \tag{4.69}$$

with:

$$D[\mu|\mu'] = S[\mu|\bar{\mu}]\,\bar{D}[\bar{\mu}|\bar{\mu}']\,S^{-1}[\bar{\mu}'|\mu'] \tag{4.70}$$

from which it follows that:

$$D[\mu|\mu']P[\mu'] = \langle T[\mu'|\sigma]\,\tilde{D}_\sigma T[\mu|\sigma]\rangle \tag{4.71}$$

as desired.

The simplest way of preserving (3.3) is by imposing the condition on the rotation matrix

$$\sum_{\{\mu\}} S[\mu|\mu'] = 1 \tag{4.72}$$

It is not difficult in perturbation theory to construct S for a given $\bar{T}$.

In this section, we have developed a general formalism for carrying out our RSDRG transformation in a way that preserves the time-independent nature of the SFO. In developing this method we have had to enlarge the class of transformation functions $T[\mu|\sigma]$ beyond the product form (3.4). In particular at first order in ϵ one finds that $T^{(1)}[\mu|\sigma]$ contains couplings between cells.

V. Applications

In this section, we discuss the application of the formalism we have developed to two different calculations. The first calculation corresponds to a direct generalization of the NvL cumulant expansion to dynamics and is carried out to first order in the intercell coupling. The second calculation builds on the first and indicates the relevant role of a dynamic variable generated in a second order calculation.

A. First Order Cumulant Expansion.

The general RSDRG procedure is very direct. Starting with a given $\tilde{D}_\sigma$ and $H[\sigma]$ one constructs $\tilde{D}_\mu$ and $H[\mu]$. From $H[\mu]$ one extracts the static parameters as in Eq. (3.12). $\tilde{D}_\mu$ and $\tilde{D}_\sigma$ will both be of the same general form given by (2.17) with the functions $W[\mu]$ being obtained from $P[\mu]$ by the same procedure used to obtain $W[\sigma]$ from $P[\sigma]$. We can then extract the functions $V[\mu]$ from D_μ. The equations relating the parameters determining $V[\mu]$ to the parameters determining $V[\sigma]$ are the dynamic recursion relations. Let us assume that our $\tilde{D}[\sigma|\sigma']$ is of the general form (2.17) with the W functions given by (2.20) in both σ and μ-space. We assume to first order in ϵ that $\tilde{D}[\sigma|\sigma']$ can be decomposed in the form

$$\tilde{D}[\sigma|\sigma'] = \tilde{D}_0[\sigma|\sigma'] + \varepsilon \tilde{D}^{(1)}[\sigma|\sigma'] + O(\varepsilon^2) \tag{5.1}$$

with $\tilde{D}_0$ acting on spins all in the same cell, and $\bar{D}^{(1)}$ coupling spins in adjacent cells only.

Using the perturbation theory results from the last section, we obtain from Eq. (4.60) and Eq. (4.66) that $D[\mu|\mu']$ is given to zeroth and first order by:

$$D^0[\mu|\mu'] P_0[\mu'] = -\sum_i \frac{\lambda_i}{2} \Lambda^{[i]}_{\mu,\mu'} P_0[\mu'] \mu_i \mu_i' \tag{5.2}$$

$$D^{(1)}[\mu|\mu'] P_0[\mu'] = \langle T_0[\mu'|\sigma] \tilde{D}^{(1)}_\sigma T_0[\mu|\sigma] \rangle_0 \; . \tag{5.3}$$

A direct calculation shows, to first order in ϵ, that

$$\langle \bar{T}[\mu|\sigma] \bar{T}[\mu'|\sigma] \rangle = \delta_{\mu,\mu'} P[\mu] \tag{5.4}$$

where

$$P[\mu] = e^{H[\mu]}/Z \tag{5.5}$$

and $H[\mu]$ is of the same form as $H[\sigma]$ given by (2.2) but with the new nearest neighbor coupling K given by Eq. (3.12). Notice (i) that we do not need $\bar{T}^{(1)}[\mu|\sigma]$ in our first order calculation and (ii) $\langle \bar{T}[\mu|\sigma]\, \bar{T}[\mu'|\sigma] \rangle$ is diagonal to first order and no rotation in μ-space is required. Detailed analysis of D_μ^{o} and $D_\mu^{(1)}$ show that they are precisely of the same form as $D[\sigma|\sigma']$ given by Eq. (2.17), with the $W[\mu]$'s appropriately determined by the new probability distribution $P[\mu]$ and with the new couplings $V[\mu]$ given up to first order by

$$V_1^{[i]}[\mu'] = -\lambda_1 + \varepsilon \langle \psi_i \tilde{D}_\sigma^{(1)} \psi_i \rangle_o \tag{5.6a}$$

$$V_2^{[i,j]}[\mu'] = \varepsilon \lambda_1 2z^2 K I_{ij} + \frac{\varepsilon}{2}\left[\langle \psi_i \tilde{D}_\sigma^{(1)} \psi_j \rangle_o + \langle \psi_j \tilde{D}_\sigma^{(1)} \psi_i \rangle_o \right] \tag{5.6b}$$

$$V_3^{[i,j]}[\mu'] = \frac{\varepsilon}{2}\left[\langle \psi_i \psi_j \tilde{D}_\sigma^{(1)} \psi_i \psi_j \rangle_o - \langle \psi_j \tilde{D}_\sigma^{(1)} \psi_j \rangle_o - \langle \psi_i \tilde{D}_\sigma^{(1)} \psi_i \rangle_o \right] \tag{5.6c}$$

$$V_4^{[i,j,k]}[\mu'] = 0 \tag{5.6d}$$

$$V_5^{[i,j,k]}[\mu'] = 0 \tag{5.6e}$$

where

$$I_{ij} = \sum_{\delta} \delta_{j,i+\delta}$$

denotes nearest neighbor interactions. To first order in the intercell coupling there are no three (or higher) spin terms generated in the dynamical operator. These results indicate how the perturbation theory method we have developed limits the number of coefficients that must be calculated, and therefore, the complexity of the initial $D(\sigma|\sigma')$ operator. We see then, as long as $\tilde{D}_\sigma$ couples only nearest neighbor cells in first order, that only three terms are generated at first order in perturbation theory. It is natural, therefore, to treat the problem where only these three parameters are included in the original operator $\tilde{D}[\sigma|\sigma']$. We start therefore with $D[\sigma|\sigma']$ given by Eq. (2.17) with

$$V_1^{[i]}[\sigma'] = -\alpha \tag{5.7a}$$

$$V_2^{[i,j]}[\sigma'] = -\beta\, I_{ij} \tag{5.7b}$$

$$V_3^{[i,j]}[\sigma'] = -\gamma\, I_{ij} \tag{5.7c}$$

and all other V's zero. We then divide $\tilde{D}[\sigma|\sigma']$ into an intracell part $\tilde{D}^{o}[\sigma|\sigma']$ and a part coupling cells. There are two types of terms coupling the cells. One coupling is through the W functions and the other through the matrix I_{ij} appearing in V_2 and V_3. The matrix

I_{ij} can be written in terms of cell variables as

$$I_{ij}(a,a') = I^{o}_{ij}(a,a') + I^{(1)}_{ij}(a,a') \tag{5.8a}$$

$$I^{o}_{ij}(a,a') = \delta_{ij}\,(\delta_{a',a+1} + \delta_{a',a-1}) \tag{5.8b}$$

$$I^{(1)}_{ij}(a,a') = \delta_{j,i+ca}(\delta_{a',a+1} + \delta_{a',a-1}) + \delta_{j,i-c(a+1)}\,\delta_{a',a+1} + \delta_{j,i-c(a-1)}\,\delta_{a',a-1} \tag{5.8c}$$

where c is the lattice constant on the μ-lattice and the notation $a+1$ stands for the vector $\vec{A}_{a+1}$. In the statistical factors like $W[\sigma_{i,a}]$, we can write, for example,

$$W[\sigma_{i,a}] = W_o[\sigma_{i,a}] + \varepsilon\, W^{(1)}[\sigma_{i,a}] + O(\varepsilon^2) \tag{5.9}$$

where

$$W_o[\sigma_{i,a}] = e^{-K\sigma_{i,a}(\sigma_{i,a+1} + \sigma_{i,a-1})} \tag{5.10}$$

$$W^{(1)}[\sigma_{i,a}] = W_o[\sigma_{i,a}]\left(-\sigma_{i,a}E^{(1)}_{i,a}(\sigma)\right) \tag{5.11}$$

and

$$E^{(1)}_{i,a}(\sigma) = \sum_j \sum_{a'} K\, I^{(1)}_{ij}(a,a')\,\sigma_{j,a'} \tag{5.12}$$

is the interaction energy on the spin at site i, a due to spins outside the ith cell. Given the explicit form for the single cell operator,

$$-\tilde{D}_o[\sigma|\sigma'] = \frac{\alpha}{2}\sum_{i,a} \sigma_{i,a}\sigma'_{i,a}\Lambda^{[i,a]}_{\sigma,\sigma'} W_o[\sigma_{i,a}] + \frac{\beta}{4}\sum_{i,a,a'} I^{o}_{ii}(a,a')\,\sigma_{i,a}\sigma'_{i,a'} \times \Lambda^{[i,a,ia']}_{\sigma,\sigma'} W_o[\sigma_{i,a},\sigma_{i,a'}] + \frac{\gamma}{4}\sum_{i,a,a'} I^{o}_{ii}(a,a')\,\sigma_{i,a}\sigma_{i,a'}\sigma'_{i,a}\sigma'_{i,a'} \times \Lambda^{[ia,ia']}_{\sigma,\sigma'} W_o[\sigma_{i,a},\sigma_{i,a'}] \tag{5.13}$$

where

$$W_o[\sigma_{i,a},\sigma_{i,a'}] = e^{-H^{i}_{o}[\sigma]}, \tag{5.14}$$

the next step is to construct the slowest odd eigenfunction and eigenvalue. One finds after a straightforward calculation that

$$\psi(\sigma) = N_1(\sigma_T - f_1\,\sigma_c) \tag{5.15}$$

and

$$\lambda_1 = \frac{1}{2}\left[(a+d) - \sqrt{(a+d)^2 - 4(ad-bc)}\right] \tag{5.16}$$

where

$$f_1 = \frac{b}{d-\lambda_1} \tag{5.17}$$

$$N_1 = \left[3(3r+1)(1-r) + (3r-f_1)^2\right]^{1/2} \tag{5.18}$$

$$r = \langle \sigma_a \sigma_{a+1} \rangle_o = (x-1)/(x+3) \tag{5.19}$$

$$a = \tilde{\alpha}(1-2g) + \tilde{\beta}(1-g+g^2) \tag{5.20a}$$

$$b = 3\tilde{\alpha} g^2 - 3g\tilde{\beta} \tag{5.20b}$$

$$c = \tilde{\alpha} g(g-2) - 6g\tilde{\gamma} \tag{5.20c}$$

$$d = 3\tilde{\alpha} + 6(1+g+g^2)\tilde{\gamma} \tag{5.20d}$$

$$\tilde{\alpha} = \alpha \operatorname{Cosh}^2 K \tag{5.21a}$$

$$\tilde{\beta} = 2\beta(1-g)\operatorname{Cosh}^3 K \tag{5.21b}$$

$$\tilde{\gamma} = \gamma(1-g)\operatorname{Cosh}^3 K \tag{5.21c}$$

and

$$g = \tanh 2K \tag{5.22}$$

$$x = e^{4K} \; . \tag{5.23}$$

We have then only to carry out the cell averages indicated in Eq . (5.6) to obtain the recursion relations, valid to first order,

$$V_1^{[i]}[\mu'] = -\lambda_1 \tag{5.24a}$$

$$V_2^{[i,j]}[\mu'] = -2\beta D^2 I_{ij} \tag{5.24b}$$

$$V_3^{[i,j]}[\mu'] = -\frac{\gamma E^2}{8} I_{ij} \tag{5.24c}$$

where

$$D = \langle W_o[\sigma_{i,a}]\, \sigma_{i,a}\, \psi_i(\sigma) \rangle_o$$
$$= \frac{2N_1}{X+3}\left[e^{2K}(1-f_1) + 1 + f_1 \right] \tag{5.25}$$

and

$$E = \langle W_o[\sigma_{i,a}]\left(\psi_i(\sigma) - \psi_i(\sigma_{a-1}, -\sigma_a, \sigma_{a+1})\right)^2 \rangle_o$$
$$= \frac{8N_1^2}{X+3}\left[(1+f_1^2)(e^{2K}-1) - 2f_1(e^{2K}+1) \right] . \tag{5.26}$$

Comparing the results on the μ-lattice given by Eq. (5.24) with Eq. (5.7), we see that the $V[\mu]$'s are of precisely the same form as the $V[\sigma]$'s. We, therefore, have the closed set of recursion relations

$$\alpha' = \lambda_1 \tag{5.27a}$$

$$\beta' = 2D^2\beta \tag{5.27b}$$

$$\gamma' = E^2\gamma/8 \tag{5.27c}$$

$$K' = 2\nu^2 K \tag{5.27d}$$

where, using Eq. (3.13) and (5.15), ν is given by

$$\nu = \frac{N_1}{X+3}\left[X(3-f_1) + 1 + f_1 \right] . \tag{5.27e}$$

In analyzing these recursion relations, we found it convenient to introduce the ratios

$$R_\beta = \beta/\alpha \tag{5.28a}$$

and

$$R_\gamma = \gamma/\alpha . \tag{5.28b}$$

We interpret the fixed points of our RG transformation in the following manner. Our original SFO is of the form

$$D_\sigma = D_\sigma(\alpha, \beta, \gamma, K) = \alpha\, D_\sigma(1, R_\beta, R_\gamma, K) . \tag{5.29}$$

We choose to extract an overall factor of α, rather than β or γ, since we expect α will scale more slowly to zero under the RSDRG than β and γ. We display the fixed points of our recursion relation in Table 1 and find that for all of the physical fixed points α is a relevant variable. R_γ is driven to zero for all of the fixed points, while there is one fixed point where R_β is finite.

TABLE I

	K	R_β	R_γ	Z	ν
1).	.212	0	0	.841	1.75
2).	.213	-.186	0	1.701	1.76

Table I. Fixed-point values for K, R_β, R_q, the dynamic critical index z and the static critical index ν.

If we are at the fixed point $K = K^*$, $R_\gamma = R_\gamma^*$ and $R_\beta = R_\beta^*$, then the SFO transforms as

$$\alpha D_\nu^*(1, R_\beta^*, R_\gamma^*, K^*) \rightarrow \alpha' D_\mu^*(1, R_\beta^*, R_\gamma^*, K^*) . \tag{5.30}$$

Since λ_1 is proportional to α we can write

$$\alpha' = \Delta_\alpha \alpha \tag{5.31}$$

and the time evolution operator transforms under our RSDRG transformation as

$$e^{D_\nu^* t} \rightarrow e^{D_\mu^* \Delta_\alpha t} \tag{5.32}$$

and dynamical scaling follows through the simple rescaling of time by

$$t' = \Delta_\alpha t \tag{5.33}$$

so

$$e^{\tilde{D}_\nu^* t} \longrightarrow e^{\tilde{D}_\mu^* t'} \quad . \tag{5.34}$$

It is then conventional to define the dynamical critical index z by

$$t' = b^{-z} t \tag{5.35}$$

where b is the spatial rescaling factor of the RG transformation ($= \sqrt{3}$ for a two dimensional triangular lattice), so

$$z = -\ln \Delta\alpha / \ln b \quad . \tag{5.36}$$

The value we find for the dynamical index z corresponding to the most stable fixed point is $z = 1.70$. This value is slightly less than the rigorous lower bound of 1.75. It is, however, not unreasonable when compared with the high temperature results of $z = 2$,[16] 2.125[17] the Monte Carlo results[18] giving $z = 1.85$.

The clear weak point in this first order calculation centers about the poor static results. The value for the transition temperature is reasonable, but the critical index ν is very poor. It is easy to trace the reason for this behavior. The parameter f_1, which characterizes the relative amount of σ_T and σ_C in Ψ, scales to very small values or zero for all of the fixed points found in this first order analysis. Recall that when one uses the majority rule for φ_i, which gives a reasonable estimate for the static exponents, one has $f_1 = 1$. Our α, β, γ, model has no mechanism in a first order theory for maintaining f_1 near 1. The fixed point eigenfunctions Ψ therefore are essentially linear.

B. The $\alpha - \delta$ Model.

The next logical step in our development is to go to the second order calculation to see if our results are substantially improved. In the other problem[15] where these methods have been used the second order calculations improved the results considerably. While we have been investigating such a complete second order calculation, the results are still incomplete. The calculation is rather involved. At second order one finds that a total of 13 new couplings are generated and one must consistently include all 19 couplings in the problem. There is also another reason we don't procede directly to the second order calculation. There is now a good deal of evidence that there is one parameter generated at second order, starting with the α, β, γ, model discussed above, which is more relevant than β or γ and which is the key parameter controlling the behavior of the nearest neighbors of a spin at site i that is involved in the flipping process.

Let us define what we will call the $\alpha - \delta$ model, a member of the class of models we have been studying thus far, by

$$\tilde{D}[\sigma|\sigma'] = -\frac{\alpha}{2}\sum_i \Lambda^{[i]}_{\sigma,\sigma'} W_i[\sigma]\,\sigma_i\sigma_i'\left(1+\frac{\delta}{2}\sum_{j,k}\Delta^{s}_{ijk}\,\sigma_j\sigma_k\right) \tag{5.37}$$

where $W[\sigma_i]$ is given by Eq. (2.15) and

$$\Delta^{s}_{ijk} = \Delta_{ijk} + \Delta_{ikj} \tag{5.38}$$

$$\Delta_{ijk} = \sum_a \delta_{j,i+ca}\left(\delta_{k,i-c(a-1)} + \delta_{k,i-c(a+1)}\right) \tag{5.39}$$

so we see that Δ_{ijk} generates the (6) nearest neighbor triangles about the ith spin. Note that the δ-term did not appear in our first order calculation and corresponds to a σ dependent contribution to $V_1^{[i]}[\sigma]$ in Eq. (2.17). Starting with the α, β, γ model a δ-term is generated in second order. We can best understand the relevance of the δ-parameter by looking at the intracell contribution to $\tilde{D}[\sigma|\sigma']$ given by:

$$\tilde{D}_c[\sigma|\sigma'] = -\frac{\alpha}{2}\sum_{i,a}\Lambda^{[i,a]}_{\sigma,\sigma'} W_0[\sigma_{i,a}]\,\sigma_{i,a}\sigma'_{i,a}\left(1+\delta\,\sigma_{i,a+1}\sigma_{i,a-1}\right). \tag{5.40}$$

We can then easily solve the eigenvalue problem

$$\tilde{D}_c[\sigma|\bar{\sigma}]\,\psi_i^n(\bar{\sigma}) = -\lambda_n\,\psi_i^n(\sigma) \tag{5.41}$$

to obtain the slowest odd eigenfunction and the corresponding eigenvalue. In Fig. 2, we have plotted the eigenvalue λ_1/α and the quantity f_1 for fixed temperature as a function of δ.

Figure 2. Plot of the eigenvalue λ_1 and coefficient f_1 defined by Eqs. (4.30) and (3.9) respectively for the operator (5.40) as a function of the parameter δ and for several values of the coupling K.

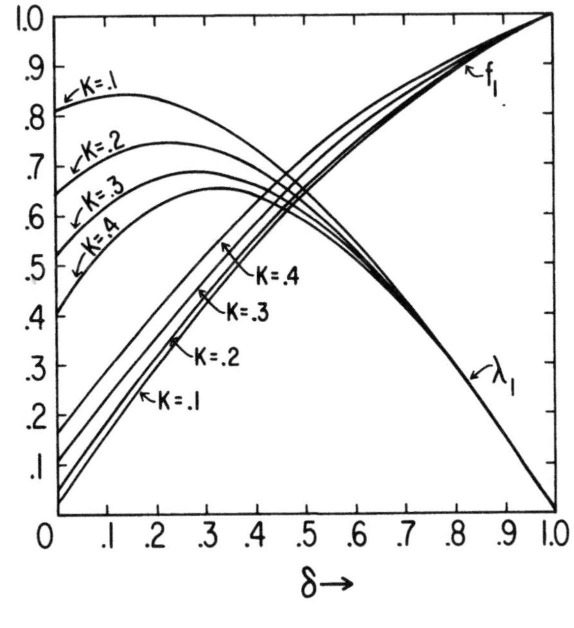

We obtain the interesting result that as $\delta \to 1$ λ_1/α approaches zero and $f_1 \to 1$. This means that for a model with $\delta = 1$, at zeroth order, the slowest mode is given by the "majority rule"

$$\psi_i'(\sigma)\Big|_{\delta=1} = \frac{1}{2}\left(\sigma_i^{\tau} - \sigma_i^{c}\right) \tag{5.42}$$

and, to lowest order, this mode doesn't decay! It should be clear that the δ-variable is important in controlling the accuracy of the static recursion relations. Armed with this information, we can set up the following perturbation theory calculation. We must, for consistency, treat the term multiplying the part of Δ^{s}_{ijk} coupling cells as being second order in ϵ. We will also treat the difference $\delta - 1$ in the intracell part of $\tilde{D}_{\sigma}$ as a second-order quantity. We can then write

$$\tilde{D}[\sigma|\sigma'] = \tilde{D}_0[\sigma|\sigma'] + \tilde{D}_I[\sigma|\sigma'] \tag{5.43}$$

where

$$\tilde{D}_0[\sigma|\sigma'] = -\frac{\alpha}{2}\sum_{i,a} \Lambda^{[i,a]}_{\sigma,\sigma'} W_0[\sigma_{i,a}]\, \sigma_{i,a}\, \sigma'_{i,a}\left(1 + \sigma_{i,a+1}\,\sigma_{i,a-1}\right). \tag{5.44}$$

We have, in this case, insured that to lowest order in ϵ, $\lambda_1 = 0$ and the majority rule choice for φ_i is an eigenfunction. This immediately gives us that

$$D_0[\mu|\mu'] = 0 \quad . \tag{5.45}$$

The first order contribution to $\tilde{D}^{(1)}[\sigma|\sigma']$ comes from the $W[\sigma_{i,a}]$ term and it is easy to show that

$$\tilde{D}^{(1)}_{\sigma}\, T_0[\mu|\sigma] = 0 \,. \tag{5.46}$$

Then we easily find that:

$$D^{(1)}[\mu|\mu']\, P_0[\mu'] = \left\langle T_0[\mu'|\sigma]\, \tilde{D}^{(1)}_{\sigma}\, T_0[\mu|\sigma]\right\rangle_0 = 0 \,. \tag{5.47}$$

Since the first order contribution to T satisfies

$$\tilde{D}_0[\sigma|\bar{\sigma}]\, \bar{T}^{(1)}[\mu|\bar{\sigma}] = 0 \tag{5.48}$$

and

$$\left\langle T_0[\mu'|\sigma]\, \bar{T}^{(1)}[\mu|\sigma]\right\rangle_0 = 0, \tag{5.49}$$

a consistent solution is

$$\bar{T}^{(1)}[\mu|\sigma] = 0 \ . \tag{5.50}$$

Going to second order, we have

$$\tilde{D}_{\sigma}^{o}\,\bar{T}^{(2)}[\mu|\sigma] + \tilde{D}_{\sigma}^{(2)}\,T_o[\mu|\sigma] = \bar{D}_{\mu}^{(2)}\,T_o[\mu|\sigma] \ . \tag{5.51}$$

Averaging over $T_o[\mu'|\sigma]$ times Eq. (5.51) gives

$$\bar{D}^{(2)}[\mu|\mu']\,P_o[\mu'] = \left\langle T_o[\mu'|\sigma]\,\tilde{D}_{\sigma}^{(2)}\,T_o[\mu|\sigma]\right\rangle_o \ . \tag{5.52}$$

Working at second order, we don't need $\bar{T}^{(2)}[\mu|\sigma]$. In this case, we find directly that

$$\begin{aligned}\tilde{G}[\mu|\mu'] &= \left\langle \bar{T}[\mu|\sigma]\,\bar{T}[\mu'|\sigma]\right\rangle \\ &= \delta_{\mu,\mu'}\left\langle T_o[\mu|\sigma]\right\rangle + O(\varepsilon^3)\end{aligned} \tag{5.53}$$

and, again we don't need to perform a rotation in μ-space. We see that at zeroth and first order only the statics are affected and one obtains the NvL statics at first order and second-order in the intercell coupling.

All that is left is the explicit evaluation of $D^{(2)}[\mu|\mu']$ using Eq. (5.52). One finds after evaluating these matrix elements that $D^{(2)}[\mu|\mu']$ is of precisely the same form as Eq. (5.37) with the new parameters.

$$\alpha' = \frac{3\alpha}{2}(1-r)\left[1+\delta(r-1)\right] \tag{5.54}$$

$$\delta' = \frac{2}{3}\,\delta v^2\left[1+\delta(r-1)\right]^{-1} \tag{5.55}$$

where, as before,

$$r = (x-1)/(x+3) \tag{5.56}$$

and, in this case,

$$v = (x+1)/(x+3) \tag{5.57}$$

with x given by (5.23). If this procedure is to be consistent than there must exist a fixed point $\delta^* \approx 1$. In particular, we find a non-trivial fixed point at

$$\delta^* = (r-1)^{-1}\,(2v^2/3-1) \ . \tag{5.58}$$

If we take this together with the second order fixed point value for $K^* = .2789$, we obtain

$$\delta^* = .94 \tag{5.59}$$

From this we obtain at this fixed point

$$\Delta_\alpha^* = (\alpha'/\alpha)^* = \frac{1-r^*}{2} \tag{5.60}$$

and

$$z = -\ln \Delta_\alpha^* / \ln b = 2.22 \tag{5.61}$$

together with the static results (due to NvL) $K_c = .251$ (.275 exact) and $\nu = .950$ (1 exact).

These calculations indicate that δ is a relevant parameter (the fixed point $\delta^* = .94$ is unstable in the same sense as $K^* = .2789$). Ultimately δ is driven by the temperature and apparently must be fixed at the critical point just as the temperature and magnetic field must be adjusted at their critical values.

VI. Discussion

We have presented a new real space dynamic renormalization group theory. This theory deals with the major problem of non-Markovian behavior associated with RG dynamical transformations. The resolution of this problem puts new constraints on the admissible RG transformations. We have seen how our formalism can be applied in two model calculations. Our results seem very encouraging and can apparently be systematically improved. We are presently investigating further applications of these ideas.

We note in closing that there are several other schemes[19] for applying the RSRG to dynamics. These methods have been more or less successful in calculating the dynamic index z for kinetic Ising models. It is very difficult at this point to compare these methods with our own since they don't address the general questions of handling non-Markovian terms and the generation of new couplings in the SFO D_σ .

REFERENCES

1. G.F. Mazenko, M. J. Nolan and O.T. Valls, Phys. Rev. Lett. 41, 500 (1978).
2. G.F. Mazenko, M. J. Nolan and O.T. Valls, unpublished.
3. K.G. Wilson, Phys. Rev. Lett. 28, 548 (1972).
4. M. Fixman, J. Chem. Phys. 36, 310 (1962).
5. K. Kawasaki, Ann. Phys. (N.Y.) 61, 1 (1970).
6. L.P. Kadanoff and J. Swift, Phys. Rev. 166, 89 (1968); R.A. Ferrell, Phys. Rev. Lett. 24, 1169 (1970); R. Zwanzig, in Statistical Mechanics, ed. by S.A. Rice, K.F. Feed, and J.C. Light (U. of Chicago Press, Chicago, 1972).
7. P. Hohenberg and B.I. Halperin, Rev. Mod. Phys. 49, 435 (1977).
8. E. Brézin, The Role of Instantons in Divergent Perturbation Series, European Particle Physics Conference, July, 1977.
9. A.N. Berker, S. Ostlund and F.A. Putnam, Phys. Rev. B17, 3650 (1978).
10. L.P. Kadanoff, Physics (NY) 2, 263 (1966).
11. T. Niemeijer and J.M.J. van Leeuwen, Phys. Rev. Lett. 31, 1411 (1973), and in Phase Transitions and Critical Phenomena, ed. by C. Domb and M.S. Green (Academic, New York, 1976), Vol. 6.
12. R. Glauber, J. Math. Phys. (NY) 4, 234 (1963).
13. K. Kawasaki, in Phase Transitions and Critical Phenomena, ed. by C. Domb and M.S. Green (Academic, NY, 1972), Vol. 2.
14. G.F. Mazenko in Correlation Functions and Quasiparticle Interactions, ed. by J.W. Halley (Plenum, NY, 1978).
15. J. Hirsch and G.F. Mazenko, to be published, Phys. Rev. B.
16. H. Yahata and M. Suzuki, J. Phys. Soc. Japan 27, 1421 (1969).
17. Z. Rácz and M.F. Collins, Phys. Rev. B13, 3074 (1976).
18. E. Stoll, K. Binder and T. Schneider, Phys. Rev. B8, 3266 (1973).
19. Y. Achiam, J. Phys. A 11, 975 (1978); Y. Achiam, J. Phys. A 11, L129 (1978); W. Kinzel, Z. Phys. B29, 361 (1978); Y. Achiam and M.J. Kosterlitz, Phys. Rev. Lett. 41, 128 (1978); M. Suzuki, K. Sogo, I. Matsuba, H. Ikada, T. Chikama and H. Tanako, Preprint S.T. Chui, G. Forgacs and H.L. Frish, Preprint.

LIGHT SCATTERING FROM HELIUM 4 NEAR THE LAMBDA POINT

T.J. Greytak

Massachusetts Institute of Technology
Cambridge, Ma. 02139/USA

LIGHT SCATTERING FROM HELIUM 4 NEAR THE LAMBDA POINT

T.J. Greytak
Massachusetts Institute of Technology
Cambridge, Ma. 02139/USA

A. Introduction

For over forty years liquid helium 4 has fascinated physicists because of the strange phenomena such as frictionless flow and the wavelike propagation of heat that occur below the superfluid transition temperature, T_λ. Recently, however, attention has been focussed on the superfluid transition itself as a testing ground for our understanding of second order phase transitions. The critical phenomena experimenter views liquid helium 4 as a homogeneous substance composed of atoms without thermally accessible internal degrees of freedom. It can be prepared with a very high degree of purity. The temperature of a sample can be made uniform, held constant in time, and measured to better than a part in 10^6. The critical phenomena theorist views superfluid helium 4 as a system of spatial dimensionality $d = 3$, possessing a two component order parameter, $n = 2$. The inconvenient fact that the order parameter is a complex scalar which is not directly observable by experiment is mitigated by the presence of two inert variables, pressure and He^3 concentration, which can be used to test the hypothesis of universality.

The critical fluctuations associated with T_λ occur at long wavelength As a consequence light, rather than neutrons, has proven to be the best scattering probe of the dynamic critical behavior. Light scattering (Brillouin scattering since one is looking at the thermal fluctuations) measures directly the dynamic structure factor, $S(q,\omega)$, for the number density. Experiments are carried out at fixed wavevector q while a spectrometer sweeps over the frequency shift ω. Unfortunately, changing q is not easily done. q is related to the laser wavelength λ, the index of refraction of the helium n_r, and the scattering angle θ by the expression

$$q = \frac{4\pi n_r}{\lambda} \sin\theta/2 \tag{1}$$

The scattering from helium is very weak and much of the important information in the spectrum is located near zero frequency shift, $\omega = 0$. Therefore, stray elastically scattered light must be eliminated or at least kept to a small, measurable amount. This is most easily done in a $\theta = 90°$ scattering geometry; thus, virtually all measurements so far have been made at this angle. This leaves λ as the only remaining way to change q. In practice λ has been fixed by the convenient laser lines at 6328 Å, 5145 Å, and 4880 Å. Tunable dye lasers could be used to vary q continuously; but, even there the range of q's available is not too useful in testing the q dependence of critical dynamics. In the future the use of discrete λ's in the ultra-violet may serve this purpose.

The latest light scattering results in liquid helium are contained in two recent, rather detailed, papers by our group at MIT[1] and by Vinen's group at Birmingham[2]. The current agreement between the results of the two groups is very satisfying; the differences, where they occur, are probably not significant. In this lecture I shall simply summarize the principal results. Before I do so, however, it will be useful to review the "expected" behavior of $S(q,\omega)$ in the region near T_λ.

The correlation length, $\xi(T)$, associated with fluctuations in the order parameter diverges as the temperature T approaches T_λ. As long as one is far enough away from T_λ so that $q\xi(T) << 1$, the equations of hydrodynamics may be used to describe the behavior of fluctuations of wavevector q. However, some of the thermodynamic parameters which are involved in the hydrodynamic equations may couple to the order parameter; in this case they may diverge or go to zero with some power law dependence on the reduced temperature, $\varepsilon \equiv |T-T_\lambda|/T_\lambda$. Very close to the transition temperature, where $q\xi(T) \gtrsim 1$, the equations of hydrodynamics no longer apply. In this "critical region" we do not yet have a set of equations which can describe the dynamics of the fluctuations. In the extreme limit $q\xi(T) >> 1$, the correlation length is so much greater than the characteristic dimension of the fluctuation, $1/q$, that its temperature dependence no longer affects the behavior of the fluctuation. In this limit the dynamics become independent of temperature.

A schematic representation of the dynamic structure factor for liquid helium 4 is shown in Fig. 1. $S(q,\omega)$ contains distinct contributions from the two hydrodynamic normal modes of the fluid: First sound and entropy fluctuations. First (or ordinary) sound is primarily a pressure fluctuation at constant entropy. It gives rise to a doublet in $S(q,\omega)$ whose shift $\pm\omega_1$ and half width at half height Γ_1 are given by

$$\omega_1 = \nu_1 q$$
$$\Gamma_1 = \alpha_1 \nu_1 \qquad (2)$$

where ν_1 and α_1 are the velocity and attenuation coefficient of first sound. First sound is not the critical mode associated with T_λ, and its properties are only weakly effected by the transition. At zero frequency ν_1 has a sharp minimum at $T = T_\lambda$; but ν_1 does not approach zero.

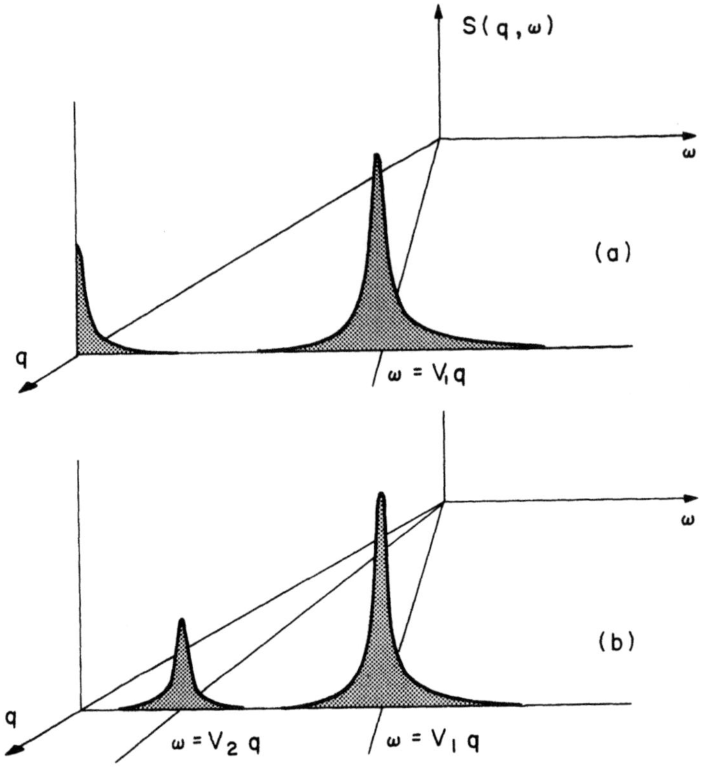

Figure 1: A schematic representation of the dynamic structure factor for liquid helium 4 above T_λ (a), and below T_λ (b). At a pressure of 23.1 Bar and 1 mK below T_λ typical spectral features are as follows: the intensity in the second sound line is about 40% of that in the first sound line, the shift of the second sound line is about 5 MHz and its half width is about 1 MHz; the shift of the first sound line is 963 MHz and its half width is 7.4 MHz.

At finite frequencies various processes associated with the phase transition will give rise to an attenuation of the sound and an associated dispersion in its velocity (an increase in the velocity over its zero frequency value). At frequencies of the order of megahertz these processes are now well understood. However, for the higher frequency sound waves studied by light scattering ($\nu_1 q/2\pi \sim 500$ to 1000 MHz) other processes give rise to the dispersion and attenuation, and these processes are not yet understood. Experimentally, then, both the attenuation and the dispersion are expected to increase as the transition is approached and exhibit maxima somewhere near T_λ.

The entropy fluctuations couple directly to the order parameter and are therefore the "critical mode" associated with the phase transition.(3,4) In the hydrodynamic region above T_λ the entropy fluctuations obey a diffusion equation and give rise to a single Lorentzian line in $S(q,\omega)$ centered at $\omega = 0$ with a half width Γ_2 given by

$$\Gamma_2 = D_{\mathcal{K}} q^2 \quad \text{for } T > T_\lambda \text{ and } q\xi(T) << 1 \tag{3}$$

Here $D_{\mathcal{K}}$ is the thermal diffusivity of the fluid, $D_{\mathcal{K}} \equiv \mathcal{K}/\rho C_p$. As T is decreased toward T_λ critical effects cause $D_{\mathcal{K}}$ to rise above its non-critical or "background" value. Dynamic scaling predicts that $D_{\mathcal{K}}$ should grow roughly as $\varepsilon^{-1/3}$. The linewidth should increase in this manner until the critical region is entered. In this region the line shape will no longer be Lorentzian and the line width will reach some limiting value. Dynamic scaling predicts that the critical contribution to this limiting spectral width should be proportional to $q^{3/2}$.

In the hydrodynamic region below T_λ the entropy fluctuations obey a wave equation and are referred to as second sound. They give rise to a doublet in $S(q,\omega)$ with a frequency shift $\pm\omega_2$ and a half width Γ_2 given by

$$\omega_2 = \nu_2 q \quad \text{for } T < T_\lambda \text{ and } q\xi(T) << 1 \tag{4}$$

$$\Gamma_2 = \left(\frac{D_{\mathcal{K}} + D_\zeta}{2}\right) q^2$$

where ν_2 is the velocity of second sound (which goes to zero as $T \to T_\lambda$), $D_{\mathcal{K}}$ is the thermal diffusivity, and D_ζ is a contribution to the damping which is intrinsic to the superfluid state. $D_{\mathcal{K}} + D_\zeta$ consists of a non-critical background contribution which is a slowly varying function of T and a critical contribution which diverges as $\varepsilon^{-1/3}$. The exact spectral shape of the entropy fluctuation part of $S(q,\omega)$ is determined by the equations of two-fluid hydrodynamics as long as $q\xi(T) << 1$. As the temperature is raised toward T_λ the splitting of the doublet will decrease and its linewidth will increase. When T becomes so close to T_λ that the critical region is entered, the width of the spectrum ceases to diverge and approaches a constant value. The exact spectral shape in this region is still a subject of theoretical speculation; but, it is unlikely that it will have the functional form characteristic of two-fluid hydrodynamics. Well inside the critical region, when $q\xi(T) >> 1$, $S(q,\omega)$ will no longer change with decreasing ε, and any parameters which might be used experimentally to describe $S(q,\omega)$ will also become independent of ε. The functional form of this limiting structure factor, which can be approached from either above or below T_λ, is of particular interest in the theory of dynamic critical phenomena.

In the hydrodynamic region the total scattered intensity is proportional to the isothermal compressibility which diverges very weakly when $T \to T_\lambda$. In the critical region the total scattering will level out and peak somewhere near $T = T_\lambda$. For a given ε, the isothermal compressibility decreases by about a factor of 2 as the pressure is raised from 0 to 25 Bar, so the total light scattered decreases by that amount. The ratio of the light scattered by second sound, I_2, to that scattered by first sound, I_1, is given to a high degree of accuracy in the hydrodynamic region by the simple relationship

$$I_2/I_1 = \gamma - 1 \tag{5}$$

where $\gamma \equiv c_p/c_v$ is the ratio of the specific heats. $\gamma - 1$ also diverges weakly as $T \to T_\lambda$. However, it is expected that I_2/I_1 will begin to fall below $\gamma - 1$ as one enters the critical region and go through a finite maximum in the vicinity of $T = T_\lambda$. $\gamma - 1$ evaluated at constant ε has a strong pressure dependence. For $\varepsilon = 5 \times 10^{-3}$, below T_λ, $\gamma - 1$ is about .0095 at .05 Bar but .28 at 25.0 Bar. Experimentally, this means that the scattering due to second sound is too weak to be measured with any precision at low pressures, but becomes comparable to the scattering from first sound at higher pressures. This is why the results of references 1 and 2 were all obtained at pressures above 20 Bar.

B. Second Sound

The frequency of the second sound observed in these experiments varies, in the hydrodynamic region, from about 5 to 30 MHz. The velocities obtained by Eq. 4 show no dispersion relative to velocities measured acoustically by Greywall and Ahlers[5] at frequencies three orders of magnitude lower. Both groups[1,2] have indicated that portions of their velocity data seemed slightly higher than the low frequency velocities; but, the increases were within the combined uncertainty of the scattering and acoustic experiments which, for our data, was about 2%.

Within the critical region Eq. 4 is not applicable. However a consistent ω_2 can be defined as a parameter in a model of the (unknown) correct structure factor. Vinen and Hurd use as such a model the hydrodynamic spectrum; we have used a pair of Lorentzian lines of equal area and equal half-width Γ_2, displaced by $\pm\omega_2$ from $\omega = 0$:

$$S_2(\omega) = \frac{I_2\Gamma_2}{2\pi}\left(\frac{1}{(\omega-\omega_2)^2 + \Gamma_2^2} + \frac{1}{(\omega+\omega_2)^2 + \Gamma_2^2}\right) \tag{6}$$

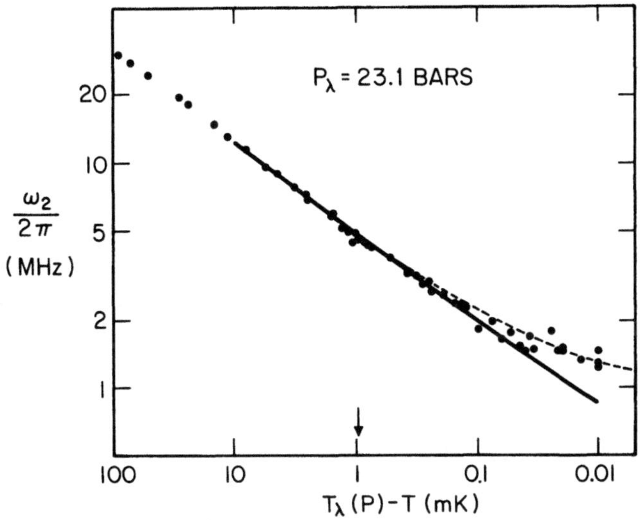

Figure 2: Second-sound frequency shift for the two-Lorentzian model. The solid line represents $q\nu_2$ with ν_2 taken from Ref. 5. The dashed curve is the prediction of the planar-spin model, Ref. 6. The arrow indicates the temperature at which $q\xi=1$.

An example of our results for ω_2 is shown in Fig. 2. Note that ω_2 does not go to zero at T_λ, as does ν_2, but it approaches a constant value. This is consistent with the concept of a limiting structure factor at $T = T_\lambda$ which is spectrally more complex than a single Lorentzian line. The data of Vinen and Hurd show a similar behavior. The arrow in Fig. 2 indicates the temperature at which $q\xi = 1$. We have used the expressions employed by Hohenberg, Siggia, and Halperin:(6)

$$\xi_t = 3.6\ \varepsilon^{-2/3}\ \text{Å}, \quad T < T_\lambda$$
$$\xi_+ = 1.4\ \varepsilon^{-2/3}\ \text{Å}, \quad T > T_\lambda \qquad (7)$$

Here the subscript t used for $T < T_\lambda$ indicates the correlation length for transverse fluctuations in the order parameter, and the subscript + indicates a correlation length defined in a different manner for $T > T_\lambda$. Notice that ω_2 deviates from $\nu_2 q$ only after ξ exceeds $1/q$. Above T_λ Vinen and Hurd switch to the two-Lorentzian model, Eq. 6, and show that ω_2 begins to decreases as ε increases and is zero to within the experimental uncertainty by the point at which ξ again falls below $1/q$. Our results show a similar behavior.

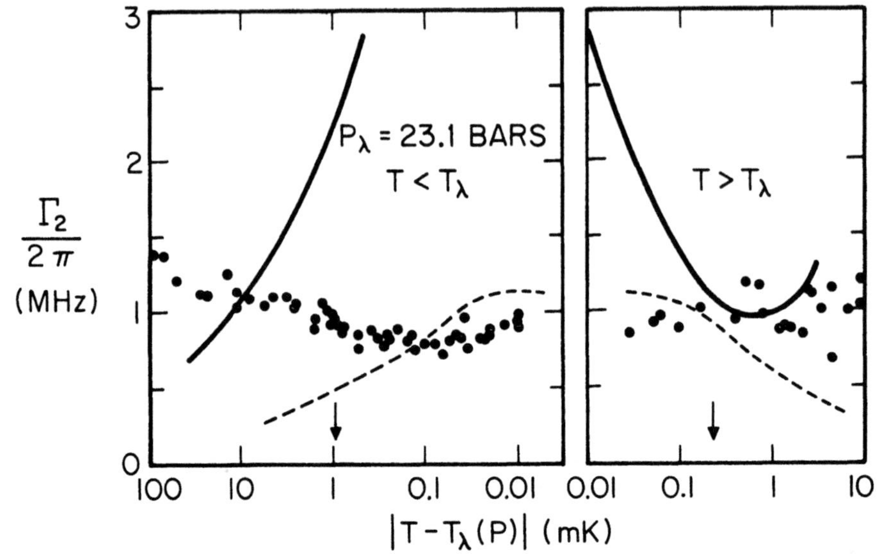

Figure 3: Half width for the two-Lorentzian model of the entropy spectrum. The solid lines correspond to Equations 3 and 4. The dashed curve is the prediction of the planar-spin model, Ref. 6. The arrow indicates the temperature at which $q\xi=1$.

The most surprising result of the light scattering measurements is that the damping of the high frequency entropy mode is almost independent of temperature over 4 orders of magnitude in ε below T_λ and 3 orders above. This is illustrated by our results at 23.1 Bar shown in Fig. 3. A particularly disturbing feature is the apparent deviation from hydrodynamic behavior in the hydrodynamic region below T_λ. Tyson[7] has measured $(D_{\varkappa} + D_\zeta)$ at low frequences by acoustic techniques. His data was taken at these temperatures but at saturated vapor pressure. The pressure dependence of the background contribution is not known; but, if it is similar to that of the background $D_{\varkappa}$ above T_λ,[8] then it is weak. The pressure dependence of the critical contribution follows from the dynamic scaling relation that $(D_{\varkappa} + D_\zeta)_{CRITICAL} = a\nu_2\xi$ where a is a numerical constant of order unity. ν_2 at a given ε decreases slightly in a known fashion[5] with increasing pressure, but it appears that ξ at a given ε is independent of pressure[9]. The solid line for $T < T_\lambda$ in Fig. 3 is obtained using Eq. 4 with Tyson's measured values of $D_{\varkappa} + D_\zeta$ multiplied by $\nu_2(23.1\text{Bar})/\nu_2(\text{SVP}) = 0.71$. Notice that the result rises significantly above the data well before $q\xi = 1$. The lowest $D_{\varkappa} + D_\zeta$ found by Tyson was 4.1×10^{-4} cm^2/sec; this corresponds to the

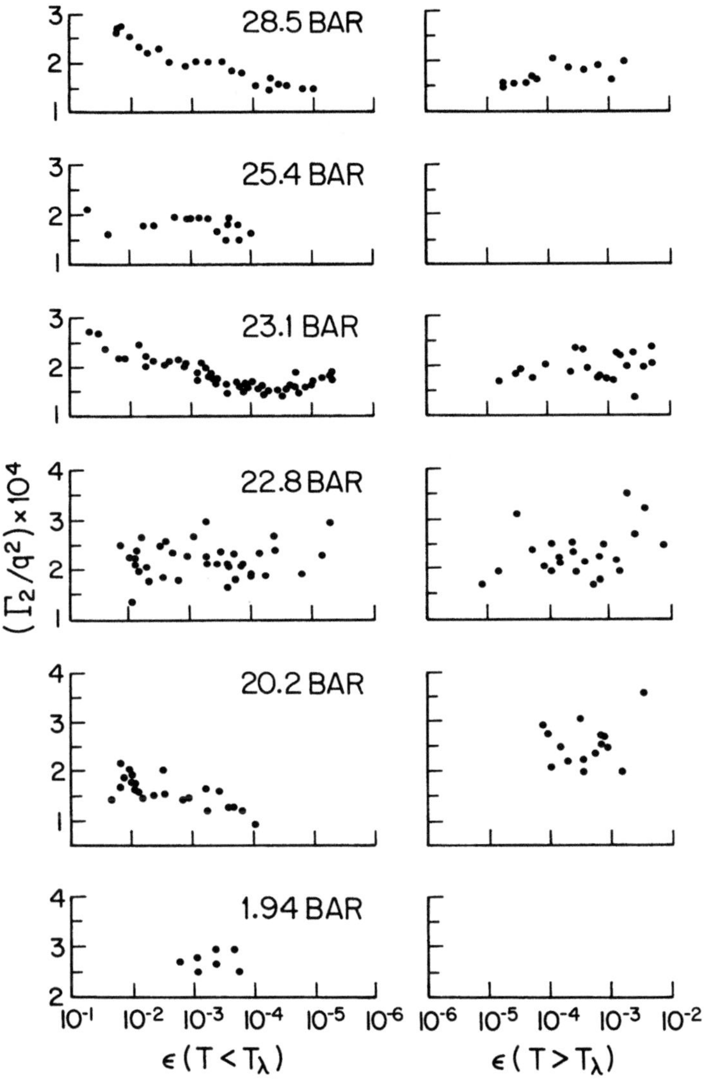

Figure 4: Damping of the entropy fluctuations at different pressures. The quantity Γ_2/q^2 is equal to $(D_\kappa+D_\zeta)/2$ in the hydrodynamic region below T_λ and to D_κ in the hydrodynamic region above T_λ. The data at 23.1 and 28.5 Bar are from Ref. 1 and were taken at $q = 1.79 \times 10^5$ cm^{-1}. The data at 20.2 and 22.8 Bar are from Ref. 2 and were taken at $q = 1.88 \times 10^5$ cm^{-1}. The data at 1.94 and 28.5 Bar were obtained by Winterling *et al.*(11,12) at $q = 1.45 \times 10^5$ cm^{-1}.

value at the minimum of the damping versus temperature curve measured by Hansen and Pellam[10] in earlier experiments. If one takes that to be the background damping and assumes that it is pressure independent, then the background $\Gamma_2/2\pi$ would be 1 MHz and the solid curve would be shifted upward by about 0.3 MHz, increasing the discrepancy even further. Above T_λ the hydrodynamic result based on Eq. 3 also rises above the experimental measurements. There is no discrepancy with hydrodynamics here, however, since the deviation does not occur until the critical region has been reached and Eq. 3 no longer applies. It is interesting, though, that in the hydrodynamic region for light scattering above T_λ the only significant contribution to D_{κ} is the non-critical one. In fact it appears as though one can cross the entire phase transition region without seeing any evidence of a critical contribution to the damping of the high wavevector entropy fluctuations.

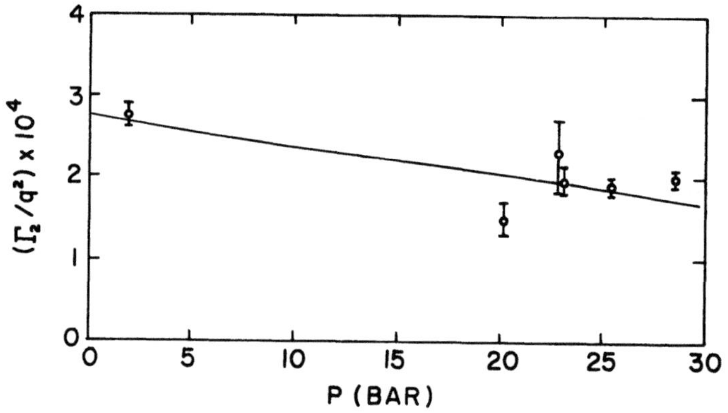

Figure 5: The pressure dependence of the second sound damping at $\varepsilon = 10^{-3}$ below T_λ. The error bars are estimated from the scatter in the data points on Fig. 4. The solid line is proportional to the velocity of second sound, v_2, and was adjusted to go through the data point at 23.1 Bar.

Similar behavior has been found in all of the light scattering measurements of the entropy damping. These results are collected in Fig. 4. Figure 5 shows the damping at a value of $\varepsilon = 10^{-3}$ below T_λ as a function of pressure. From this it appears that the substantial increase in the damping between 20.2 Bar and 22.8 Bar found by Vinen and Hurd may be spurious.

The constancy of the entropy damping, in particular the discrepancy with hydrodynamics below T_λ, has been quite puzzling to theorists.[13] Recently, however, two new ideas have emerged which may explain the results. First recall that the entropy, s, is a quadradic function of the order parameter, ψ.[3,4] Then a fluctuation in the entropy at a wavevector $\vec{q}$, $s(\vec{q})$, must be represented by a sum over products of order parameter fluctuation $\psi(\vec{k})\ \psi(\vec{k}')$ whose wavevectors are related to $\vec{q}$ by the expression

$$\vec{q} = \vec{k} + \vec{k}' \tag{8}$$

Dynamic scaling arguments apply primarily to the order parameter. In particular the critical region boundary should be $k\xi = 1$ rather than $q\xi = 1$. The distinction is not important, however, as long as the magnitudes of the k's in Eq. 8 are close to the magnitude of the resultant $\vec{q}$; and this had been assumed to be the case. But recently Ferrell, Dohm, and Bhattacharjee[14] pointed out that a model can be made for liquid helium in which the k's entering Eq. 8 are an order of magnitude larger than the resultant $\vec{q}$. This is associated with the assumption that the fluctuations in the order parameter may relax an order of magnitude more slowly than fluctuations in the entropy. Now if $k >> q$ then the condition $k\xi = 1$ occurs for a value of ξ which is smaller than that for which $q\xi = 1$. This means that the critical region will begin farther away from T_λ than had been assumed. The calculations show that the ε at which hydrodynamics breaks down could be an order of magnitude larger than we have indicated. This would explain why the second sound damping does not follow the divergent $D_{\mathfrak{K}} + D_\zeta$: the critical region is entered before the rise is appreciable.

Ferrell and Bhattacharjee[15] have now extended their calculations to take into account the non-critical background contribution to diffusivities. They find that these terms do not simply contribute in an additive fashion to the total damping. Rather, in a subtle way they also decrease the magnitude of the critical contribution to the damping. Above T_λ, their calculations can reproduce the magnitude and temperature dependence of the data shown in Fig. 3. They have also explained quantitatively the deviations from the scaling prediction found in the precise thermal conductivity measurements made by Ahlers[9] very close to T_λ. The theory seems less successful below T_λ. Although it gives the correct magnitude for the damping data in Fig. 3 near T_λ, it predicts that Γ_2 decreases slightly as T is lowered, whereas the measured Γ_2 rises. A more serious problem is that the background corrections should also lower the critical damping contribution to $D_{\mathfrak{K}} + D_\zeta$ at short wavevectors to a value much smaller than that which was actually observed by Tyson.

Vinen and Hurd[2] have developed a phenomenological theory for the critical dynamics. They are able to use their theory to account for the anomalous behavior of the second-sound linewidth in the hydrodynamic region, although it gives predictions for other measurable quantities (such as the dispersion in ν_2 and the deviation of I_2/I_1 from $\gamma - 1$) which are not in agreement with the data. It is interesting to note, however, that their theory shares with the microscopic theories(14,15) the idea that the dominant order-parameter fluctuations relax more slowly than the fluctuations in the thermodynamic variables, such as the entropy.

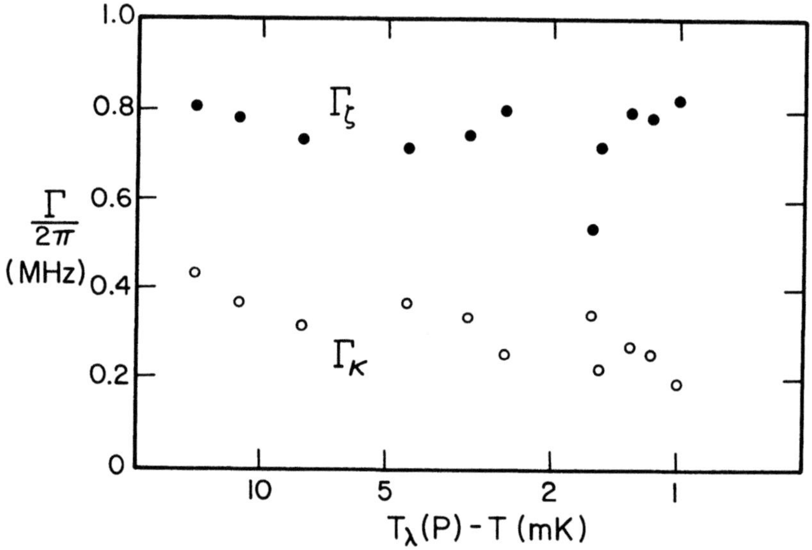

Figure 6: The two separate contributions to the hydrodynamic linewidth at a pressure of 23.1 Bar. $\Gamma_\zeta/2\pi = D_\zeta q^2/4\pi$ and $\Gamma_\mathcal{K}/2\pi = D_\mathcal{K} q^2/4\pi$.

The precise spectral shape for the entropy fluctuations in the hydrodynamic region below T_λ depends on only three parameters: ν_2, $D_\mathcal{K} + D_\zeta$, and $D_\zeta/D_\mathcal{K}$. They can be roughly thought of as determining respectively the shift, width and asymmetry of the lines. The ratio $D_\zeta/D_\mathcal{K}$ is difficult to measure accurately in these experiments. At low temperatures and large shifts the lines are close to Lorentzian in shape and insensitive to $D_\zeta/D_\mathcal{K}$. Close to T_λ where the shift and width become comparable, and $D_\zeta/D_\mathcal{K}$ has a stronger influence on the spectral shape, one enters the critical region where the hydrodyanmic form may no longer be applicable. Fig. 6 shows our results for the separate contribution of $D_\mathcal{K}$ and D_ζ to the total linewidth. They were determined from fits of the data to the exact hydrodynamic spectral shape. They indicate that

D_ζ is about 2 or 2.5 times as large as $D_\mathcal{K}$. Vinen and Hurd find that this ratio is of the order of 1 or 2, a result which they find extends into the critical region when they use the hydrodynamic spectrum to model the true critical one. Although it is clear that light scattering does not give a precise value for $D_\zeta/D_\mathcal{K}$, the information that is obtained is helpful. For example, a theory of the dynamics based on the planar-spin model predicts that $D_\zeta/D_\mathcal{K} = 0.36$.[6]

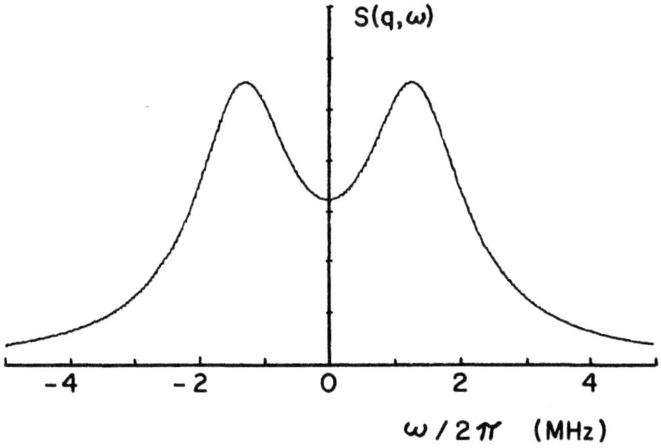

Figure 7: The two-Lorentzian model of the limiting structure factor as suggested by the data from Figures 2 and 3. The shift $\omega_2/2\pi$ was taken to be 1.3 MHz and the half-width $\Gamma_2/2\pi$ was taken to be 0.9 MHz. This model was used as a convenience in parametrizing the data. No one has seriously suggested that the true limiting structure factor should have this form.

Near T_λ the shifts and linewidths that are used to experimentally parametrize the unknown spectral shape are comparable to or less than the instrumental linewidth of the spectrometer. Under these circumstances we are not able to determine the limiting structure factor. However, we can make some general statements about its form. Both groups find that the limiting structure factor cannot be approximated by a single Lorentzian line. Fits by Vinen and Hurd using the hydrodynamic spectrum as a model "suggest strongly that $S(q,\omega)$ still has a double peak at the λ point"[2]. Our two-Lorentzian model indicates the same result, as is shown in Fig. 7. However, the data is not conclusive on this point. Hohenberg, Siggia, and Halperin[6] have calculated $S(q,\omega)$ in the symmetric planar-spin model using no adjustable parameters. Spectra calculated from this theory were convolved with our instrumental profile and are compared with our experimental traces in Fig. 8. This theory does predict

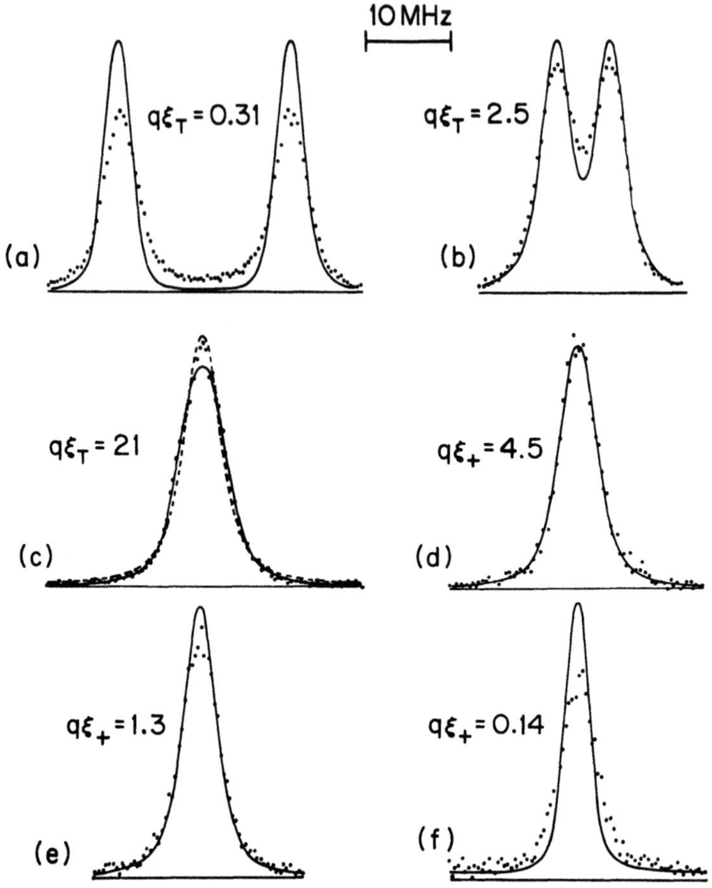

Figure 8: Central portion of spectra recorded at P_λ=23.1 Bars. The contributions of first sound, stray light, and dark count have been removed from the spectra. In order, the temperatures are ($T-T_\lambda$): -5.5, -0.25, -0.01, 0.025, 0.16, 4.5 mK. The solid lines are theoretical spectra predicted by the planar-spin model convolved with the line shape of our instrument. The dashed line in (c) is the fit to a single Lorentzian.

a two-peaked limiting structure factor, but the dip in the function at $\omega = 0$ is only a few percent of its peak value. This theoretical spectrum seems to fit our data at $T = T_\lambda$ as well as the spectrum in Fig. 7. The planar-spin model predicts the limiting spectral extent at T_λ very well; however, as shown in Figures 3 and 8, it predicts linewidths which are too small in the hydrodynamic regions.

C. First Sound

Brillouin scattering allows the simultaneous measurement of the velocity and attenuation of high frequency sound waves, even though the attenuation may be quite large. The technique has the added advantage that one is studying sound waves which are present in the medium due to its thermal fluctuations. There is never a question of perturbations due to a transducer or of finite amplitude effects. The frequency of the sound wave studied is $\nu_1 = v_1 q/2\pi$. The wavevector q is fixed in a scattering experiment, so the exact frequency of the sound wave varies somewhat as the sound velocity changes with the temperature. Figures 9 and 10 show some of the data obtained by Vidal, Tarvin, and Greytak[16] on sound waves at about 960 MHz. The results cover the critical region as well as the hydrodynamic regions above and below T_λ.

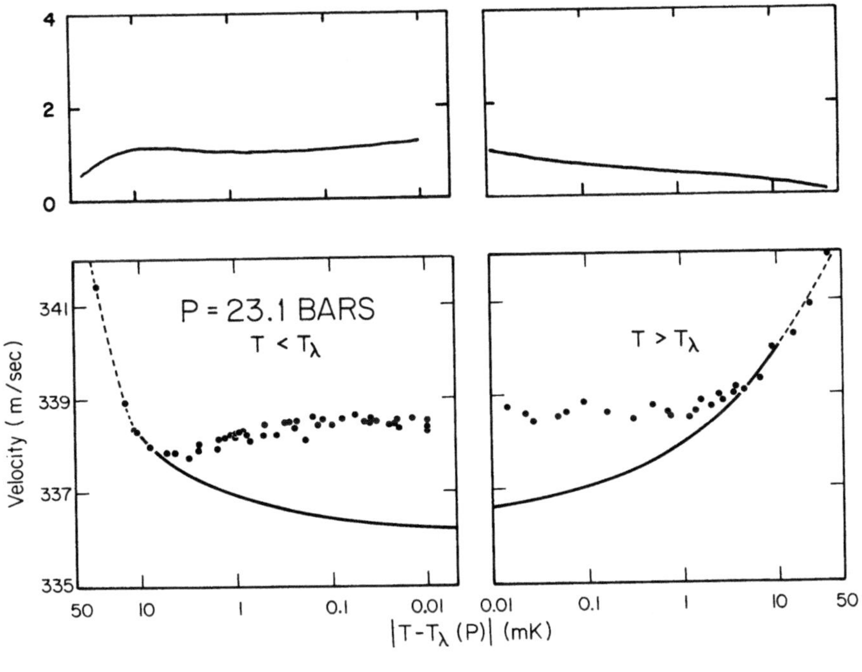

Figure 9: Lower graph: The velocity of first sound at a frequency of about 960 MHz. The solid lines are the zero frequency velocity whose absolute values are matched to agree with the data 10 mK below T_λ. Upper graph: The dispersion of first sound predicted by the low frequency model of Ref. 18.

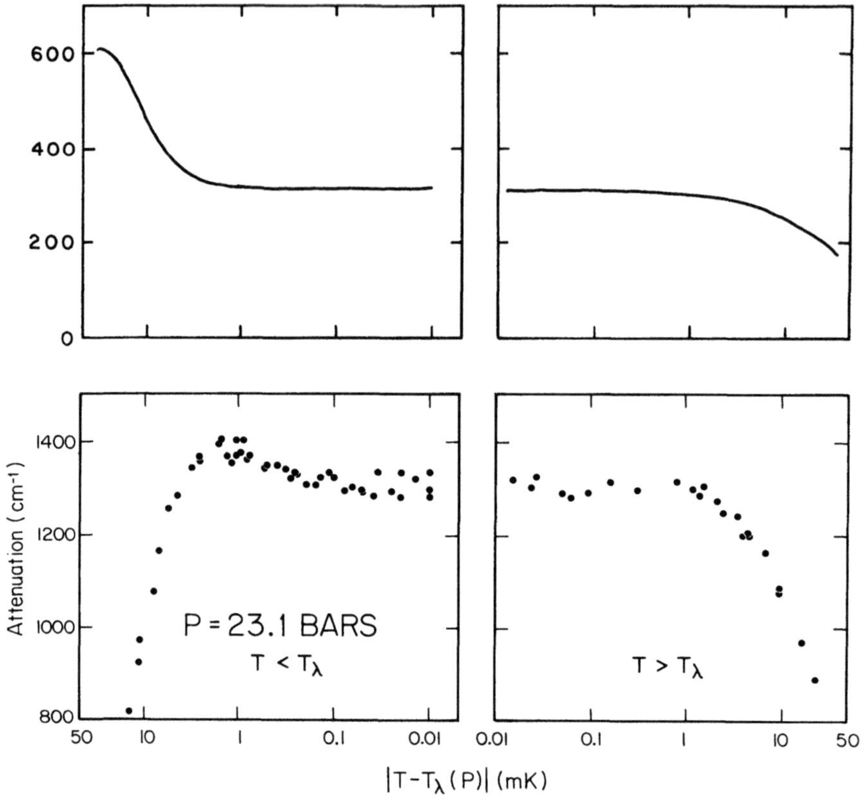

Figure 10: Lower graph: The attenuation of first sound at a frequency of about 960 MHz. Upper graph: The attenuation predicted by the low frequency model of Ref. 18.

Two mechanisms have been proposed to explain the critical behavior of low frequency sound in the hydrodynamic regions.(4) The first is a relaxation process associated with the mean value of the order parameter $\langle\psi\rangle$. In many ways $\langle\psi\rangle$ can be treated as an internal thermodynamic degree of freedom associated with the superfluid. $\langle\psi\rangle$ adjusts itself to the local conditions of pressure and temperature. However $\langle\psi\rangle$ can only follow sudden changes in these conditions with a characteristic relaxation time τ_2 which can be shown to be close to ξ/v_2. It follows that the temperature dependence of τ_2 would be roughly proportional to ε^{-1}. When the frequency of sound is in the vicinity of $1/\tau_2$, the contributions to the dispersion and the attenuation normally associated with a relaxation arise. Since $\langle\psi\rangle$ is finite only for $T < T_\lambda$, the dispersion and absorption associated with this process occur only below T_λ. The second process thought to effect the sound waves is a coupling to the fluctuations in the order parameter. These fluctuations, which become large near T_λ, interfere with the normal propagation of the sound wave. This process is not completely understood, but the attenuation and dispersion it produces are assumed to be symmetric about the transition temperature. These concepts were first applied to the analysis of sound propagation data near T_λ by Williams and Rudnick.[17] They showed that these simple concepts could fully explain their results at SVP. Recent extensive experiments by Pobell and his coworkers[18,19] have shown that the two mechanisms also completely describe data taken at various pressures and He^3 concentrations. As a result the pressure and concentration dependence of the parameters entering the theories are now known.

Unfortunately, these two processes which are so successful in describing the critical behavior of low frequency sound do not explain the results which we find at high frequency. The upper graphs in Figures 9 and 10 show the dispersion and the attenuation predicted by these theories. The vertical scales are the same as those of the data shown in the lower graphs. The predicted dispersion is generally less than that which is observed, yet at low temperatures it is finite where no dispersion is found experimentally. The largest attenuation predicted does not even come up to the value at the lower end of the scale of measured attenuations. It is not surprising that basically hydrodynamic theories do not represent the high frequency data in the critical region. Even in the hydrodynamic region there are reasons to believe that these particular models will break down at high frequencies. As yet, however, there are no theories which can explain the observed critical behavior of high frequency sound.

D. Intensities

Both the total scattered intensity I_1+I_2 [2] and the intensity ratio I_2/I_1 [1,2] fall below their divergent hydrodynamic values inside the critical region, as was expected. In both cases the maximum occurs below T_λ in the vicinity of the temperature where $q\xi = 1$. There is as yet no theoretical prediction for the complete behavior of I_1 or I_2 in the critical region. However, Bray[20] has shown that the energy density-energy density correlation function, which is related to these intensities, must peak below rather than above T_λ. He also points out that the slope of these intensities as a function of temperature at $T = T_\lambda$, and the q dependence of the position of the maximum in the intensities, contain important information about the critical behavior. The experimental data are not yet complete enough to investigate this interesting suggestion.

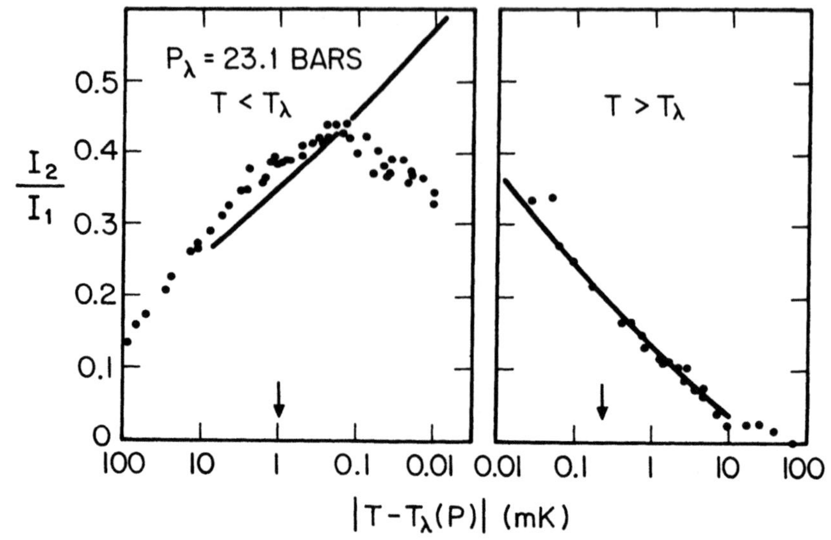

Figure 11: The intensity ratio I_2/I_1. The solid line gives values of $\gamma - 1$ calculated from the thermal-expansion coefficient. The arrows indicate the temperature at which $q\xi = 1$.

Figure 11 shows I_2/I_1 at 23.1 Bar and illustrates an interesting phenomenon. I_2/I_1 appears to rise above the predicted value of $\gamma - 1$ before the critical region is entered below T_λ. This feature of I_2/I_1 was thought to be of spurious origin when it first occured in earlier measurements of Winterling *et al.*[11] but not in measurements at a lower pressure by O'Connor *et al.*[21] Now, however, both groups[1,2] find a rise in I_2/I_1 above $\gamma - 1$ for data taken at pressures of 20.2 Bar and above. The data on I_2/I_1 were extracted from high resolution Fabry-Perot interferometer traces in which some overlapping of the spectral features of first and second sound necessarily occured. This usually gives rise to no particular difficulty as long as the spectra can be fit to a model which includes all the spectral features. There is a remote chance, however, that there is some spectral feature which has been overlooked. Low resolution traces should be taken in the future with the sole purpose of obtaining precise data on this ratio.

The work described here which was done at MIT was supported by the NSF under grants DMR7803017 and DMR7680895.

References

1. J.A. Tarvin, F. Vidal, and T.J. Greytak, Phys. Rev. B15, 4193 (1977)
2. W.F. Vinen and D.L. Hurd, Advances in Physics 27, 533 (1978)
3. R.A. Ferrell, N. Menyhard, H. Schmidt, F. Schwabl, and P. Szepfalusy, Ann. Phys. (N.Y.) 47, 565 (1968)
4. P.C. Hohenberg, in Critical Phenomena, Edited by M.S. Green (Academic, N.Y. 1971)
5. D.S. Greywall and G. Ahlers, Phys. Rev. A7, 2145 (1973)
6. P.C. Hohenberg, E.D. Siggia, and B.I. Halperin, Phys. Rev. B14, 2865 (1976)
7. J.A. Tyson, Phys. Rev. Lett. 21, 1235 (1968)
8. J. Kerrisk and W.E. Keller, Phys. Rev. 177, 341 (1969)
9. See the review article by G. Ahlers, in The Physics of Liquid and Solid Helium, edited by K.H. Bennemann and J.B. Ketterson (Wiley, N.Y. 1976)
10. W.B. Hanson and J.R. Pellam, Phys. Rev. 95, 321 (1954)
11. G. Winterling, F.S. Holmes, and T.J. Greytak, Phys. Rev. Lett. 30, 427 (1973)
12. G. Winterling, J. Miller, and T.J. Greytak, Physics Lett. 48A, 343 (1974)
13. P.C. Hohenberg and B.I. Halperin, Rev. Mod. Phys. 49, 435 (1977)
14. R.A. Ferrell, V. Dohm, and J.K. Bhattacharjee, Phys. Rev. Lett. 41, 1818 (1978)
15. R.A. Ferrell and J.K. Bhattacharjee, to be published.
16. F. Vidal, J.A. Tarvin, and T.J. Greytak, to be published.
17. R.D. Williams and I. Rudnick, Phys. Rev. Lett. 25, 276 (1970)
18. R. Carey, C. Buchal, and F. Pobell, Phys. Rev. B16, 3133 (1977)
19. C. Buchal and F. Pobell, Phys. Rev. B14, 1103 (1976)
20. A. Bray, to be published.
21. J.T. O'Connor, C.J. Palin, and W.F. Vinen, J. Phys. C 8, 101 (1974)

DYNAMIC SCALING NEAR THE LAMBDA POINT OF LIQUID He^4

R.A. Ferrell and J.K. Bhattacharjee

Institute for Physical Science and Technology
and Department of Physics and Astronomy
University of Maryland
COLLEGE PARK, Maryland 20742, USA

DYNAMIC SCALING NEAR THE LAMBDA POINT OF LIQUID He^4

Richard A. Ferrell
and
Jayanta K. Bhattacharjee
Institute for Physical Science and Technology
and Department of Physics and Astronomy
University of Maryland, College Park, Maryland 20742

The original versions[1,2] of the dynamic scaling theory were limited to the extreme long wave lengths or equivalently the very small wave vectors. The striking predictions of the theory were the divergences in the thermal conductivity and the second-sound damping when the lambda point is approached from above and below, respectively. Both of these quantities were predicted to diverge as $t^{-1/3}$ where $t = |T - T_\lambda|/T_\lambda$. The early experiments of Ahlers[3] on thermal conductivity and of Tyson[4] on second-sound damping were in agreement with these predictions (when the critical behavior of the specific heat was properly accounted for).

Renormalization group was introduced in critical dynamics by Halperin, Hohenberg and Ma.[5] The equations of motion for the order parameter, and any secondary variable that it may couple to, determine the dynamic universality class. Halperin, Hohenberg and Siggia[6] described liquid Helium in terms of the planar spin model. Application[6] of the renormalization group to this model yielded the familiar dynamic scaling exponents. The amplitudes of thermal conductivity and second-sound damping were expressed in terms of certain universal ratios.[6] Exponents for the correction to scaling terms were calculated. A single-loop self-consistent calculation of the density-correlation function[7] was also carried out.

Turning now to the experimental side, high precision measurements on thermal conductivity as a function of t were carried out by Ahlers[8] at different pressures. The result was a small but definite (about 20%) deviation from the dynamic scaling prediction. The experimental divergence was stronger than that expected from the theory. Light scattering experiments were carried out by Tarvin, Vidal and Greytak[9] at various temperatures above and below the lambda point. The light scattering directly probes the density correlation function which, above the lambda point, is expected to be nearly Lorentzian with a half-width proportional to the thermal conductivity. Thus the spectrum above the λ-point should shrink, as one leaves the critical region, corresponding to the $t^{-1/3}$ fall-off of the conductivity. Below the lambda point it is the width of second-sound peaks which should decrease as $t^{-1/3}$ outside the critical region. Tarvin et al[9] found that, above the lambda point, the width hardly changes and below, the width of the second-sound peak falls at first but then rises slowly. The measurements of Tyson[4] on the second-sound damping gave an amplitude which was five times higher than that obtained from the universal ratio of Hohenberg, Siggia and Halperin.[7] Thus, there were significant discrepancies between the dynamic scaling predictions and the experimental observations.

A possibility of violation of dynamic scaling was discovered by DeDominicis and Peliti[10] when they applied the field theoretic renormalization group in the study of critical dynamics. It is convenient to picture this in terms of the SSS[11] model which has an n-component, non-conserved order parameter field (ψ) with O(n) symmetry and a n(n-1)/2 component conserved generator of rotations (S). The Fourier components have the equations of motion

$$\dot{\psi}_i(k) = \sum_{j\neq i} \sum_q S_{ij}(k-q)\psi_j(q) \tag{1a}$$

$$\dot{S}_{ij}(k) = \sum_{p+p'=k} (p^2 - p'^2)\psi_i(p)\psi_j(p') \tag{1b}$$

where the dot denotes differentiation with respect to time. Thus, each component of the order parameter is coupled to and perturbed by n-1 components of the tensorial generator field. As n is decreased in this model, the perturbations are weakened, resulting in less Brownian motion in the order parameter. The increased value of the relaxation time γ_ψ^{-1} produces more noise in the generator field, thereby increasing its relaxation rate γ_S and consequently shortening the duration of the Langevin pulses in order parameter dynamics. This feedback situation results in a kind of catastrophe that leads to a breakdown of dynamic scaling at a finite value of n-1 for a given value of D, the dimensionality of space. The curve along which this breakdown occurs in the n-D plane was calculated to two-loop accuracy by DeDominicis and Peliti,[12] Dohm and Ferrell[13] and the present authors.[14] It was found that liquid Helium is on the "scaling" side of the boundary but close to it. Consequently it is possible to characterize Helium by a small value of w, the ratio of the relaxation rates of the order parameter and the entropy. Dohm[15] and the present authors[14] found this ratio to be O(0.1).

A consequence of the small value of w is the existence of a "slow transient" i.e., a correction to scaling term which has a small exponent of O(w). This correction term persists deep down into the critical region and hence the true scaling domain becomes practically inaccessible. It will be seen below that a further consequence of a small w is to give this slow transient a strong amplitude. This is related to the existence of a noncritical background for the thermal conductivity and the Onsager coefficient for the order parameter.

We now turn to a study of this noncritical background, the importance of which in understanding critical phenomena was first pointed out by Sengers and Keyes[16] in their study of the thermal conductivity of a single component fluid. Using simple kinetic theory, we have obtained an expression[17] for the thermal conductivity which has both the critical and noncritical parts. A straightforward linearized Boltzmann equation treatment of the drift of the Helium atoms of mass m, density n, velocity v and energy $\varepsilon = mv^2/2$ in a temperature gradient leads to the standard picture of the drift of the "hot" atoms opposite to the temperature gradient. These carry the heat, while the "cold" atoms constitute a counterflow in the reverse direction such that the net mass current vanishes. The resulting thermal conductivity is[17]

$$\lambda = \frac{n}{mT} \langle \tau(\delta\varepsilon)^2 \rangle \tag{2a}$$

where

$$\delta\varepsilon = \varepsilon - \langle\varepsilon\rangle \tag{2b}$$

and τ is the relaxation time. The angular brackets indicate an "average value" weighted according to the distribution function

$$F(\varepsilon) = -\frac{2}{3}\,\varepsilon\, g'(\varepsilon) \quad . \tag{3}$$

$g(\varepsilon)$ is the Bose-Einstein distribution function

$$g(\varepsilon) = \frac{1}{e^{\beta(\varepsilon-\mu)} - 1} \tag{4}$$

where μ is the chemical potential and β^{-1} is T times Boltzmann's constant. Thus, in this notation the average value $\langle Q\rangle$ of some quantity Q is the velocity-space integral

$$\langle Q\rangle = -\frac{2}{3n}\left(\frac{m}{h}\right)^3 \int d^3 v\; F(\varepsilon) Q \tag{5}$$

where h is Planck's constant.

Now, what is special about the λ-point of Helium is that $\tau(\varepsilon)$ is a rapidly varying function of ε with a divergence in the low energy limit $\varepsilon \to 0$. It is this diverging lifetime that produces the critical divergence in the thermal conductivity. In order to decompose Eq. (2a) into its noncritical and critical parts, we assume that the strong energy dependence of τ is restricted to the very small energies. For the remaining energy range we assume that τ is some smooth function of ε, say τ_B. By extrapolating τ_B back into the low energy critical range, we obtain the background conductivity

$$\lambda_B = \frac{n}{mT} \langle \tau_B(\delta\varepsilon)^2 \rangle \tag{6}$$

What is left is the critical component

$$\lambda_c = \frac{n}{mT} \langle (\tau - \tau_B)\; (\delta\varepsilon)^2 \rangle \tag{7}$$

By assumption, $\tau - \tau_B$ approaches zero rapidly and $\tau - \tau_B > 0$ only in the very small energy range $\varepsilon \ll \langle\varepsilon\rangle$. We can make the further approximation

$$\langle\varepsilon\rangle = Ts + \mu \approx Ts \quad , \tag{8}$$

because the approach to criticality requires a very small μ. Equation (8) gives $(\delta\varepsilon)^2 \approx T^2 s^2$, so

$$\begin{aligned} \lambda_c &= \frac{ns^2T}{m} \langle (\tau - \tau_B) \rangle \\ &\approx \frac{ns^2T}{m} \langle \tau \rangle \quad , \end{aligned} \tag{9}$$

proportional to the weighted mean of the relaxation time. The simplified version of Eq. (9) in the second line can be used if the function that is substituted for $\tau(\varepsilon)$ leads to a convergent integral.

In order to apply Eq. (9) quantitatively, we introduce the Bose-Einstein distribution from Eq. (4) and note that, in the range $\varepsilon << k_BT$, it is equivalent to the standard Ornstein-Zernike form for the order parameter correlation function at wave number $p = mv/\hbar$,

$$g(\varepsilon) \approx \frac{k_BT}{\varepsilon - \mu} = \frac{2m\ k_BT/\hbar^2}{p^2 + \kappa^2} \quad . \tag{10}$$

This depends on identifying the square of the inverse correlation length as $\kappa^2 = -2m\mu/\hbar^2$. The derivative of the B.E. distribution is

$$g'(\varepsilon) \simeq -\frac{k_BT}{(\varepsilon - \mu)^2} \simeq -\beta g(\varepsilon)^2$$

$$\approx -\frac{4m^2 k_BT/\hbar^4}{(p^2+\kappa^2)^2} \tag{11}$$

Introducing the order-parameter relaxation rate, $\gamma_\psi = (2\tau)^{-1}$, and expressing the velocity in terms of the wave number, we get

$$\lambda_c = \frac{2}{3} k_B^{-1} \left(\frac{sk_BT}{\hbar}\right)^2 \frac{1}{(2\pi)^3} \int d^3p \frac{p^2}{(p^2 + \kappa^2)^2} \frac{1}{\gamma_\psi} \tag{12}$$

This picture of a critical variation imposed on a noncritical background is borne out by the experimental data presented in Fig. 1. It will be seen that, very close to

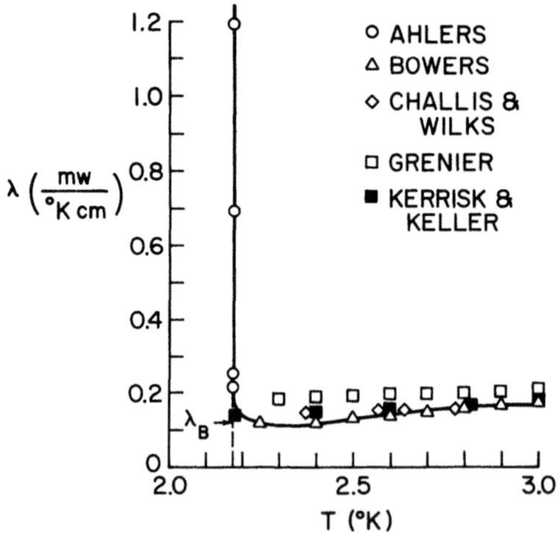

Fig. 1 Thermal conductivity near the λ-point as a function of temperature.

the lambda point, there is the critical divergence. Away from the critical point, the behavior is fairly flat with a small rise farther away. The background λ_B is estimated to be 0.12 mw/cm°K from the following analysis of Ahlers' data.

In what follows, we discuss Ahlers' data at 22 Bars, since the light scattering work of Tarvin et al. was carried out at a similar pressure and the two experiments can be linked up.[17] The first departure from noncritical behavior can be obtained by carrying out the integration indicated in Eq. (12) with $\gamma_\psi = B_\psi(p^2 + \kappa^2)$, where B_ψ is the constant noncritical background value for the Onsager coefficient of the order parameter. This yields

$$\lambda_c = \frac{k_B^{-1}}{16\pi}\left(\frac{sk_BT_\lambda}{\hbar}\right)^2 \frac{\kappa^{-1}}{B_\psi} = \frac{k_B^{-1}}{16\pi}\left(\frac{sk_BT_\lambda}{\hbar}\right)^2 \frac{\kappa_o^{-1}}{B_\psi}\, t^{-2/3} \tag{13}$$

where, in the second step, $\kappa = \kappa_o t^{2/3}$ has been used. Thus a plot of the thermal conductivity against $t^{-2/3}$ should be linear when Eq. (13) is valid, i.e., for large values of t. The intercept on the λ-axis will give λ_B and the slope will provide B_ψ. Fig-

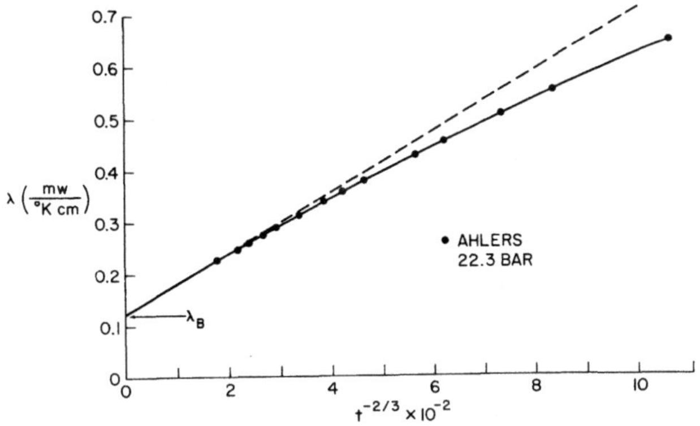

Fig. 2 Thermal conductivity as a function of $t^{-2/3}$.

ure 2 shows such a plot. For small $t^{-2/3}$ it is indeed linear, giving $\lambda_B = 0.12$ mw/cm °K and $B_\lambda = 1.35 \times 10^{-4}$ cm^2/sec. As $t^{-2/3}$ is increased, i.e., as the lambda point is approached, the data deviates more and more from the linear behavior as expected, but a parabolic form

$$\lambda = \lambda_B + (0.587\ x - 0.0086\ x^2) \times 10^{-4}\ \frac{w}{\text{cm °K}} \tag{14}$$

where $x = (10^3\ t)^{-2/3}$ gives a satisfactory fit to the data over a fairly wide range. The solid curve in Fig. 2 represents this parabola.

In the scaling region Eq. (14) is not the correct behavior for λ. We shall return to the study of the scaling solution and corrections to it, but for now turn to Eq. (12) to see what functional form for γ_ψ would produce a given λ_c. We assume that γ_ψ is a function of the variable $r = \sqrt{p^2 + \kappa^2}$ alone, whence Eq. (12) yields

$$\begin{aligned}
\lambda_c &\propto \int_0^\infty dp \left(\frac{p^2}{p^2 + \kappa^2}\right)^2 \gamma_\psi^{-1}(r) \\
&= \int_\kappa^\infty dr \left(1 - \frac{\kappa^2}{r^2}\right)^{3/2} \gamma_\psi^{-1}(r) \\
&= \int_{C\kappa}^\infty dr\ \gamma_\psi^{-1}(r) \quad ,
\end{aligned} \tag{15}$$

where $C\kappa$ is an effective cut-off that allows the factor $(1 - \kappa^2/r^2)^{3/2}$ to be replaced by 1. Differentiating Eq. (15)

$$\gamma_\psi^{-1}(C\kappa) \propto \frac{d\lambda_c}{d\kappa} \quad . \tag{16}$$

Thus, given $\lambda_c(\kappa)$, Eq. (16) determines the order parameter relaxation rate that would produce such a $\lambda_c(\kappa)$. For large values of κ, λ_c can be obtained from Eq. (14). Substitution of this in Eq. (16), and finally the replacement of κ by the wave number k, yields for the nonlocal Onsager coefficient of the order parameter a constant background value and a 1/k rise. The important factor is the fairly large background value which, coupled with the small w, makes the critical region more inaccessible.

The small w causes the scaling solution of the order parameter Onsager coefficient to be small. The background value on the other hand is large. The consequence is that the leading correction to scaling -- the slow transient -- will have to be very strong to lift the scaling value above the background and then allow it to merge into it. The correction to scaling being so large, the true scaling region recedes further away. The lowest order ε-expansion shows[14] that the amplitude of the slow transient has to be equal and opposite for the entropy and the order parameter. Thus the slow transient decreases the true scaling solution for the entropy relaxation rate and the thermal conductivity. This is the key idea in explaining the observed deviation of the thermal conductivity data from the dynamic scaling prediction.

According to dynamic scaling a fluctuation of wave number k is characterized by the frequency $D(k)k^2$. The nonlocal "diffusion coefficient" is predicted to scale as

$D(k) = ak^{-1/2}$ (the slowly varying logarithmic specific heat has been included in the definition of a). From our mode-coupling calculation to two-loop accuracy, we find

$$a = \frac{1}{\pi}\sqrt{\frac{J}{2}}\frac{\gamma_o s}{\sqrt{nc_p}} \qquad (17)$$

$J = J_1 + J_2$, where $J_{1,2}$ are one- and two-loop convolution integrals that have to be evaluated numerically. n is the Helium atom density. s and c_p are the mean entropy and constant pressure specific heat per particle, both in units of k_B. $\gamma_o = k_B T_\lambda/\hbar$ (where $2\pi\hbar$ is Planck's constant) is the basic reference frequency of the system. The origin of γ_o can be traced to the equation of motion for the phase of the order parameter. (The coupling to the temperature fluctuations is via the ratio $k_B s/\hbar$.)

We now introduce the order-parameter and entropy relaxation rates, γ_ψ and γ_S, respectively, as $\gamma_{\psi,S} = a_{\psi,S}\, k^{3/2}$, where $a_\psi/a_S = w$ and $a_\psi a_S = a^2$. Thus, $a_\psi = a\sqrt{w}$ and $a_S = a/\sqrt{w}$. Dividing the rates by k^2 yields $D_{\psi,S}(k) = a_{\psi,S}\, k^{-1/2}$ which is exhibited in Fig. 3 by the lowermost and uppermost curves, labeled "SCALING." The horizontal line in the figure represents the noncritical background B_ψ and it is quite clear that the transients must lift the scaling solution to above the background value in the small k-range. The transient solutions can be described as the replacement of $a_{\psi,S}$ by $a_{\psi,S}\left(1 \pm b_s k^{\omega_s} + b_f k^{\omega_f}\right)$. As explained earlier, the slow transient enters

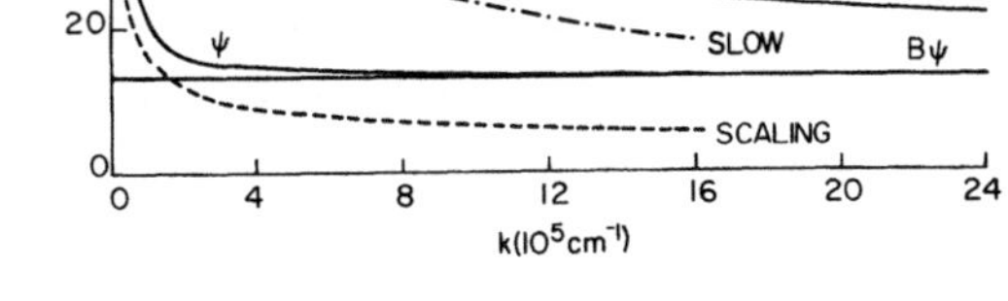

Fig. 3 The Onsager coefficients D_S and D_ψ as functions of the wave vector.

with equal and opposite amplitudes for the order parameter and entropy. The fast transient, according to the lowest order ε-expansion, has the same amplitude in the two cases. The exponents ω_s and ω_f are universal. $\omega_s \simeq w$ and for ω_f it is reasonable to use the $w = 0$ value, $\varepsilon - 0.038\,\varepsilon^2 \simeq 0.96$. In the small k-range k^{ω_f} is vanishingly small and the slow transient suppresses D_S according to

$$D_S(k) = a_S k^{-1/2}\left(1 - b_S k^{\omega_s}\right) . \tag{18}$$

This is illustrated by the middle (dot-dash) curve in Fig. 3. The label "SLOW" signifies that only the slow transient has been included in the computation of this curve. The logarithmic derivative of the correction factor in Eq. (18) corresponds to an effective incremental exponent $-\Delta z$, where

$$\Delta z = \omega_s \frac{b_S k^{\omega_s}}{1 - b_S k^{\omega_s}} \simeq w\,\frac{\text{CORR}}{\text{NET}} . \tag{19}$$

With $w = 0.1$, we see that a ratio of unity between the correction and the net amount raises the scaling exponent $z = 0.5$ by 20%. The k-dependent thermal conductivity $\lambda(k)$ is obtained from $\lambda = nc_p D_S$. For $\kappa \gg k$, the k-dependence goes over smoothly to a κ-dependence. With $\kappa = \kappa_o\, t^{2/3}$ (where we use the conventional value $\kappa_o = 0.7$ Å^{-1}) we can arrive at λ as a function of t. The plot of $\lambda(t)$ vs. t undergoes the same 20% steepening described above and accounts satisfactorily for the discrepancy in the scaling exponent reported by Ahlers. A quantitative application of the above ideas gives the excellent fit[22] to Ahlers' data shown in Fig. 4 with

$$\lambda = 21.6\,\sqrt{c_p}\; t^{-1/3}\,(1 - 1.18\, t^{0.067} + 26.1\, t^{0.64}) \tag{20}$$

in μw/cm °K. This fit has been used to obtain D_S as a function of k which is shown by the curve labeled S in Fig. 3. The curve marked ψ represents the complete order-parameter Onsager coefficient and was obtained by using Eq. (16) and the substitution $\kappa = k$ in the final step.

Writing Eq. (19) as

$$\Delta z = w\,\frac{\text{CORR}}{\text{NET}} = w\left(\frac{\text{SCALING}}{\text{NET}} - 1\right) \tag{21}$$

we can arrive at the following uniqueness theorem. For a given Δz, Eq. (21) determines a unique $w = w_o$ such that CORR/NET = 1 or equivalently SCALING/NET = 2. Let w' be some other value of w satisfying Eq. (21). Using the fact that the scaling solution is proportional to $w^{-1/2}$, we have

$$w_o = w'\left(2\sqrt{\frac{w_o}{w'}} - 1\right) \tag{22}$$

leading to

$$\left(\sqrt{\frac{w_o}{w'}} - 1\right)^2 = 0$$

or

$$w' = w_o \tag{23}$$

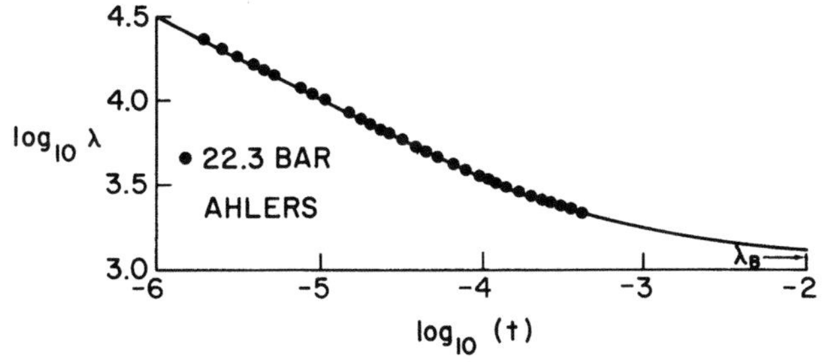

Fig. 4 Thermal conductivity λ vs. reduced temperature t.

The above proof depends on two assumptions: i) Helium lies on the scaling side of the boundary and ii) $\omega_s = w$. The first assumption has been consistently supported by theoretical considerations. The approximation $\omega_s = w$ is fairly accurate for small w, as can be checked from general expressions given in Refs. 12, 14, and 15.

To make contact with the light scattering experiments, we need to compute the fluctuation spectrum, which is the real part of the Green's function

$$G_S = \frac{1}{-i\omega + \gamma_S(\omega,k,\kappa)} \quad . \tag{24}$$

Clearly we need the frequency- and temperature-dependent relaxation rate for this computation. A frequency-dependent thermal conductivity would give the necessary information and so, following Ref. 17, we present a derivation of Eq. (12) for λ_c that allows for frequency-dependent generalizations.

The critical dynamic properties are associated with the thermodynamic fluctuations

$$\delta\mu = -s\delta T + \frac{1}{n}\,\delta P \tag{25}$$

These fluctuations of the chemical potential are produced by the fluctuations δT and δP in the temperature and pressure, respectively. At the low frequencies of interest here, δP can be neglected. According to the time-dependent Schrödinger equation, the time-derivative of the phase of ψ is

$$\dot{\phi} = \frac{d\phi}{dt} = -\frac{1}{\hbar}\,\delta\mu \quad . \tag{26}$$

As we are interested in the response of liquid Helium to an applied thermal gradient, we can write

$$\delta\mu = \underset{\sim}{r}\cdot\text{grad}\ \mu = -\underset{\sim}{r}\cdot\underset{\sim}{F} \tag{27}$$

where $\underset{\sim}{F} = -\text{grad}\ \mu$ is a constant vector force field, independent of the spatial coordinate $\underset{\sim}{r}$. After the lapse of a short time Δt, Eqs. (26) and (27) produce the phase change

$$\Delta\phi(\underset{\sim}{r}) = \frac{1}{\hbar}\,\Delta t\ \underset{\sim}{F}\cdot\underset{\sim}{r} = \frac{1}{\hbar}\,\Delta\underset{\sim}{p}\cdot\underset{\sim}{r} \tag{28}$$

where

$$\Delta\underset{\sim}{p} = \underset{\sim}{F}\ \Delta t$$

or

$$\dot{\underset{\sim}{p}} = \underset{\sim}{F} \quad . \tag{29}$$

Equation (29) is nothing other than the standard superfluid equation of motion applied to the fluctuations in the condensate wave-function. It is now convenient to switch from wave mechanical to classical particle language, and to regard Eq. (29) as simply Newton's second law describing the acceleration of Helium atoms of momentum $\underset{\sim}{p}$.

The steady-state linearized Boltzmann equation in the relaxation time approximation is

$$\underset{\sim}{F}\cdot\underset{\sim}{v}\ g'(\varepsilon) = -2\gamma_{\psi}\,(k,\kappa)\Delta g(k,\kappa) \tag{30}$$

where $\Delta g(k,\kappa)$ is the perturbation in the distribution function, $\underset{\sim}{v} = \underset{\sim}{p}/m$ is the particle velocity, and $(2\gamma_{\psi})^{-1}$ is the mean particle relaxation time. Solving Eq. (30) and using Eq. (11) for $g'(\varepsilon)$ gives

$$\Delta g = \frac{4m^2 k_B T_\lambda}{\hbar^4}\ \frac{\underset{\sim}{F}\cdot\underset{\sim}{v}}{(p^2+\kappa^2)}\ \frac{1}{2\gamma_{\psi}} \quad . \tag{31}$$

The critical mass current density is now found by multiplying Eq. (31) by $\underset{\sim}{v}$ and integrating over all momentum values. Taking $\underset{\sim}{v}$ in the direction of $\underset{\sim}{F}$ and performing angular averages yields

$$\underset{\sim}{J}{}^{c}_{m} = \underset{\sim}{F}\ \frac{4}{3}\ \frac{k_B T_\lambda}{\hbar^2}\ \frac{m}{(2\pi)^3} \int \frac{d^3p}{(p^2+\kappa^2)}\ \frac{p^2}{2\gamma_{\psi}} \tag{32}$$

We now invoke the basic feature of the two-fluid model -- namely the counterflow of normal fluid that must take place, equal and opposite to Eq. (32) in order that no net mass flow occur. This counterflow carries entropy at the rate of $k_B s$ per particle, and hence heat at the rate of $k_B T_\lambda s$ per Helium atom. The density of heat current is consequently

$$\underset{\sim}{Q} = -k_B T_\lambda s\ \underset{\sim}{J}_{at} \quad . \tag{33}$$

The force can be written in terms of the applied temperature gradient as

$$\underset{\sim}{F} = -\text{grad}\ \mu = k_B s\ \text{grad}\ T \quad . \tag{34}$$

Substituting Eqs. (32) and (34) into Eq. (33) enables one to identify the conductivity from

$$\underset{\sim}{Q} = -\lambda \text{ grad } T \tag{35}$$

and recover Eq. (12).

To find the response to an oscillatory but spatially uniform temperature gradient at angular velocity ω, we need to include $\partial(\Delta g)/\partial t$ in Eq. (30). This is equivalent to replacing $-2\gamma_\psi$ by $-2\gamma_\psi + i\omega$ and hence

$$\lambda(\omega,\kappa) = \frac{4}{3}(k_B)^{-1}\left(\frac{k_B T_\lambda s}{\hbar}\right)^2 \frac{1}{8\pi^3} \int \frac{d^3p}{(p^2+\kappa^2)^2} \frac{p^2}{-i\omega+2\gamma_\psi(p,\kappa)} , \tag{36}$$

which is identical to the single-loop expression in the standard approach to critical dynamics for $\kappa >> k$. The two-loop corrections can be obtained from the technique presented here by considering the effect of persistence of velocities.[23]

The frequency dependence of Eq. (36) has already been studied in Ref. 24, where it was found that

$$\frac{\lambda(\omega,\kappa)}{\lambda(0,\kappa)} = \left(1 - i\frac{\omega}{\sigma'}\right)^{-1/3} . \tag{37}$$

$$\sigma' = \frac{20}{21}\sqrt{\pi}\,\frac{\Gamma(1/4)}{\Gamma(3/4)} \cdot a_\psi \kappa^{3/2}$$

$$\simeq 5.0\, a_\psi \kappa^{3/2} \tag{38}$$

differs somewhat from the definition of σ in that work. $a_\psi \kappa^{3/2}$ is the local order-parameter relaxation rate. As $\lambda(0,\kappa) \propto \kappa^{-1/2} \propto \sigma'^{-1/3}$, it follows from Eq. (37) that $\lambda(\omega,\kappa)$ is obtained from $\lambda(0,\kappa)$ by replacing σ' by

$$Z' = \sigma' - i\omega . \tag{39}$$

Therefore it is not necessary to regard $\lambda(\omega,\kappa)$ as a function of the two separate variables ω,κ. The two aspects are described by one combined scaling variable Z'. This makes it possible to extract the frequency dependence necessary for the analysis of the entropy fluctuation spectrum, with a minimum amount of formal theoretical manipulation. In fact, the frequency dependence is fixed by the critical temperature dependence found by Ahlers.[8]

Equation (20), which gives λ as a function of t, will give the necessary frequency dependence when we eliminate t in favor of σ' and make the substitution $\sigma' \to Z'$. The transformation $t \to \sigma'$ involves the use of $\kappa = \kappa_o t^{2/3}$ and Eq. (38). It should be noted that Eq. (37) is not an exact representation of the integral in Eq. (36). Reference 24 estimated the maximum deviation to be 20% in the very high frequency range. A detailed study in Ref. 17 showed that, in the frequency range of interest, the deviation is less than 10% and can be neglected. The single-loop frequency dependence to $O(\varepsilon)$, without the restriction $\kappa >> k$, was carried out by Dohm[25] using the field theoretic renormalization group and the results were qualitatively similar to those in Ref. 24. Dividing

$\lambda(Z')$ by the specific heat gives a "universal" critical entropy diffusion coefficient $D_S(Z')$, containing both temperature and frequency dependence. $D_S(Z')$ takes on the background value $B_S = 2.0 \times 10^{-4}$ cm^2/sec as Z' becomes large. We note that there is a basic diffusion constant in the problem, namely $\hbar/m$, which is equal to 1.58×10^{-4} cm^2/sec. It is reassuring to see that both B_S and B_ψ are close to this number. At this stage the entropy relaxation rate γ_S can be obtained by multiplying D_S by the square of $k = 1.79 \times 10^5$ cm^{-1}, the wave number in the Tarvin et al. experiment.

It is convenient to redefine the scaling variable Z' in the dimensionless form

$$Z = Z'/(a_\psi k^{3/2}) = \sigma - i\Omega \tag{40}$$

where

$$\Omega = \omega/(a_\psi k^{3/2}) \tag{41}$$

and

$$\sigma = 5.0 \left(\frac{\kappa}{k}\right)^{3/2} . \tag{42}$$

$a_\psi k^{3/2}$ is the scaling value of the order parameter relaxation at $\kappa = 0$. From the above equations we find, in units of MHz,

$$\frac{\gamma_S}{2\pi} = \frac{10.35}{Z^{1/3}} \sqrt{\frac{k_B}{c_P}}\,(1 - 0.582\, Z^{0.067} + 0.0234\, Z^{0.64}) \tag{43}$$

where c_P is the specific heat per particle in units of k_B. Over the intermediate range of Z-values, c_P can be fitted sufficiently accurately by a power law, which puts Eq. (43) in the convenient form

$$\frac{\gamma_S}{2\pi} = \frac{5.192}{Z^{0.263}} - \frac{3.022}{Z^{0.196}} + 0.122\, Z^{0.381} \tag{44}$$

again in MHz.

$(\gamma_S - B_S k^2)/2\pi$ is shown in Fig. 5. As explained in Ref. 17, the $\kappa = 0$ λ-point value of σ is estimated at $\sigma_\lambda = 0.34$, which is illustrated in Fig. 5 by the downwards-pointing arrow close to the origin. This gives a zero-frequency rate of 3.1 MHz, as indicated at the top of the figure by the upwards-pointing arrow. The second downwards-pointing arrow at the bottom left corner of the figure indicates the frequency at which the maximum in the spectrum of Re G_S occurs (infinite resolution case, or $\Gamma_0 = 0$). As is evident, γ_S drops rapidly with increasing Z. This can result from a rise in temperature or from a change in the frequency variable. A finite resolution adds the width Γ_0 to Z, so that for a given temperature (i.e., σ-value), the whole frequency range is shifted to the right. This is illustrated in Fig. 5 by the two vertical dashed lines for $\kappa = 0$ and $\kappa = k$. In each of these cases the structure of γ_S to the left of the dashed line is inaccessible to observation. As the sharpest structure occurs at the smallest value of Z, it is clear why even a small value of Γ_0 can have a big effect in washing out the structure in the spectrum. This picture applies, of course, also to the temperature dependence. A rise in temperature moves the accessible range of Z to the right, causing a loss of structure, with the spectrum

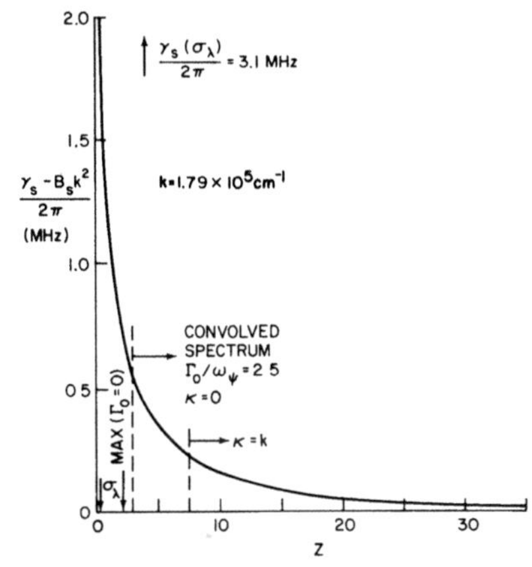

Fig. 5 The entropy relaxation rate above the background value vs. scaling variable Z for $k = 1.79 \times 10^5$ cm^{-1}.

soon becoming Lorentzian.

The self-energy function γ_S of Eq. (44) is used in Eq. (24) to obtain the fluctuation spectrum. To allow for the finite resolution of the instrument, a convolution of Re $G_S(\omega)$ with the resolution function $R(\omega)$ has to be carried out. The effect of different resolution functions has been demonstrated in Ref. 17. A detailed comparison with the spectra of Tarvin, Vidal and Greytak has also been carried out there. Tarvin et al fitted their spectra at different temperatures to double Lorentzian shapes and reported the results of their measurements in the form of two fitting parameters ω_2 (frequency shift) and Γ_2 (damping) as functions of temperature. The spectra in Ref. 17 were consequently fitted the same way and the temperature dependence of ω_2 and Γ_2 predicted. The comparison of Γ_2 above the λ-point is shown in Fig. 6.

The open circles are the experimental Γ_2 data points. The dashed curve is the prediction of the planar spin model calculation of Hohenberg, Siggia and Halperin.[7] The reason why their curve falls off so rapidly lies in their neglecting the fairly strong thermal conductivity background, as well as their use of too large a value for w. Theirs is the standard dynamic scaling prediction, which would lead to a vanishing half width at large values of κ. The solid curve represents the theory with the back-

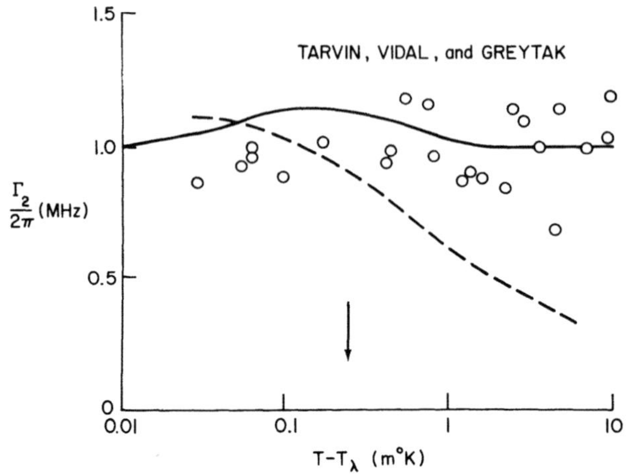

Fig. 6 Γ_2 vs. $T - T_\lambda$ from experiment and theories. The open circles are the experimental points. The dashed curve is from Ref. 7, while the solid curve is from Ref. 17.

ground effects built in. It uses the self-energy given by Eq. (44), and the parameters Γ_2 and ω_2 are determined by matching the central value and curvature of the actual convolved spectrum with the similarly convolved double Lorentzian. The improvement obtained by incorporating the background effects is evident. It is also shown in Ref. 17 that the predicted and experimentally inferred values of ω_2 at the lambda point are in good agreement.

It is worth exhibiting the predicted lambda point spectra at two different scattering wave numbers. Figure 7 shows the spectrum for $k = 1.79 \times 10^5$ cm^{-1} (the wave number used by Tarvin et al.) for three different resolutions. The infinite resolution spectrum, shown by a solid curve, has a pronounced structure. Almost all of the structure is lost for a resolution of 0.75 MHz (dashed curve), while for $\Gamma_o/2\pi = 1.5$ MHz the structure disappears completely. The latter, however, is the resolution of Tarvin et al. and the width of the curve allows for a satisfactory comparison[17] of the absolute scale of the spectral width in theory and experiment.

Figure 8 shows the predicted λ-point spectrum for $k = 3.58 \times 10^5$ cm^{-1} and the three different resolutions, $\Gamma_o/2\pi = 0$, 0.75 MHz and 1.5 MHz. The $\Gamma_o = 0$ spectrum is more filled-in than in Fig. 7 because of the larger value of σ_λ. There is no structure at finite resolutions. Though not useful for the purpose of detecting structure in the spectrum, this frequency doubling can demonstrate the validity of the considerations involving the background effects. Experiments at this higher frequency should again lead to a flat Γ_2 for $\kappa >> k$, with a value four times as large as the present value of

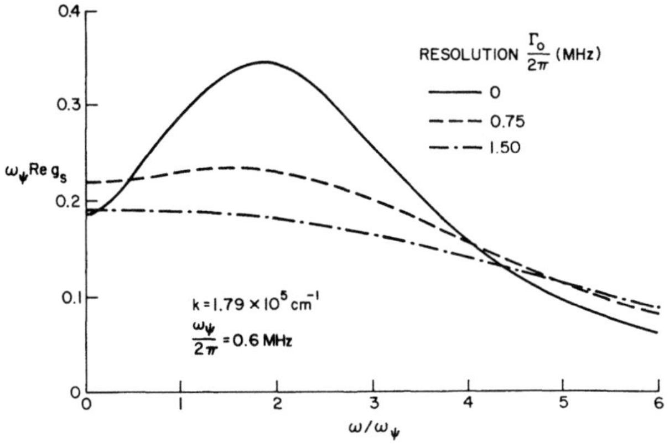

Fig. 7 λ-point spectrum for different resolutions at $k = 1.79 \times 10^5\ \mathrm{cm}^{-1}$.

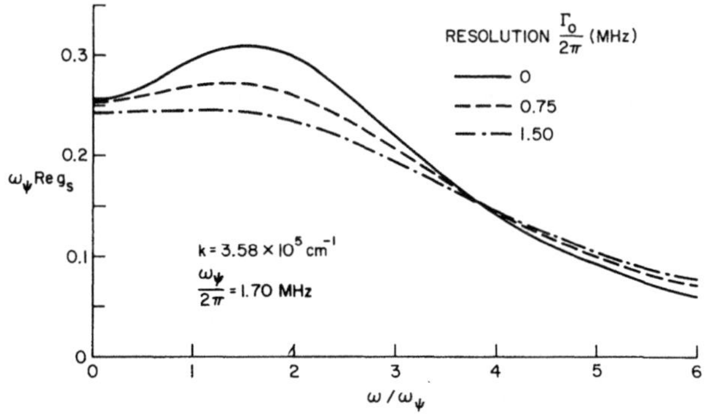

Fig. 8 λ-point spectrum for different resolutions at $k = 3.58 \times 10^5\ \mathrm{cm}^{-1}$.

1 MHz. The conventional dynamic scaling would predict an increase by a factor of $2\sqrt{2}$ only.

The light scattering below T_λ is also fairly well understood on similar lines[22] with the slow rise in the parameter Γ_2 being caused by the dropping specific heat. However, the large amplitude of the second-sound damping that was observed by Tyson still remains unexplained.

As a final item we shall demonstrate how the light scattering experiments can determine the constant κ_o in $\kappa = \kappa_o t^{2/3}$ with some knowledge of the scaling funcion for the specific heat. The value of κ_o used in all experimental and theoretical works is the number obtained by Hohenberg et al.[26] They used the series estimate for the universal amplitude ratio connecting the correlation length to the singular part of the specific heat and the measured specific heat and the mean specific heat to obtain $\kappa_o = 0.7$ Å^{-1}. To show how this alternate scheme works, we proceed as follows.

The temperature dependence of the specific heat (C) near the lambda point is established to be logarithmic. We adjust the constants in such a way that, in certain arbitrary units,

$$C = -\ln \kappa \qquad \text{for} \qquad \kappa \gg k \quad . \tag{45}$$

According to the usual scaling idea, the κ-dependence will smoothly go over to a k-dependence, and in the critical region,

$$C = -\ln(k/k_o) \quad . \tag{46}$$

k_o is a constant that will be determined by the general scaling function F(y), in terms of which

$$C = -\ln k + F(y) \qquad \text{with} \qquad y = \kappa/k \quad . \tag{47}$$

Equations (45) and (46) are the necessary ingredients for a determination of κ_o. The experiments are performed at a fixed wave vector k, and the temperature (t) dependence of C for that value of k can be extracted from the intensity measurements. Normalizing the large κ behavior to that expressed by Eq. (45), one can extrapolate to see where the value given by Eq. (46) is obtained. At this t, the value of κ is clearly k/k_o and hence determines κ_o.

The basic task of the theory is to provide the number k_o. To this end, one requires the scaling function F(y). For D = 4 the scaling function has been determined exactly by Bray[27] as

$$F(y) = 1 - \ln y - (1 + 4y^2)^{1/2} \ln\left[\frac{1}{2y} + \left(1 + \frac{1}{4y^2}\right)^{1/2}\right] \tag{48}$$

The limiting values are

$$F(y) = \begin{cases} 1 & y \ll 1 \\ -\ln y & y \gg 1 \end{cases} \quad . \tag{49}$$

Using these limiting values to obtain the hydrodynamic and critical behavior of C corresponding to Eqs. (45) and (46) leads to the identification

$$k_o = e = 2.718 \cdots \qquad (50)$$

This scaling function is shown in Fig. 9. The dashed curve represents the simple

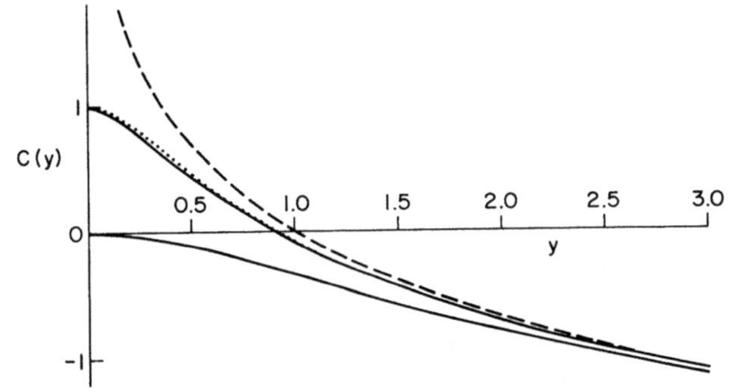

Fig. 9 Scaling function for specific heat vs. $y = \kappa/k$ at $D = 4$.

logarithmic behavior, $-\ln y$ valid for $y \gg 1$. The lower solid curve is a naive expectation for the scaling behavior -- namely $-\frac{1}{2}\ln(1 + y^2)$ -- which implies that going from hydrodynamic to critical region involves the replacement of κ by k. The upper solid curve is the actual function F(y) of Eq. (48), and the dotted curve is the simple scaling function $f(y) = -\frac{1}{2}\ln(y^2 + e^{-2})$, suggested by Eq. (50).

The calculations for D = 3 are more involved. Ornstein-Zernike-like functions have been used by Stephen.[28] A dispersion theory approach to the scaling function has been studied by Kroll[29] and more recently by Nicoll.[30] Preliminary results[30] indicate that the constant k_o is equal to π. This is exactly true for n = -2 and in the spherical limit. The deviations for other values of n are expected to be of $O(\eta)$.

The authors would like to thank Dr. Volker Dohm and Dr. Jeffrey Nicoll for many fruitful conversations and their collaborations during various parts of this work. In addition, the senior author (R. A. Ferrell) would like to thank Profs. N. Ményhard, H. Schmidt, F. Schwabl, and P. Szépfalusy for their collaboration during the early part of the work on dynamic scaling.

REFERENCES

1. R. A. Ferrell, N. Ményhard, H. Schmidt, F. Schwabl and P. Szépfalusy, Phys. Rev. Lett. 18, 891 (1967); Ann. Phys. (N.Y.) 47, 565 (1968).
2. P. C. Hohenberg and B. I. Halperin, Phys. Rev. 177, 952 (1969).
3. G. Ahlers, Phys. Rev. Lett. 21, 1169 (1968).
4. J. A. Tyson, Phys. Rev. Lett. 21, 1235 (1968).
5. B. I. Halperin, P. C. Hohenberg and S. Ma, Phys. Rev. B10, 139 (1974).
6. B. I. Halperin, P. C. Hohenberg and E. D. Siggia, Phys. Rev. B13, 1299 (1976).
7. P. C. Hohenberg, E. D. Siggia and B. I. Halperin, Phys. Rev. B14, 2865 (1976).
8. G. Ahlers in "The Physics of Liquid and Solid Helium," ed. by K. H. Benneman and J. B. Ketterson (Wiley, New York, 1976) Vol. I, Chap. II.
9. J. A. Tarvin, F. Vidal and T. J. Greytak, Phys. Rev. B15, 4193 (1977).
10. C. DeDominicis and L. Peliti, Phys. Rev. Lett. 38, 505 (1977).
11. L. Sasvàri, F. Schwabl and P. Szépfalusy, Physica 81A, 108 (1975).
12. C. DeDominicis and L. Peliti, Phys. Rev. B18, 353 (1978).
13. V. Dohm and R. A. Ferrell, Phys. Lett. A67, 387 (1978).
14. R. A. Ferrell and J. K. Bhattacharjee, J. of Low Temp. Phys. 36 (1979).
15. V. Dohm, Z. Physik B31, 327 (1978).
16. J. V. Sengers and P. H. Keyes, Phys. Rev. Lett. 26, 70 (1971).
17. R. A. Ferrell and J. K. Bhattacharjee, Univ. of Md. Tech. Report #79-099.
18. R. Bowers, Proc. Phys. Soc. (London), A65, 511 (1952).
19. C. Grenier, Phys. Rev. 83, 598 (1951).
20. J. F. Kerrisk and Wm. F. Keller, Phys. Rev. 177, 341 (1969).
21. L. J. Challis and J. Wilks, Proc. Roy. Soc. (London), 231, 515 (1955).
22. R. A. Ferrell and J. K. Bhattacharjee, Univ. of Md. Tech. Report #79-075.
23. S. Chapman and T. G. Cowling, "The Mathematical Theory of Non-Uniform Gases," Cambridge Univ. Press, 1953.
24. R. A. Ferrell, V. Dohm and J. K. Bhattacharjee, Phys. Rev. Lett. 41, 1818 (1978).
25. V. Dohm, Z. Physik B (1979).
26. P. C. Hohenberg, A. Aharony, B. I. Halperin and E. D. Siggia, Phys. Rev. B13, 2986 (1976).
27. A. J. Bray, preprint, 1978.
28. M. J. Stephen in "The Physics of Liquid and Solid Helium," ed. by K. H. Benneman and J. B. Ketterson (Wiley, New York, 1976), Vol. I., Chap. IV.
29. D. M. Kroll, Z. Physik B31, 309 (1978).
30. J. F. Nicoll, Univ. of Md. IPST Tech. Report #BN-903.

TRANSPORT PROPERTIES NEAR THE SUPERFLUID TRANSITION AND NEAR THE TRICRITICAL POINT OF He^3-He^4 MIXTURES

H. Meyer, G. Ruppeiner and M. Ryschkewitsch

Department of Physics, Duke University
Durham, NC 27706, USA

TRANSPORT PROPERTIES NEAR THE SUPERFLUID TRANSITION AND NEAR THE TRICRITICAL POINT OF He^3-He^4 MIXTURES

H. Meyer, G. Ruppeiner and M. Ryschkewitsch
Department of Physics, Duke University
Durham, NC 27706/USA

A short survey is presented on transport properties in liquid He^3-He^4 mixtures under saturated vapor pressure. We discuss the mass diffusion D, the thermal diffusion ratio k_T, and the effective thermal conductivity κ. The theoretical predictions of singular behavior near the superfluid transition and near the tricritical point are reviewed. New measurements of κ and k_T for dilute mixtures ($\leq$30% He^3) and for concentrated mixtures (between 60 and 70% He^3), carried out at Tokyo and at Duke Universities, are presented. These experiments show the predicted features, in particular the weak singularity in k_T near the superfluid transition and the strong one near the tricritical point. Results of relaxation times obtained in the course of these measurements are discussed in terms of the coupled diffusion modes associated with transport properties.

I. INTRODUCTION and THERMODYNAMIC RELATIONS

In his 1976 review article on experiments near the superfluid transition in helium, Ahlers[1)] remarked that the experimental investigations of the transport properties of He^3-He^4 mixtures are far from complete. Among the experiments described in detail in Ahlers' review, were measurements of the thermal conductivity of a X=0.15 He^3-He^4 mixture[2)] and of the mass diffusion D for X=0.10, 0.21, and 0.40 mixtures [3)]. Here, X is the He^3 number concentration. The thermal conductivity in the normal phase shows a singular behavior that abruptly stops at T_λ, the superfluid transition temperature. In the superfluid phase, the temperature dependence is much smaller. Within the experimental scatter, D was found to diverge with a simple power law in $(T-T_\lambda)$ as T_λ was approached. Ahlers also pointed out that, although further theoretical progress was needed, the hydrodynamics of mixtures is well known[4-6)]. Predictions on transport coefficients by mode coupling theories had been made a few years before[7,8)].

In the last three years, attention has been focussed in more detail on transport properties in dilute mixtures and near the tricritical point. The tricritical point terminates the critical (second-order) line of the superfluid transition and starts the (first-order) line that defines the coexistence region of two separated phases, one He^4-rich and superfluid, the other He^3-rich and normal. For convenience, a phase diagram of the liquid mixtures under saturated vapor pressure is presented in Fig. 1. We refer to the review by Ahlers[1)] for a detailed presentation of the static and dynamic properties of liquid helium. Therefore, the transport experiments presented there

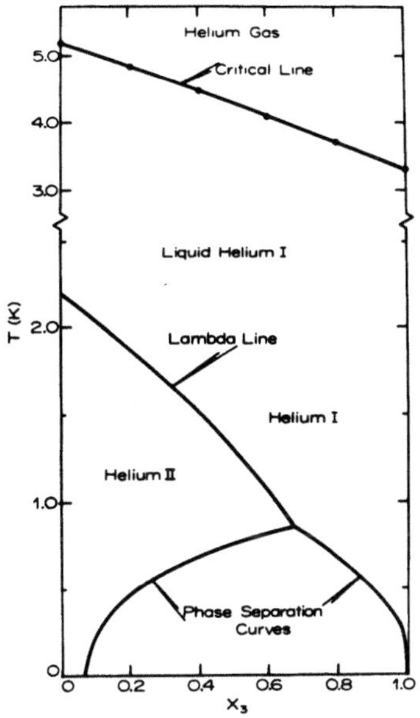

Fig. 1. The phase diagram for He^3-He^4 mixtures at saturated vapor pressure.

will only be briefly mentioned here.

This article attempts to give a short review of experiments that lead to information on transport properties. It is organized as follows. In Section II, several transport coefficients will be defined and appropriate thermodynamic relations will be quoted. Section III deals with the predicted singularities of quantities relevant to this article and reviews some recent experiments. The hydrodynamic modes in the normal phase are described in Section IV. Finally, in Section V, the most recent experiments on transport properties - most of them unpublished - are presented. These experiments are measurements of the thermal conductivity and the thermal diffusion ratio k_T in dilute

$(0.1 \leq X \leq 0.33)$ and in concentrated $(0.6 \leq X \leq 0.7)$ mixtures performed respectively at Tokyo and at Duke Universities. Relaxation time measurements are also discussed. Because the analysis of the Duke results is not final, this article serves as a progress report.

II. DEFINITIONS and RELATIONS

The notation of Khalatnikov[5] and of Landau and Lifshitz[4] is used for the thermal conductivity κ. The expressions below are in terms of X rather than the mass concentration $c = XM_3/\overline{M}$ used in Refs. 4 and 5. Here

$$\overline{M} = (XM_3 + (1-X)M_4) \tag{1}$$

where M_3 and M_4 are the isotopic molar masses of He^3 and He^4 respectively. The transport coefficients which are dealt with in this paper are:

1. The mass diffusion D.
2. The thermal diffusion ratio k_T. In the normal phase, under steady state conditions (no mass flow, $\underset{\sim}{i}=0$), k_T is defined by[4,5]

$$\frac{k_T}{T} = -\frac{\nabla X}{\nabla T}\frac{M_4 M_3}{\overline{M}^2} \quad . \tag{2}$$

We consider the case only of one dimension where ∇X is the concentration gradient and ∇T the corresponding temperature gradient. When a temperature gradient is established, the He^3 concentration becomes enriched near the colder part, hence $k_T > 0$.

The situation in the superfluid is more complicated since there is also a superfluid current from the cold to the warm end of the cell. Khalatnikov[5] shows that if we assume that the non-diffusive part of the He^3 mass flux is carried along by the normal component of the He^4, which travels with velocity V_n,

$$\frac{\nabla X}{\nabla T}\frac{M_4 M_3}{\overline{M}^2} = -\frac{k_T}{T} + \frac{V_n}{\nabla T}\frac{XM_3}{D\overline{M}} \tag{3}$$

in the steady state where $\underset{\sim}{i} + \rho c \underset{\sim}{V}_n = 0$ (see Eqs. 24-52 and 24-55 of Ref. 5). By making several hydrodynamic assumptions in the superfluid, it can be shown that[5]

$$\frac{\nabla X}{\nabla T} = X\left(\frac{\partial X}{\partial \Delta}\right)_{T,P}\left[\frac{\partial(S/X)}{\partial X}\right]_{T,P} \tag{4}$$

where $(\partial X/\partial\Delta)_{T,P}$ is the molar concentration susceptibility, with $\Delta\equiv\mu_3-\mu_4$ the isotopic chemical potential difference. In the limit of dilute mixtures, Eq. 4 becomes[5)]

$$-\frac{\nabla X}{\nabla T} = \frac{XM_3}{\bar{M}\,T}\left[1 + \frac{\sigma_o\bar{M}}{R\,X}\right] \tag{5}$$

where R is the gas constant and σ_o the entropy of liquid He^4 per gram. In particular at $T=T_\lambda$, where Eq. 2 is valid (see also below), k_T for dilute mixtures is independent of X. Using σ_o=1.56 Joules/gr[1,10)], one obtains

$$\lim_{X\to 0} k_T = 0.57 \tag{6}$$

a value which compares well with experiment, as we shall see in Section V.

3a. <u>The thermal conductivity κ in the normal phase</u>. Here we mean the thermal conductivity measured under steady state conditions, $\underset{\sim}{i}=0$, in the presence of a concentration gradient.

3b. <u>The conductivity κ_s in the normal phase in the absence of a concentration gradient</u>. Of theoretical interest[7-9)] is the flux $J_T = Q - \frac{1}{\bar{M}}\left(\Delta - T\left(\frac{\partial\Delta}{\partial T}\right)_{P,X}\right)\underset{\sim}{i}$, where $\underset{\sim}{Q}$ is the heat flux. In the absence of a concentration gradient, $\underset{\sim}{J}_T$ is given by

$$\underset{\sim}{J}_T = -\kappa_s \underset{\sim}{\nabla} T \tag{7a}$$

where

$$\kappa_s = \kappa + \left(\frac{\partial\Delta}{\partial X}\right)_{T,P} \frac{\bar{M}^4}{M_3^2 M_4^2}\, k_T^2 D(VT)^{-1} \quad . \tag{7b}$$

κ_s is not easily accessible to observation.

3c. <u>The measured thermal conductivity κ_{eff} in the superfluid phase</u> which, according to Khalatnikov[5)], is related to κ via

$$\kappa_{eff} = \kappa + V^{-1}DT(\partial X/\partial\Delta)_T\left\{X\,\frac{\partial}{\partial X}\left(\frac{S}{X}\right)_{T,P} + \frac{k_T}{T}\,\frac{\bar{M}^2}{M_3M_4}\left(\frac{\partial\Delta}{\partial X}\right)_{T,P}\right\}^2 \quad . \tag{8}$$

As Papoular[9)] has discussed, the term $(\kappa_{eff}-\kappa)$ in Eq. 8 is to be regarded as associated with the extra heat flux proportional to the normal fluid velocity V_n and is to vanish as the superfluid transition and the tricritical point are approached. Hence, near T_λ or T_t, and in the superfluid phase

$$\lim_{T\to T_{\lambda,t}} \frac{k_T}{T} = -\,\frac{M_3M_4}{\bar{M}^2}\,X\left(\frac{\partial X}{\partial\Delta}\right)_{T,P}\left[\frac{\partial(S/X)}{\partial X}\right]_{T,P} \quad . \tag{9}$$

III. PREDICTION OF SINGULARITIES and REVIEW OF EARLY EXPERIMENTS

We briefly recapitulate here the prediction of the singularities in the susceptibility and in the transport coefficients. We compare these predictions with recent experiments giving k_T, κ, and D.

1) The susceptibility $(\partial X/\partial \Delta)_{T,P}$ is predicted[11] to show a weak singularity along the critical line, similar to that of C_p for pure He^4 at T_λ. As the tricritical point is approached, a cross-over to a strong divergence should be observed[12]. The predictions are in good agreement with experiments[13-16].

2) For the thermal diffusion ratio k_T, the dynamic scaling (DS)[7], mode coupling (MC)[8] and renormalization group (RG)[17] theories predict that the singularity of k_T above T_λ is the same as that for $(\partial X/\partial \Delta)_{T,P}$. Hence, a weak singularity with a peak is predicted for k_T in mixtures at the lambda line. There should be a cross-over into a strong divergence as the concentration becomes close to the tricritical one, X_t=0.675, namely

$$\text{Critical:} \quad k_T \propto (\partial X/\partial \Delta)_{T,P} \propto (\varepsilon_\lambda)^{-\alpha} \qquad \alpha \approx -0.02 \qquad (10a)$$

and

$$\text{Tricritical:} \quad k_T \propto (\partial X/\partial \Delta)_{T,P} \propto \varepsilon_t^{-1} \qquad (10b)$$

upon approaching along a path $X=X_t$ or along the coexistence curve σ_+ and σ_-. Here

$$\varepsilon_\lambda = (T-T_\lambda)/T_\lambda \quad , \quad \varepsilon_t = (T-T_t)/T_t$$

and the subscripts + and - denote the He^3-rich and He^3-poor branches of the coexistence curve.

The behavior of k_T below T_λ is predicted from Eq. 9. For concentrated mixtures, numerical calculations using calorimetric data[18] show $\left[\partial(S/X)/\partial X\right]_{T,P}$ to be a slow varying function, and hence $k_T \propto (\partial X/\partial \Delta)_T$ should exhibit a sharp peak. For very dilute mixtures, however, the temperature dependence below T_λ is calculated with Eq. 5 to be much smaller than that above T_λ.

The earliest experiments on k_T were performed under non-stationary conditions. The method adopted by Lucas and Tyler[19] utilized the convection velocity in a vertical column of a X=0.15 mixture. The NMR signal of He^3 was the probe for the concentration. These measurements showed conclusively the expected singularity in the normal phase, and a milder temperature dependence in the superfluid phase. Unfortunately they did not yield absolute values for k_T. Roe and Meyer[20] derived k_T from the concentration relaxation of mixtures in the range $0.6<X<0.7$. They obtained all the data in the same arbitrary units and were able to normalize the results for one mixture using Eq. 4. This procedure yielded k_T for the other mixtures. The results showed the weak singular behavior expected for the superfluid transition and also the strong divergence for the tricritical mixture X-X_t as indicated by Eq. 10b.

3) <u>Mass diffusion</u>. The predictions from both DS and RG theories[7,17] for the divergence along the critical line are consistent, namely

$$D \propto \varepsilon_\lambda^{-1/3} \tag{11}$$

and this behavior has been observed in the isothermal concentration-relaxation measurements by Ahlers and Pobell[3]. There is a difference between the tricritical singularity expected from the DS[7] and the RG[17] approach,

$$D \propto \varepsilon_t^{\frac{1}{2}} \qquad \text{(DS)} \tag{12a}$$

$$D \propto \varepsilon_t^{1/3} \qquad \text{(RG)} \quad . \tag{12b}$$

The later prediction appears to be confirmed by the analysis of acoustic attenuation experiments[21], which in the "low frequency" limit (hydrodynamic regime) give

$$D \propto \varepsilon_t^{0.32\pm0.1} \tag{13}$$

Furthermore, the attenuation data analysis shows the cross-over of D from a divergent critical behavior to the tricritical confluent singularity of Eq. 12b.

4) <u>Thermal conductivity</u>. Predictions are for κ and κ_{eff} to be finite both in the critical and the tricritical regions. A microscopic calculation by Kawasaki and Gunton[7] shows that κ_s diverges. However, its combination with the second term in Eq. 7b leads to a cancellation of divergences in the first order so that κ is finite. As Ahlers has remarked, "the theory seems to make no predictions about higher order singular contributions which must evidently be responsible for the remaining singularity that is observed experimentally in κ above T_λ"[1].

For very dilute mixtures, two recent theories[22,23] predict $\kappa(T_\lambda)$ as a function of X. Obviously for pure He^4(I) the conductivity must tend to infinity as T_λ is approached. The dynamic scaling argument by Kawasaki[22] predicts $\kappa(T_\lambda)\propto X^{-0.33}$, while RG calculations by Siggia[23] give $\kappa(T_\lambda)\propto X^{-\frac{1}{2}}$. Experiments by Tanaka and Ikushima[22] can be represented by the expression

$$\kappa(T_\lambda)^{-1} = 50.5X + 0.63X^{0.37}. \qquad (14)$$

While the authors claim good agreement with Ahlers' data for X=0.15[2], there is a disagreement for κ in pure He^4 [24]. The exponent 0.37 quoted above is closer to Kawasaki's prediction[22] than Siggia's[23]. However, if the analysis is restricted to concentrations less than 0.5%, the possibility that the data can be represented by a power law $\kappa(T_\lambda)\propto X^{-\frac{1}{2}}$ within the experimental scatter is not excluded. Hence, this situation still needs to be clarified.

Light scattering[25] and thermal relaxation experiments[26] in the superfluid phase near T_t have also shown κ_{eff} to be finite. From their relaxation time observations during second sound measurements, Ahlers and Greywall[26] deduce $\kappa(T_t)\simeq 47$ µWatt/cm deg. The Rayleigh linewidth[25] could be expressed by the experimental relation

$$\Gamma/q^2 = 1.5\text{x}10^{-4}\varepsilon_t^{\,0.95\pm0.07}\ \text{cm}^2\text{sec}^{-1} \qquad (15)$$

where q is the momentum transfer. This slow diffusive mode is given in the limit $\varepsilon_t \to 0$ by[9,25]

$$\Gamma_o^s/q^2 = \frac{V\kappa\ (\partial\Delta/\partial X)_{T,P}}{TX^2\left[\partial(S/X)/\partial X\right]^2_{T,P}} \quad . \qquad (16)$$

Because $(\partial\Delta/\partial X)_{T,P}\propto\varepsilon_t$ its temperature dependence is consistent with the one observed for Γ_o^s and hence κ is constant. Upon using Eq. 4, we obtain

$$\lim_{\varepsilon_t\to 0}\Gamma_o^s/q^2 \approx \frac{\kappa D}{\kappa_s} \approx V\kappa\left[\left(\frac{\partial\Delta}{\partial X}\right)_T \frac{k_T^2}{T}\,\frac{\bar{M}^4}{M_3^2M_4^2}\right]^{-1} \quad . \qquad (17)$$

Using the value of $\kappa(T_t)$ from Ref. 26, one finds Γ_o/q^2=1.3x10^{-4}cm^{2}sec^{-1}. This excellent agreement with the experimental value is somewhat fortuitous, however, as will be discussed in Section V.

IV. HYDRODYNAMIC MODES IN THE NORMAL PHASE

In He^3-He^4 mixtures, the dynamics of concentration and thermal fluctuations is given by two coupled differential equations[4]. The solution of these equations yields two well known dispersion relations[6,9] which we write, following Tanaka and Ikushima[27], as

$$\frac{\Gamma_o^n}{q^2} = D_o \quad \text{with} \quad D_o = \frac{D_T + D - \left[(D_T+D)^2 - 4DD_{eff}\right]^{1/2}}{2} \tag{18}$$

$$\frac{\Gamma_2^n}{q^2} = D_2 \quad \text{with} \quad D_2 = \frac{D_T + D + \left[(D_T+D)^2 - 4DD_{eff}\right]^{1/2}}{2} ,$$

where D has the same meaning as before, and

$$D_T = \kappa_s/\rho C_{P,X} \quad , \quad D_{eff} = \kappa/\rho C_{P,X} \quad . \tag{19}$$

Since D_o and D_2 are real, the eigenmodes are dissipative. Griffin[6] has discussed these two modes and concluded that in the superfluid phase, D_o is again related to diffusion and is associated with the Rayleigh linewidth mentioned before, while D_2 determines the Brillouin linewidth of the second sound. In the case of slow relaxation of the fluid from an initial thermal gradient to a final zero gradient, the boundary value problem must be solved to determine the relative amplitude of the modes. The formal solution of this problem for well established experimental conditions has been worked out by Behringer[28], but is too lengthy to be presented here.

Let us consider just the limiting cases for the two dispersion relations.

a) Close to the tricritical point, where k_T and D have singularities as given by Eqs. 10b and 12b, $(\partial\Delta/\partial X)_{T,P} k_T^2 D$ diverges. Numerical estimations based on experiments show that for $\varepsilon_t<10^{-3}$, $\kappa_s>>\kappa$, and $D_T>>D_{eff}>>D$. Therefore

$$D_o \approx \frac{D_{eff} D}{D} \approx \frac{\kappa D}{\kappa_s} \tag{20a}$$

$$\approx V\kappa \left[\left(\frac{\partial\Delta}{\partial X}\right)_{T,P} \frac{k_T^2}{T} \frac{\overline{M}^4}{M_3^2 M_4^2}\right]^{-1} \propto \varepsilon_t \tag{20b}$$

which is the same result as that for the slow mode Γ_o^s/q^2 quoted before in Eq. 16. The fast mode is characterized by

$$D_2 \approx D_T \approx \frac{\overline{M}^4}{M_3^2 M_4^2} \left(\frac{\partial\Delta}{\partial X}\right)_T \frac{k_T^2}{T} D \propto \varepsilon_t^{-0.68} \quad . \tag{20c}$$

The situation in dilute mixtures near T_λ is quite different. D_o is predicted to stay roughly constant and D_2 is expected to diverge approximately like $D \propto \varepsilon_\lambda^{-1/3}$, which is the only nonviscous transport coefficient having a strong divergence.

b) Far away from the tricritical point and from the critical line, $k_T \to 0$ and $\kappa_s \to \kappa$, therefore

$$\Gamma_o^n/q^2 = D \quad , \; \Gamma_2^n/q^2 = D_{eff} \tag{21}$$

The thermal and mass diffusion modes are uncoupled.

V. RECENT EXPERIMENTS ON THERMAL TRANSPORT PROPERTIES

We now describe two very similar series of experiments that give simultaneously the thermally induced concentration gradient $\nabla X/\nabla T$ and the effective conductivity κ or κ_{eff} under stationary conditions. Experiments on dilute mixtures ($0.1 \leq X \leq 0.33$) were performed at Tokyo University[27,29] and those on concentrated mixtures ($0.6 \leq X \leq 0.7$) by the present authors[30-32].

The design of the experimental cells used both at Duke and Tokyo is based on the same principle. The cells consist of two horizontal copper plates, separated by thin-walled stainless steel spacers. To each of these plates is attached a horizontal capacitor made of two thin perforated stainless steel discs. These capacitors, are used to measure the dielectric constant of the fluid at the top and bottom of the cell. Changes in X are calculated with the Clausius Mossotti relation and from the known change of the molar volume V with X. Matched thermometers inserted in the sides of the top and bottom plates allow measurements of the temperature gradient established when a small heater at the bottom of the cell is turned on. In Fig. 2, we present first the results for dilute mixtures at three molefractions[29]. It can be seen that k_T diverges at most weakly, which is consistent with the expectations, Eq. 10a). Furthermore, the limiting value at T_λ is only weakly concentration-dependent and $k_T(T_\lambda) \simeq 0.6$, which is in good agreement with the prediction by Eq. 6. In Fig. 3, the results for κ in several mixtures are shown, and these exhibit a singularity similar to that first reported by Ahlers[2] for a X=0.15 mixture . By using the data of Ahlers and Pobell[3] for D and their own results for k_T and κ, Tanaka and Ikushima[27] have shown that the diffusivity D_o, defined by Eq. 18, does not diverge as T_λ is approached. D_2, however, diverges like $(T-T_\lambda)^{-1/3}$ irrespective of concentration, at least up to X=0.3. These results are consistent with the expectations mentioned above. The tricritical behavior is quite different, as we will discuss below.

Fig. 4 shows the results[30-32] for concentrated mixtures in the normal and in the superfluid phase, represented respectively by k_T and by k_T^*, the latter defined by

$$-\frac{k_T^*}{T} \equiv \frac{\nabla X}{\nabla T}\,\frac{M_4 M_3}{\overline{M}^2} = -\frac{k_T}{T} + \frac{V_n X M_3}{\nabla T D \overline{M}} \;, \qquad (22)$$

since it is $\nabla X/\nabla T$ that is actually measured. From our previous discussion, it follows that $k_T^* \to k_T$ as T_t or T_λ is approached. For clarity, results for a number of other mixtures have not been shown. We note the sharp peak at T_λ, which is consistent with expectations, and the steady increase of k_T^* as the phase separation is approached. These results can be compared with those in Fig. 3 of Ref. 20 that show the early data obtained from a transient method. The improvement achieved since then is quite striking. In contrast to the case for very dilute mixtures, the conductivity κ shows only a weak variation with

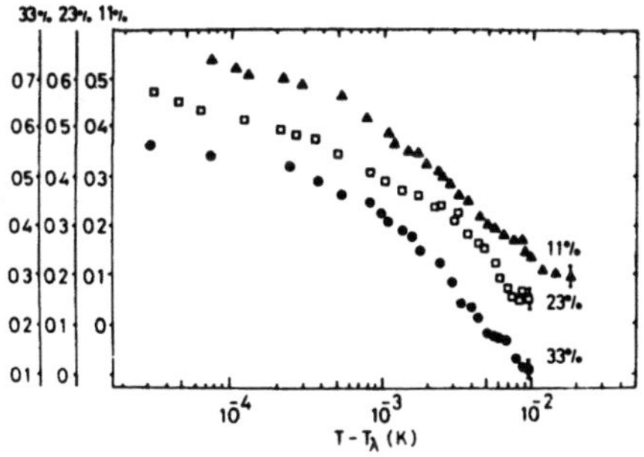

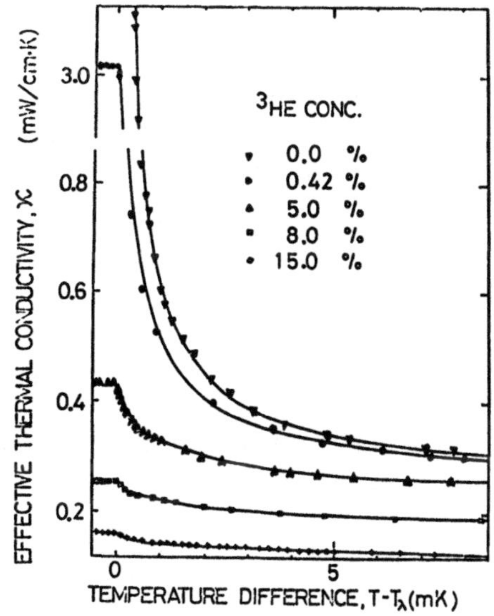

Fig. 2. The thermal diffusion ratio k_T versus $T-T_\lambda$ for three mixtures, on a semi-logarithmic scale. After Ref. 29.

Fig. 3. The effective thermal conductivity κ plotted versus $T-T_\lambda$ in mixtures of various concentrations. After Ref. 22.

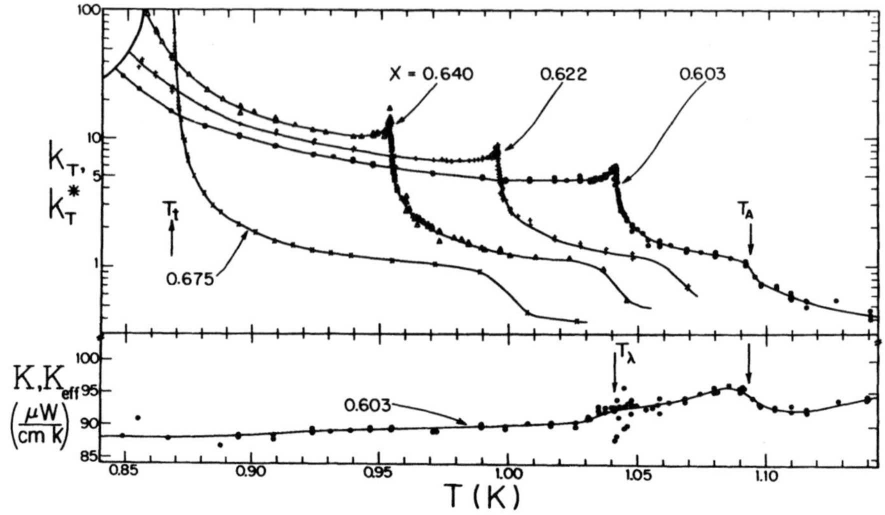

Fig. 4. The thermal diffusion ratio k_T and the effective conductivity κ for $T>T_\lambda$ and the corresponding properties k_T^* and κ_{eff} for $T<T_\lambda$ in concentrated mixtures. The conductivity units are μWatt/cm deg.

temperature near the transition. Only one representative curve for κ is shown, since very similar values have been obtained for the other mixtures. We note that for the tricritical mixture, we find $\kappa(T_t)=80\pm5$ μW/cmK, a value that is in disagreement with that inferred from relaxation time measurements[26], $\kappa(T_t)\simeq47$ μW/cmK.

Let us next examine the results for the "tricritical" mixture with $X\cong X_t=0.675$. The logarithmic plot in Fig. 5 shows that k_T can be taken as

$$k_T = k_{T,sing} + k_{T,reg} \tag{23}$$

where the subscripts denote "singular" and "regular". A preliminary analysis gives

$$\begin{aligned} k_{T,sing} &= 5.0\times10^{-2}\times\varepsilon_t^{-1.0} \\ k_{T,reg} &= 0.6 \end{aligned} \tag{24}$$

In the same Figure, data are shown for mixtures ($X>0.673$) at the coexistence curve σ_+ on the normal side. Here again, the same power law for the singular term is obtained, in excellent agreement with theoretical predictions[7,17].

In Fig. 4, we note that for each mixture there is a shoulder in the k_T curve above T_λ and above T_t. At the temperature of this anomaly, marked by T_A for X=0.603, the conductivity shows a slight anomalous temperature variation. The locus of $T_A(X)$ for all the mixtures coincides with the temperature below which Gearhard and Zimmermann found the onset of superfluid flow in films through porous glass[33].

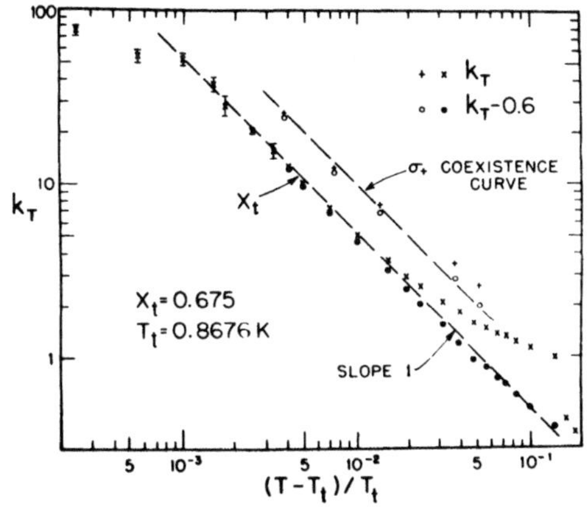

Fig. 5. The thermal diffusion ratio as a function of the distance $(T-T_t)/T_t$ from the tricritical point for the mixture $X \simeq X_t$ and along the coexistence curve σ_+ on the He^3-rich side.

Hence, we attribute the shoulder in k_T to this phenomenon, and suggest the following mechanism: During an applied temperature gradient, there is a continuous flow of superfluid He^4 along the stainless steel wall of the cell towards the warmer region. Hence, the concentration gradient ∇X is enhanced. Steady state conditions are reached by a continuous diffusion of He^4 back toward the colder end through the center of the cell, and as a result the net vertical flow of He^4 is zero. The enhancement of k_T due to film flow is the regular term in Eq. 24, as can be seen by inspection of the data in Fig. 5 for $\varepsilon > 10^{-1}$.

We now examine the weak singularity of k_T or k_T^* at T_λ. Fig. 6 shows a comparison between k_T and the susceptibility[16] $(\partial X/\partial \Delta)_{T,P}$ close to T_λ for the mixture with X=0.622[31]. As expected from predictions, the behavior is very similar, although the k_T peak tends to be less sharp. This rounding may well be caused by the smearing out of the transition by the non-zero thermal gradient in the mixture during the measurements. In Fig. 6b, we show the ratio $(\nabla X/\nabla T)/(\partial X/\partial \Delta)_{T,P}$

from experimental data and compare it with the quantity $X[\partial(S/X)/\partial X]_{T,P}$ calculated from the calorimetric measurements[18]. Strictly speaking, this comparison is only valid for the superfluid phase. The error

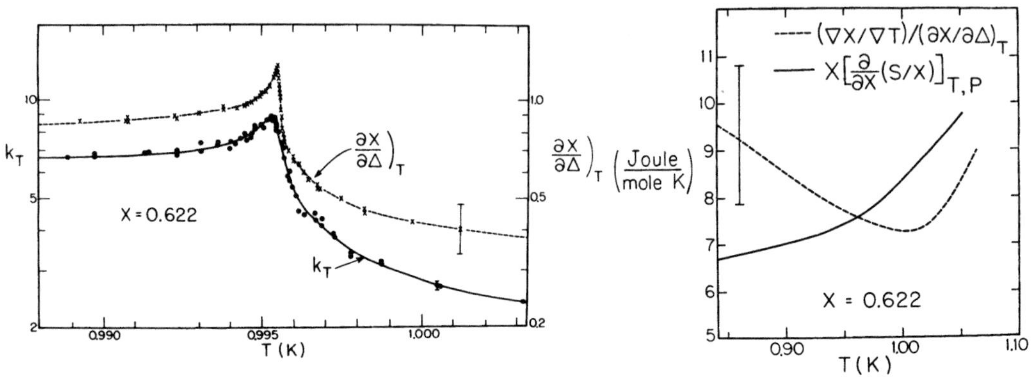

Fig. 6. a) Comparison between k_T and the susceptibility[16] $(\partial X/\partial\Delta)_{T,P}$ in units of (mole/Joule) for the mixture X=0.622.
b) Test of the Khalatnikov – relation, Eq. 4, with data for the mixture X=0.622.

bars show an estimation of possible systematic geometric uncertainties in both susceptibility and $(\nabla X/\nabla T)$ measurements. According to Khalatnikov's theory[5], the two curves in Fig. 6b should coincide. Considering the assumptions made in the theory and the possibilities of systematic errors in the various experiments, the agreement is considered to be quite satisfactory. However, there is considerable disagreement for $(\partial X/\partial\Delta)_T$ near the phase separation curve σ_- between the results from various groups[13,14,16], and this matter still needs to be clarified.

We note from Fig. 4 that our k_T^* data along σ_- can be represented roughly by

$$k_T^* = (1.0 \pm 0.2)|\varepsilon_t|^{-1} . \tag{25}$$

Let us briefly come back to the light scattering experiment by Leiderer et al[25]. If in their analysis we use our value for κ=80 μW/(cm K) then, with $X[\partial(S/X)/\partial X]_{T,P} \simeq 6.5$ Joules/mole K, $(\partial\Delta/\partial X)_T = 3.1|\varepsilon_t|$ (from Ref. 14) inserted into Eq. 16, we obtain $\Gamma_o/q^2 \simeq 2.2\times10^{-4}\varepsilon_t$ cm^2/sec in fair agreement with the experimental value, Eq. 15. But if we use Eq. 17 together with Eq. 25, where we assume

$k^*_{T_2} \to k_T$ for $\varepsilon_t \to 0$ and the same $(\partial X/\partial\Delta)_T$ as before, then $\Gamma^s_o/q^2 \approx 8\times10^{-4}\varepsilon_t$ cm^2/sec, in disagreement with experiment.

Finally, we discuss the relaxation times $\tau(\nabla T)$ and $\tau(\nabla X)$ that are measured by observing the establishment or the decay of ∇T and ∇X after switching the heat current on or off. In Fig. 7, we show the results for the tricritical mixture $X \simeq X_t = 0.675$. Both temperature and concentration gradients relax with the same time constants, which indicates strong coupling via k_T (see Section III). We have used the available data for the quantities entering in Eq. 18 to calculate the inverse diffusion rate D_o^{-1} and D_2^{-1}. Because in the limit of small ε_t, $D_o \propto \varepsilon_t$ and $D_2 \propto \varepsilon_t^{-0.68}$, it is obvious that the slow mode is observed. We have estimated the corresponding relaxation times by setting $q=\pi/d$, namely

$$\tau = \left(\frac{d}{\pi}\right)^2 D_o^{-1} \quad , \tag{26}$$

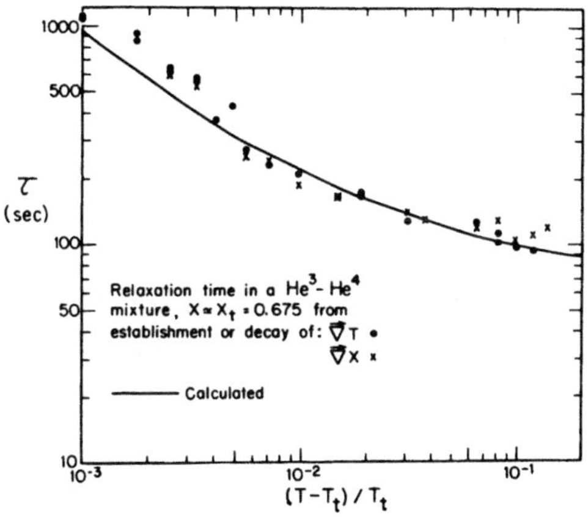

Fig. 7. The relaxation times $\tau(\nabla T)$ and $\tau(\nabla X)$ characterizing the establishment or decay of the concentration and temperature gradients, respectively. The mixture has a concentration $X \simeq X_t$ and the solid line is calculated from Eq. 18 as explained in the text, and normalized at $\varepsilon_t = 10^{-1}$ to the data.

where d=0.28 cm is the height of the conductivity cell, and we find $\tau(calc.)/\tau(obs.) \approx 1.5$ which is a good agreement considering the geometric uncertainties. Hence, we feel justified in normalizing the calculated curve to the experiment, and this is done arbitrarily at $\varepsilon_t = 10^{-1}$. The result is shown in Fig. 7 and the agreement over the whole temperature range is very satisfactory.

An interesting situation that illustrates the coupling by k_T of

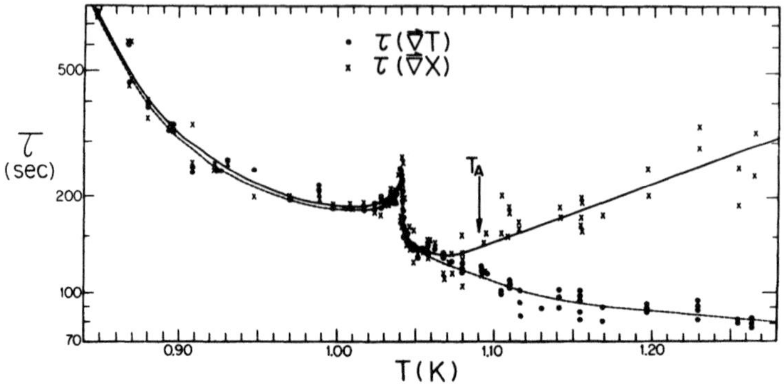

Fig. 8. The same relaxation times for the mixture X=0.603 showing the decoupling of the thermal- and concentration diffusion mechanisms above T_λ=1.09 K. The locus of the k_T "shoulder" is indicated by T_A. The solid and dashed lines are simply a guide to the eye.

the diffusive modes is shown in Fig. 8 for the mixture X=0.60. For sufficiently large k_T, both thermal and mass diffusion are coupled and show one common relaxation time. Here τ is an increasing function of k_T since it reflects qualitatively the same temperature dependence. However, above T_λ at the same temperature where the "shoulder" in k_T is observed, k_T becomes small enough for the relaxation processes to decouple, giving a fast thermal- and a slow mass diffusion. The analysis of these processes, including the calculation of the amplitudes for the slow and the fast modes will be reported elsewhere in collaboration with R. P. Behringer.

In conclusion, there has been some increasing activity on the thermal transport phenomena in mixtures, with exception of the viscosity. In general, the preliminary analysis for experiments near the tricritical point shows good agreement with expectations from theory, but a number of intriguing questions remain, and more will surely come up as a more careful analysis of the results is carried out.

ACKNOWLEDGMENTS

The authors are grateful to Mr. Octavio Diaz and Christopher Meyer for help in the analysis of the data. One of us (H.M.) appreciates stimulating conversations with Drs. P. Hohenberg, M. Papoular and G. Ahlers. The research has been supported by a grant from the National Science Foundation and from the Air Force Office of Scientific Research.

REFERENCES

1. G. Ahlers in The Physics of Liquid and Solid Helium, Part I, edited by K.H. Bennemann and J.B. Ketterson (Wiley, N.Y.,1976).
2. G. Ahlers, Phys. Rev. Lett. 24, 1333 (1970).
3. G. Ahlers and F. Pobell, Phys. Rev. Lett. 32, 144 (1974).
4. L.D. Landau and I.M. Lifshitz, Fluid Mechanics (Pergamon Press, London, 1959), Chapter VI.
5. I.M. Khalatnikov, Introduction to the Theory of Superfluidity (Benjamin, N.Y.,1965), Section 24.
6. A. Griffin, Can. J. Phys. 47, 429 (1969).
7. K. Kawasaki and J.D. Gunton, Phys. Rev. Lett. 29, 1661 (1972).
8. M.K. Grover and J. Swift, J. Low Temp. Phys. 11, 751 (1973).
9. M. Papoular, J. Low Temp. Phys. 24, 105 (1976).
10. J. Wilks, Liquid and Solid Helium, (Oxford University Press, 1967) Appendix A1.
11. R.B. Griffiths and J.C. Wheeler, Phys. Rev. A2, 1047 (1970).
12. E.K. Riedel, Phys. Rev. Lett. 28, 675 (1972).
13. G. Goellner and H. Meyer, Phys. Rev. Lett. 26, 1534 (1971); G. Goellner, R.P. Behringer and H. Meyer, J. Low Temp. Phys. 13, 113 (1973).
14. P. Leiderer, D.R. Watts and W.W. Webb, Phys. Rev. Lett. 33, 483 (1974).
15. C.A. Gearhart, Jr. and W. Zimmermann, Jr., to be published.

16. M. Ryschkewitsch, T. Doiron, M. Chan and H. Meyer, Phys. Lett. 64A, 219 (1977); M. Ryschkewitsch and H. Meyer, J. Low Temp. Phys. 35, 103 (1979).
17. E.D. Siggia and D.R. Nelson, Phys. Rev. B15, 1427 (1977).
18. S.T. Islander and W. Zimmermann, Jr., Phys. Rev. A7, 188 (1973).
19. P. Lucas and A. Tyler, J. Low Temp. Phys. 27, 281 (1977).
20. D. Roe and H. Meyer, J. Low Temp. Phys. 28, 349 (1977).
21. D. Roe, G. Ruppeiner and H. Meyer, J. Low Temp. Phys. 27, 747 (1977).
22. M. Tanaka, A. Ikushima and K. Kawasaki, Phys. Lett. 61A, 119 (1977).
23. E.D. Siggia, Phys. Rev. B15, 2830 (1977).
24. G. Ahlers, Phys. Rev. Lett. 21, 1159 (1968); G. Ahlers in Proceedings of the 12th International Conference on Low Temperature Physics, edited by E. Kanda (Keigaku, Tokyo, 1971) p. 21.
25. P. Leiderer, D.R. Nelson, D.R. Watts and W.W. Webb. Phys. Rev. Lett. 34, 1080 (1975).
26. G. Ahlers and D. Greywall in Low Temperature Physics-LT 13, edited by K.D. Timmerhaus, W.J. O'Sullivan and E.F. Hammel (Plenum, N.Y.,1974), Vol. 1, p. 586; G. Ahlers (private communication).
27. M. Tanaka and A. Ikushima (to be published by J. Low Temp. Phys. 1979).
28. R.P. Behringer, private communication.
29. M. Tanaka and A. Ikushima, Phys. Lett. 64A, 402 (1978).
30. M. Ryschkewitsch, G. Ruppeiner and H. Meyer, J. de Physique C-6, Vol. 1, 186 (1978).
31. G. Ruppeiner and H. Meyer, Phys. Lett. A 70, (1979).
32. G. Ruppeiner, M. Ryschkewitsch and H. Meyer (to be published).
33. C.A. Gearhart and W. Zimmermann, Jr., Phys. Lett. 48A, 49 (1974).

RENORMALIZATION GROUP CALCULATIONS FOR CRITICAL AND TRICRITICAL DYNAMICS APPLIED TO HELIUM

L. Peliti

GNSM-CNR, Unità di Roma*

Istituto di Fisica, Università di Camerino, Camerino (MC), Italy

* Mailing address : Istituto di Fisica "G. Marconi",
Piazzale delle Scienze 5, 00185 Roma (Italy)

RENORMALIZATION GROUP CALCULATIONS FOR CRITICAL AND TRICRITICAL DYNAMICS APPLIED TO HELIUM

L.Peliti

GNSM-CNR, Unità di Roma*

Istituto di Fisica, Università di Camerino, Camerino (MC), Italy

Abstract Renormalization group calculations for the critical dynamics of liquid Helium and of He^3-He^4 mixtures are reviewed. The role of different fixed points in determining the asymptotic behaviour is discussed. Emphasis is laid upon ambiguities in extrapolating results of $d=4-\epsilon$ calculations down to d=3.

* * *

The aim of my talk is to review the successes of renormalization group (RG) calculations on the critical dynamics of Helium and to point out that some problems still remain to be settled, which will keep us busy for some time still.

The very first work on critical dynamics via RG by Halperin, Hohenberg and Ma [1] showed that RG could indeed be applied to the problem of critical dynamics, provided that a suitable model evolution equation be chosen. This equation should be able to reproduce the hydrodynamical features of the physical system under consideration, which mode-coupling theories and dynamic scaling theories had shown essential to determine the dynamic behaviour [2] . Such a model equation should be able to satisfy the following requirements:

- it should explicitly contain all physical quantities ("fields") which have such a slow evolution to be able to contribute to the critical slowing down; this includes the order parameter and the conserved quantities;
- it should represent the effect of non-Gaussian equilibrium fluctuations of these quantities ("static coupling") as well as the effect of Poisson-bracket relations among them ("dynamic reversible couplings") and of whatever off-diagonal Onsager coefficients are allowed for by symmetry requirements.

Halperin , Hohenberg and Siggia [3] (HHS) were able to introduce on the basis of these requirements a model evolution equation which still stands as the basis of our understanding of this problem. This equation describes the universality class to which liquid Helium at the λ-line and a planar antiferromagnet in an external normal magnetic field belong. It contains explicitly the dynamics of the order parameter ψ , a complex quantity (the condensate wavefunction for Helium, the staggered magnetization in the easy plane for the antiferromagnet), and of a scalar conserved quantity m (the normal magnetization for the antiferromagnet, a linear combination of energy and mass density for Helium); the conservation of energy is not taken into account.

The equations are [3]

$$\frac{\partial \psi}{\partial t} = -\Gamma_0 (1+ib_0) \frac{\delta \mathcal{H}}{\delta \psi^*} - ig_0 \psi \frac{\delta \mathcal{H}}{\delta m} + \vartheta, \tag{1}$$

$$\frac{\partial m}{\partial t} = \Lambda_0 \nabla^2 \frac{\delta \mathcal{H}}{\delta m} - ig_0 \left[\psi^* \frac{\delta \mathcal{H}}{\delta \psi^*} - \psi \frac{\delta \mathcal{H}}{\delta \psi} \right] + \zeta, \tag{2}$$

where

$$\mathcal{H} = \int d^d x \left[|\nabla \psi|^2 + r_0 |\psi|^2 + \frac{u_0}{3!} |\psi|^4 + \frac{1}{2} m^2 + \gamma_0 m |\psi|^2 \right] \tag{3}$$

is the Landau-Ginzburg functional, and the Gaussian noise fields ϑ, ζ have zero mean and correlation functions:

$$\langle \vartheta^*(\underset{\sim}{x}, t)\, \vartheta(\underset{\sim}{x}', t') \rangle = 2\Gamma_0\, \delta(\underset{\sim}{x} - \underset{\sim}{x}')\, \delta(t-t'), \tag{4}$$

$$\langle \zeta(\underset{\sim}{x}, t)\, \zeta(\underset{\sim}{x}', t') \rangle = -2\Lambda_0 \nabla^2 \delta(\underset{\sim}{x} - \underset{\sim}{x}')\, \delta(t-t'), \tag{5}$$

The quantities Γ_0 , Λ_0 are "bare" transport coefficients and their appearance in both eqs. (1)-(2) and (4)-(5) guarantees that the equilibrium properties of this model are those derived from the Landau-Ginzburg functional $\mathcal{H}$ via purely static methods. The first terms in eqs.(1) (2) represent the dynamic effect of the static couplings of ψ with itself (proportional to u_0) and of ψ and m (proportional to γ_0). If they were alone the model would reduce to a purely relaxational model: model C of Halperin, Hohenberg and Ma [1] . This is a rather pathological model (see also Brézin,De Dominicis [4]) whose pathologies we shall not be completely able to avoid later.

The second term in eqs. (1)-(2) represents the effect of the Poisson bracket relation between ψ and m: in Helium, this corresponds to the precession of the condensate wavefunction ψ in the complex plane at a speed proportional to the chemical potential μ , conjugate to the conserved quantity m. This appears physically as the Josephson effect: the phase of ψ precesses according to

$$\frac{d\phi}{dt} = \mu/\hbar\, . \tag{6}$$

The appearance of an imaginary part, ib_o, in the kinetic coefficient of eq.(1) is a reminder of the Poisson bracket relation between ψ and its complex conjugate.

It is easy to see that eqs.(1)-(5) contain the relevant hydrodynamic features of liquid Helium [5, 6]: in particular they contain below T_c a propagation mode ("second sound") with frequency

$$\omega_-(\underset{\sim}{k}) = c\,k, \tag{7}$$

$$c^2 = g_o^2\,\rho_s / \chi_m, \tag{8}$$

where χ_m is the static susceptibility of the m field (related to the specific heat and the compressibility in Helium) and ρ_s is the stiffness constant, related to the coherence length below T_c, ξ_-,

$$\xi_- = \rho_s^{(2-d)^{-1}} \,. \tag{9}$$

Above T_c, there are two relaxing modes: the conserved field m has a diffusive mode with frequency

$$\omega_m(\underset{\sim}{k}) = (\Lambda_{eff} / \chi_m)\, k^2, \tag{10}$$

the order parameter relaxes to its equilibrium value with a frequency

$$\omega_\psi(\underset{\sim}{k}) = \Gamma_{eff} / \chi_\psi, \tag{11}$$

which is finite as $k \to 0$.

The fact that the second sound velocity c is given (eq.(8)) in terms only of g_o and of static quantities has an important consequence: if the frequency-dependent correlation functions obey dynamic scaling around T_c, then the critical exponents for Γ_{eff}, Λ_{eff} should be given only in terms of static critical exponents [6,7]. One can define a dynamic critical exponent z via

$$\omega(\underset{\sim}{k}, T-T_c) = k^z\,\Omega(k\,\xi(T-T_c)), \tag{12}$$

where ω is the characteristic frequency of any fluctuation and ξ is the coherence length; then

$$z = \frac{d}{2} + \frac{\tilde{\alpha}}{2\nu}, \tag{13}$$

where $\tilde{\alpha} = \max(\alpha, 0)$ is the exponent for the divergence of the spe cific heat. At $d=4-\epsilon$ one has therefore

$$\Gamma_{eff} \propto \xi^{\frac{\epsilon}{2}-\eta-\frac{\alpha}{2\nu}}, \tag{14}$$

$$\Lambda_{eff} \propto \xi^{\frac{\epsilon}{2}+\frac{\alpha}{2\nu}} \tag{15}$$

Halperin, Hohenberg and Siggia [3] first showed how these results could be obtained within the Wilson renormalization group framework and how the difficult problem of handling the corrections to asymptotic behaviour, essential for an understanding of experimental data, could be handled. Since the Martin-Siggia-Rose (MSR) [8] formalism allows us to cast the dynamics expressed in eqs. (1)-(5) in field theoretical form, the whole apparatus of "standard" renormalization theory becomes available [9, 10]. This allows for a systematic treatment of higher orders in the ϵ expansion, which appears crucial for the interpretation of the behaviour of helium in three dimensions.

The formalism may be simply expounded, although its justification and the corresponding computations are far from trivial. The most convenient framework is the use of dimensional regularization and minimal subtraction procedures [11], which simplify the computations by allowing for an automatic separation of static and dynamic properties of the renormalized quantities and by reducing the need of computing complicated momentum-dependent integrals.

In this framework an arbitrary normalization momentum μ is introduced, which allows for the definition of dimensionless renormalized couplings:

$$K_d u_0 = \mu^{\epsilon} u \, (Z_u / Z_\psi^2), \tag{16}$$

$$K_d g_0^2/(\Lambda_0 \Gamma_0) = \mu^{\epsilon} f \, Z_g^2 Z_\Lambda Z_\Gamma, \tag{17}$$

$$K_d \gamma_0^2 = \mu^{\epsilon} v \, Z_\gamma^2/(Z_\psi^2 Z_m), \tag{18}$$

where the renormalization factors Z are functions of ϵ and of renormalized quantities, singular as $\epsilon \to 0$, in order to remove the logarithmic singularities which appear in a four-dimensional calculation of the correlation functions of interest without a cutoff [11]. The factor $K_d = 2\,(\pi)^{d/2}(2\pi)^{-d}/\Gamma(\frac{1}{2}d)$ is introduced for convenience. To remove fully these singularities it is also necessary to define renormalized transport coefficients and fields according to:

$$\Gamma_o^{-1} = Z_\Gamma \quad , \tag{19}$$

$$\Lambda_o^{-1} = w \, Z_\Lambda , \tag{20}$$

$$b_o = b \, Z_b \; ; \tag{21}$$

and

$$\psi = Z_\psi^{1/2} \, \psi_R \; , \tag{22}$$

$$m = Z_m^{1/2} \, m_R \; , \tag{23}$$

plus the corresponding ones for the Martin-Siggia-Rose conjugate fields.

The arbitrariness of the normalization momentum μ entails a Callan -Symanzik equation for each of the "observable" correlation or response functions of interest. If we consider for instance the Fourier transform of the response function for the order parameter, R_ψ :

$$R_\psi(\underset{\sim}{x}-\underset{\sim}{x}', t-t') = \delta\langle\psi(\underset{\sim}{x},t)\rangle / \delta h(\underset{\sim}{x}', t') , \tag{24}$$

where h(x,t) is a field which appears in a term $-\int d^d x \; h(\underset{\sim}{x},t)\psi^*(\underset{\sim}{x},t)$ in the Landau-Ginzburg functional, then we may define the renormalized response function R_ψ^R via:

$$R_\psi(\underset{\sim}{k}, -\frac{i\omega}{\Gamma_o}; u_o, \gamma_o, g_o, \Gamma_o/\Lambda_o, b_o) =$$

$$= Z_\psi \; R_\psi^R(\underset{\sim}{k}, -i\omega; u, v, f, w, b; \mu) \; ; \tag{25}$$

and R_ψ^R satisfies

$$\left(\mu\frac{\partial}{\partial\mu} + \sum_\ell W_\ell \frac{\partial}{\partial\ell} - \eta_\Gamma \, \omega \frac{\partial}{\partial\omega} - \eta_\psi\right) R_\psi^R = 0 . \tag{26}$$

We have here introduced the <u>exponent functions</u> η_ℓ, $(\ell = \psi, m, \Gamma, \Lambda)$ as

$$\eta_\ell = \mu \frac{d}{d\mu} \ln Z_\ell \quad , \tag{27}$$

and the <u>Wilson functions</u>. W_ℓ (ℓ = u, f, v, w, b) as

$$W_\ell = \mu \frac{d}{d\mu} \ell \quad ; \tag{28}$$

the derivatives are understood to be taken at fixed bare parameters.

As it is well known, the asymptotic critical behaviour is determined by the stable fixed points of the RG flows: i.e. by those values ℓ^* of the renormalized parameters ℓ = u,f,v,w,b, for which

$$W_\ell = 0, \quad (\ell = u, f, v, w, b), \tag{29}$$

and the "stability matrix", i.e. the matrix

$$\| W_{\ell,m} \| = \| \partial W_\ell / \partial m \|, \quad (\ell, m = u, f, v, w, b), \tag{30}$$

has all its eigenvalues ω_ℓ positive. These eigenvalues are the <u>subcritical</u> <u>exponents</u> and govern the corrections to the asymptotic behaviour each being related to a term vanishing like $\xi^{-\omega_\ell}$.

One readily obtains [4] that the quantity $\tilde{u} = u - 3v$ goes to the same fixed point value as in the n=2 Landau-Wilson model for the statics:

$$\tilde{u}^* = \frac{3}{5} \epsilon + O(\epsilon^2) ; \tag{31}$$

whereas v goes to two different fixed points depending on the sign of the specific heat exponent α :

$$v^* = \begin{cases} 0 & \text{if } \alpha < 0, \\ \neq 0 & \text{if } \alpha > 0. \end{cases} \tag{32}$$

If $v^* \neq 0$ one readily obtains $\eta_m^* = -\alpha/\nu$. Since it is known experimentally [12] that α is negative (though very small) for Helium, the relevant fixed point value appears to be $v^* = 0$; the smallness of α/ν means however that this asymptotic value is reached exceedingly slowly, so that a careful treatment of the corresponding corrections is made necessary [3] . The <u>dynamic</u> Wilson functions are

$$W_f = -f(\epsilon + \eta_r + \eta_\lambda + \eta_m), \tag{33}$$

$$W_w = w(\eta_r - \eta_\lambda), \tag{34}$$

and W_b, which for $v^* = 0$ yields a stable value of b at $b^* = 0$. The fact that only η_r, η_λ and η_m appear in the expression for W_f is forced upon by the Josephson relation (6) which may be interpreted to mean that

$$Z_g = Z_m^{1/2} , \tag{35}$$

from which (17) may be seen to imply (33). It is then seen that $W_f=0$ implies either $f^* = 0$-which is ruled out by stability requirements - or

$$\eta_\Gamma^* + \eta_\Lambda^* + \eta_m^* = -\epsilon , \tag{36}$$

where η_ℓ^* are the fixed point values of the exponent functions.

The second Wilson function (34) leaves us with two possibilities for the fixed point value of w :

(a) $w^* \neq 0$, which implies $\eta_\Gamma^* = \eta_\Lambda^*$;

(b) $w^* = 0$.

The two prossibilities correspond to two different asymptotic behaviours of the kinetic coefficients. The Callan-Symanzik equation implies that the asymptotic behaviour of the kinetic coefficients for $T > T_c$ is given by:

$$\Gamma_{eff} \propto \xi^{-\eta_\Gamma^* - \eta} , \tag{37}$$

$$\Lambda_{eff} \propto \xi^{-\eta_\Lambda^* - \eta_m^*} , \tag{38}$$

where η is the critical exponent of the (static) correlation function.

In case (a) we have therefore, for $\alpha < 0$,

$$\eta_\Gamma^* = \eta_\Lambda^* = -\frac{\epsilon}{2} , \tag{39}$$

which implies

$$\Gamma_{eff} \propto \xi^{\frac{\epsilon}{2} - \eta} , \tag{40}$$

$$\Lambda_{eff} \propto \xi^{\frac{\epsilon}{2}} , \tag{41}$$

also obtained from the phenomenological dynamic scaling theory [6,7]. It is therefore warranted to call this fixed point the <u>dynamic scaling</u> fixed point.

In case (b) we may define an exponent ω_w by:

$$\omega_w = \eta_\Gamma^* - \eta_\Lambda^* , \tag{42}$$

and we obtain instead of (40)

$$\Gamma_{eff} \propto \xi^{\frac{\epsilon}{2} - \eta - \omega_w/2} , \tag{43}$$

and instead of (41)

$$\Lambda_{eff} \propto \xi^{\frac{\epsilon}{2} + \omega_w/2} . \tag{44}$$

In this case there appears to be a breakdown of the full homogeneity of frequency-dependent correlation functions near the critical point: the corresponding fixed point is referred to as the <u>weak scaling</u> one. It is interesting to note that the asymptotic behaviour for the second sound velocity obtained from dynamic scaling,

$$c \propto \xi^{\frac{\epsilon}{2} - 1} \tag{45}$$

holds good in <u>both</u> cases.

Since it appeared experimentally [12, 13] that the divergence of Λ_{eff} was indeed stronger than expected from dynamic scaling, attention was drawn on the possibility that the weak scaling fixed point could be the relevant one for Helium (De Dominicis and Peliti [10]).

In order to see which of the two possibilities is the actual one, one has to discuss the stability of the point (b), which is related to the sign of the difference $\eta_\Gamma - \eta_\Lambda$. For the model we are discussing we obtain to one loop order

$$\eta_\Gamma = - f/(1+w), \tag{46}$$

$$\eta_\Lambda = - f/2 \tag{47}$$

One has therefore at $w^* = 0$, to lowest order,

$$\eta_\Gamma - \eta_\Lambda = - f/2 < 0 , \tag{48}$$

and it may seem that this point is definitely instable. If we now take $w^* \neq 0$, eq. (34) implies

$$\eta_\Gamma^* = \eta_\Lambda^* , \tag{49}$$

which together with (33) yields

$$\eta_r^* = \eta_\lambda^* = -\frac{\epsilon}{2} - \frac{1}{2}\eta_m^* = -\frac{\epsilon}{2} + \frac{\tilde{\alpha}}{2\nu} \quad , \tag{50}$$

which may be seen to be equivalent to (13). One may then exploit the knowledge of the fixed point values of f, w to compute universal combination of coefficients [3], shape functions [14, 15], and other universal quantities.

One may however suspect that the weak-scaling fixed point $w^* = 0$ has been disposed of with undue haste. A similar thing happens in the statics when one computes the specific heat exponent α out of the ϵ-expansion. To lowest order, for one model, $\alpha = 0.1\,\epsilon$, which is positive; to next order $\alpha = 0.1\epsilon - 0.2\,\epsilon^2$ and the estimation of the actual value of α becomes extremely dependent on the way we treat the series. In fact we know that α is negative, but very small, and that it is difficult to hit near the actual value by just manipulating a few terms of the ϵ-expansion.

In this case it is more convenient to generalize the model to a n-dependent one and to exploit all the evidence we may get from this n-dependence to understand what happens for the real system. There is an excellent discussion by Dohm [15] of how this can be made in the present case. We set v=0 from the outset, since as we discussed earlier this appears to be the relevant fixed point value for Helium. We then have at least two possible generalizations of our model to an n-dependent one:

(a) the antiferromagnetic (AF) model of Sasvàri, Schwabl, Szépfalusy [16,14]: ψ is a n-dimensional vector, m a n-rank antisymmetric tensor, whose (α,β) component generates rotations of ψ in the (α,β) plane;

(b) the "Helium" model (He) in which ψ is a n/2 dimensional complex vector, and m is a scalar which generates rotations of each component of ψ in the complex plane.

Another generalization, due to Halperin [17], introduces a field which generates rotations which mix up complex components of ψ : but it has less relevance for our problem.

If we were able to perform exact calculations, all results obtained for these two models should coincide for our case (n=2). We may take therefore any discrepancies as an indication of the inherent incertainties of the extrapolation of the ϵ-expansion.

A two loop calculation of the relevant exponent functions yields [10,15]:

(AF): $$\eta_\Gamma = -f\,\frac{n-1}{1+w} + f^2 G_{AF}(w) + \epsilon^2 \eta_\Gamma^A , \tag{51}$$

$$\eta_\Lambda = -\frac{1}{2} f + f^2 L_{AF}(w), \tag{52}$$

(He): $$\eta_\Gamma = -f/(1+w) + f^2 G(w) + \epsilon^2 \eta_\Gamma^A , \tag{53}$$

$$\eta_\Lambda = -n/4 \cdot f + f^2 L(w), \tag{54}$$

where

$$\eta_\Gamma^A = \frac{n+2}{2(n+8)^2}\left(6 \ln\frac{4}{3} - 1\right), \tag{55}$$

and

$$G_{AF}(w) = \frac{n-1}{2(1+w)^3}\Big\{[1+(2-n)w](1+2w)\ln\frac{2(1+w)}{1+2w} + (1+nw)\ln\frac{1+w}{2} + \frac{1}{2}(27\ln 4/3 - 6)(1+w) + (n-1)w\Big\}, \tag{56}$$

$$L_{AF}(w) = -\frac{1}{4(1+w)}\left\{w(2+w)(n-2-w)\ln\frac{(1+w)^2}{w(2+w)} + w + \frac{3}{2} - \frac{n}{2}\right\}, \tag{57}$$

$$G(w) = \frac{1}{8(1+w)^2}\left\{4(1+2w)\ln\frac{(1+w)^2}{w(2+w)} + 9(4+n)\left(\ln\frac{4}{3}\right)(1+w) - (4+2n)w - (8+2n)\right\}, \tag{58}$$

$$L(w) = \frac{n}{8(1+w)}\left\{w^2(2+w)\ln\frac{(1+w)^2}{w(2+w)} - w - \frac{1}{2}\right\}. \tag{59}$$

In order to understand whether $w^* = 0$ or not for $\epsilon = 1$; $n=2$, we look for the boundaries $n(\epsilon)$ which separate the stability region of the <u>weak scaling</u> fixed point $w^* = 0$, from that of the <u>dynamic scaling</u> fixed point $w^* \neq 0$. For the "Helium" models one readily obtains [10,15] that

$$n(\epsilon) = 4 - \left(19 \ln\frac{4}{3} - \frac{11}{3}\right)\epsilon \simeq 4 - 1.80\,\epsilon \tag{60}$$

is, to first order in ϵ, the common boundary crossing which in the sense of decreasing n's the weak-scaling point destabilizes and the dynamic scaling fixed point takes over.

For the AF models, due to the presence of a term proportional to

$w \ln w$ in L_{AF}, the situation is less clear [18] . There appear to be two boundaries: the first one

$$n^{(1)}(\epsilon) = \frac{3}{2} + p\epsilon + O(\epsilon^2), \tag{61}$$

$$p = \left(\frac{27}{8} + \frac{42}{361}\right) \ln\frac{4}{3} - \left(\frac{9}{16} + \frac{7}{361}\right) = 0.42 \tag{62}$$

is the line crossing which towards higher values of n the weak-scaling fixed point destabilizes, the other,

$$n^{(2)}(\epsilon) = n^{(1)}(\epsilon) - \frac{\epsilon}{2}\left[2 - n^{(1)}(\epsilon)\right] \exp\left(-\frac{c}{\epsilon}\right), \tag{63}$$

is the line, crossing which the dynamic scaling fixed point becomes stable. The two leave a region, exponentially small as $\epsilon \to 0$, in which both points are stable: which one is chosen by the actual system will depend on the details of the original interaction. Moreover is we take (63) seriously for large values of ϵ we see that for $n^{(1)}(\epsilon) > 2$ there will be a gap between the two stability regions (corresponding perhaps to a region where exponents depend continuously on the details of the interaction?). The point $n^{(1)}(\epsilon) = 2$ appears as it were a "tricritical point" in the (ϵ , n) plane, where the nature of the stability boundary changes qualitatively: from (61), (62) we obtain that it lies at n=2, d=2.81, i.e. very near the point representing real Helium.

The two lines, $n(\epsilon)$ and $n^{(1)}(\epsilon)$ should cross on the n=2 line. In fact the values $\bar{\epsilon}$ of ϵ at which they cross the n=2 line mismatch by $\Delta\epsilon = 0.08$ which, however small, is of the same order of magnitude as the distance of the physical point from the boundaries. Thus even if the physical point appears to be safely on the scaling side of the boundaries, the uncertainties pertaining to the extrapolation up to finite ϵ are sufficient to cloud quantitative conclusions.

If we try to estimate the fixed point values of the parameters f, w we face similar difficulties. Eqs. (36),(39) imply for the dynamic scaling fixed point at n=2

$$f^* = \epsilon + \epsilon^2 L(w^*, n=2) + O(\epsilon^3), \tag{64}$$

where w^* is given as a solution of an implicit equation, of the form

$$w^* = w^\circ(n) + \epsilon\, H(w^*, n) + O(\epsilon^2)\,, \tag{65}$$

where $w^\circ(n)$ is the lowest order estimate of w^* ((He): $w^\circ(n) = (4-n)/n$; (AF): $w^\circ(n) = 2n-3$), and the function H is simply related to L,G, η_Γ^A [10, 15] . One may try to solve (65) by setting

$$w^* = w^\circ(n) + \epsilon\, w^1(n) + O(\epsilon^2), \tag{66}$$

obtaining therefore

$$w^1(n) = H(w^\circ(n), n), \tag{67}$$

but this procedure is not warranted, for <u>any</u> value of ϵ , if we are near the boundaries $n(\epsilon)$ or $n^{(1)}(\epsilon)$: in this case w° and ϵw^1 are of the same order of magnitude. Since the physical point lies very near the boundary, it is easy to make an error of some magnitude when we use an extrapolation procedure which does not take care of the boundary. But the boundary itself has been estimated just to first order in ϵ ... Dohm [15] introduces a suitable extrapolation procedure, which amounts to solving (65) and the analogous equation for AF models with respect to n and then expanding with respect to ϵ , obtaining, in lieu of (65), the equation

$$n = \frac{4}{1+w^*} - \frac{\epsilon}{1+w^*} H\left(w^*, \frac{4}{1+w^*}\right) + O(\epsilon^2) \tag{68}$$

This yields for the fixed point value w^* at n=2, ϵ =1 the result

$$\text{(He)}: \quad w^* = 0.06, \tag{69}$$
$$\text{(AF)}: \quad w^* = 0.15$$

which shows the large uncertainty in the result, but also that w^* is quite small however it is estimated. Ferrel, Dohm and Bhattacharjee [19] and Dohm [15] have discussed the importance of a small value of w^* in the interpretation of critical point experiments both on the thermal conductivity [13] and on light scattering [20] . The main point is that the spectrum of entropy fluctuations has a very different shape at T_c depending on the size of w^* : if w^* is small it shows a substantial <u>dip</u> at small frequencies, which eventually reaches zero at $w^* = 0$. As w^* increases, the dip begins to fill in, leaving the outer "wings" essentially unchanged. Hohenberg, Siggia and Halperin [21] computed the shape function for a value of w^* for which the central dip had almost filled in, leaving only a characteristic "shoulder" as a remainder. A similar phenomenon appears when we move away from T_c at fixed wevenumber $\underset{\sim}{k}$ in either direction. First the central valley of the spectrum fills in, only after do the outer wings feel the distance from the critical point and start behaving as expected from dynamic scaling. The net effect is a widening of the critical region for entropy fluctuations up to temperatures for which the coherence length ξ satisfies

$$k\xi \lesssim \sigma_\psi^{+2/d} , \tag{70}$$

where σ_ψ , the ratio between order parameter and entropy fluctuation characteristic frequencies, is a number which becomes small with w^* . All the more important is a proposed test [15] which allows one to estimate the critical value w^* from the actual shape function of entropy fluctuations.

Let us also observe that a small value of w^* would imply a small value for the subcritical exponent ω_w which should vanish on the boundaries. The estimation of this exponent runs parallel to the estimation of w^* . A straightforward ϵ-expansion yields

$$\omega_w = \epsilon/4 - 0.108\,\epsilon^2 , \tag{71}$$

whereas a more careful extrapolation, which takes into account the boundaries, yields at n=2, ϵ=1 [15]

$$\text{He} : \quad \omega_w = 0.048, \tag{72}$$

$$\text{AF} : \quad \omega_w = 0.098, \tag{73}$$

i.e. a rather small exponent, known with an uncertainty of a factor two. This means that there are rather slow corrections which mar the asymptotic behaviour and must be taken into account in the interpretation of experimental data.

Ferrell [19] has discussed at length the implications of the different fixed points as applied to an understanding of experimental properties of pure Helium. What I tried to point out is the care needed in interpreting the results of a ϵ-expansion: one should investigate the whole (ϵ, n) plane and try to understand from all available information the location of the physical point with respect to the relevant physical boundaries. If one sticks to a straighforward extrapolation; one may easily be led in error. Sometimes it will also be necessary to receive a hint from experiment to understand which of the different possible asymptotic behaviours has actually been chosen by Nature.

The renormalization group analysis of the dynamic critical behaviour of He^3-He^4 mixtures is essentially due to Nelson and Siggia [22, 23]. Although their work has found substantial experimental confirmation [24] there still are some open problems about the applicability of their analysis in d=3. What follows is a pedestrian account of their work.

As soon as one introduces some He^3 in He^4 one must consider the presence of two conserved fields coupled to the order parameter: an entropy-like field q

$$q(\underset{\sim}{x}) = \rho_0 \sigma(\underset{\sim}{x}) + \rho_0 \left.\frac{\partial \mu}{\partial T}\right|_{c,p} c(\underset{\sim}{x}), \tag{79}$$

where ρ_0 is the λ point mass density, $\sigma(\underset{\sim}{x})$ the entropy density, $c(\underset{\sim}{x})$ the He^3 concentration (both intended as deviation from equilibrium values) and μ its conjugate chemical potential. As the concentration c_o of He^3 reaches zero, q eventually becomes the entropy-like mode we have previously called m; for any $c_o > 0$ the q fluctuations are asymptotically decoupled from order parameter fluctuations as one approaches the λ line, whereas they diverge (for $\alpha > 0$) in pure Helium. On the contrary the c fluctuations yield a divergent susceptibility.

$$\chi_c = \frac{1}{\rho_0} \left.\frac{\partial c}{\partial \mu}\right|_{T,p} , \tag{80}$$

if the λ line is approached at constant μ corresponding to a nonzero concentration. While one should in principle study a model in which both fields, q and c, appear statically coupled to the order parameter, one may simplify the task if one is interested on either asymptotic behaviour, namely:

- if one is interested on the neighbourhood of the tricritical point, the range where the q field susceptibily is finite is sufficiently large to take a model in which its fluctuations are Gaussian [22];
- if one is interested on dilute solutions, the quasidivergence of q susceptibility dominates over most of the accessible temperature range, thus calling for the introduction of a model where q fluctuations are coupled to the order parameter, and c ones are Gaussian [23].

I shall dwell on the first model, as the one relevant for tricritical behaviour. It reads:

$$\frac{\partial}{\partial t}\psi = -\Gamma_0 (1+ib_0)\frac{\delta \mathcal{H}}{\delta \psi^*} - i g_{1,0}\psi \frac{\delta \mathcal{H}}{\delta q} - i g_{2,0}\psi \frac{\delta \mathcal{H}}{\delta c} + \zeta, \tag{81}$$

$$\frac{\partial}{\partial t} q = K_0 \nabla^2 \frac{\delta \mathcal{H}}{\delta q} + L_0 \nabla^2 \frac{\delta \mathcal{H}}{\delta c} + 2 g_{1,0} \,\mathrm{Im}\left[\psi^* \frac{\delta \mathcal{H}}{\delta \psi^*}\right] + \vartheta, \tag{82}$$

$$\frac{\partial}{\partial t} c = \lambda_0 \nabla^2 \frac{\delta \mathcal{H}}{\delta c} + L_0 \nabla^2 \frac{\delta \mathcal{H}}{\delta q} + 2 g_{2,0} \,\mathrm{Im}\left[\psi^* \frac{\delta \mathcal{H}}{\delta \psi^*}\right] + \varphi, \tag{83}$$

where the Landau-Ginburg functional $\mathcal{H}$ is given by

$$\mathcal{H} = \int d^d x \left[|\nabla\psi|^2 + r_0 |\psi|^2 + \frac{u_0}{3!}|\psi|^4 + \frac{1}{2}c^2 + \gamma_0 c|\psi|^2 + \frac{1}{2}q^2 \right], \tag{84}$$

and the correlations of the noise sources, ζ , ϑ , φ , satisfy obvious Einstein relations with the kinetic coefficients Γ_0 , λ_0, L_0, K_0. The reversible couplings $g_{1,0}$ and $g_{2,0}$ are related to experimentally accessible quantities. The term proportional to γ_0 in eq.(84) represents the <u>static</u> coupling between fluctuations of c and of ψ .

The off-diagonal transport coefficient L_0 , coupling the conserved fields q and c, makes the perturbation expansion more complicated. The essential features of the renormalization scheme may nevertheless be written down almost by inspection. We have besides the usual static renormalization-with a nontrivial factor Z_c forced upon by the static coupling γ_0 , as in eq. (35) - the dynamic renormalizations:

$$K_d\, g_{1,0}^2/(\Gamma_0 K_0) = \mu^\epsilon f_1 Z_1^2 Z_\Gamma Z_K, \tag{85}$$

$$K_d\, g_{2,0}^2/(\Gamma_0 \lambda_0) = \mu^\epsilon f_2 Z_2^2 Z_\Gamma Z_\lambda, \tag{86}$$

$$\Gamma_0/K_0 = w_1 Z_K/Z_\Gamma, \tag{87}$$

$$\Gamma_0/\lambda_0 = w_2 Z_\lambda/Z_\Gamma, \tag{88}$$

$$L_0^2/(\lambda_0 K_0) = w_3 Z_K Z_\lambda/Z_L^2, \tag{89}$$

$$b_0 = b Z_b \tag{90}$$

Moreover a Ward identity implied by the reversible nature of the couplings $g_{1,0}$, $g_{2,0}$ entails

$$Z_1 = 1, \tag{91}$$

$$Z_2 = Z_c^{1/2} . \tag{92}$$

One obtains therefore the following Wilson functions:

$$W_{f_1} = -f_1 (\epsilon + \eta_r + \eta_K), \tag{93}$$

$$W_{f_2} = -f_2 (\epsilon + \eta_r + \eta_\lambda + \eta_c), \tag{94}$$

$$W_{w_1} = w_1 (\eta_r - \eta_K), \tag{95}$$

$$W_{w_2} = w_2 (\eta_r - \eta_\lambda), \tag{96}$$

in addition to W_{w_3}, W_b and to the static ones. Of these we shall need the one relative to v, defined by an equation similar to (18):

$$W_v = -v (\epsilon + 2 \hat{\eta}(\tilde{u}) + \eta_c), \tag{97}$$

where $\hat{\eta}(\tilde{u})$ is the exponent function of the ψ^2 insertions [4]. From (97) one may readily obtain the static crossover. One has to lowest order in ϵ the following fixed points:

(a) the critical fixed point $\tilde{u}^* = \frac{3}{5}\epsilon$, $v^* = \frac{\epsilon}{5}$, at which $\eta_c^* = -\alpha/\nu$;

(b) an instable critical fixed point $\tilde{u}^* = \frac{3}{5}\epsilon$, $v^* = 0$;

(c) the tricritical fixed point $\tilde{u}^* = 0$, $v^* = \epsilon$, at which $\eta_c^* = -\alpha_t/\nu_t = -\epsilon$. Here the suffix "t" refers to the tricritical exponents.

As ϵ increases, the fixed points (a) and (b) approach each other: they cross on the line $\alpha = 0$ and exchange stability. On the λ- line in three dimensions the relevant fixed point is therefore (b), but the slow vanishing of v introduces important corrections, as in [3].

The analysis of the flow given by eqs. (93)-(96) appears rather difficult but simplifies owing to the presence of fast transients. One may observe that f_1 and f_2 settle down rapidly to a common value f, driving w_3 to the value:

$$w_3^* = 1 \tag{98}$$

One then notices that as soon as $f_1 = f_2$ one has

$$\eta_\lambda = \eta_K - \eta_c , \tag{99}$$

and for nonzero v the extra term of η_c appearing in (96) is enough to drive w_2 to the value:

$$w_2^* = \infty . \tag{100}$$

In this limit b settles down to zero and the remaining exponent functions, η_r and η_K , simplify drastically, yielding:

$$\eta_r = - \frac{f}{1+w_1} + v, \tag{101}$$

$$\eta_K = - \frac{1}{2} f, \tag{102}$$

$$\eta_c = - v. \tag{103}$$

At any dynamic fixed point with $f^* \neq 0$ one must have

$$\epsilon + \eta_r^* + \eta_K^* = 0, \tag{104}$$

which together with (98) (99) implies the following asymptotic behaviour for the effective transport coefficients:

$$\Gamma_{eff} K_{eff} \sim \Gamma_{eff} \lambda_{eff} \sim \Gamma_{eff} L_{eff} \sim \xi^{\epsilon} . \tag{105}$$

On the other hand, depending on the value of v^*, we have different fixed point values of the dynamic parameters:

at the <u>critical</u> fixed point (a) the stable fixed point is at $w_1^* \neq 0$, implying

$$\eta_r^* = \eta_K^* = - \frac{\epsilon}{2} ; \tag{106}$$

therefore <u>all</u> transport coefficients diverge like $\xi^{\epsilon/2}$, in contrast with pure Helium behaviour, where factors of $\xi^{\alpha/2\nu}$ appear;

at the <u>tricritical</u> fixed point (c) the stable value is $w_1^* = 0$, yielding

$$\begin{aligned} f^* &= \frac{4}{3} \epsilon , \\ \eta_r^* &= - \frac{\epsilon}{3} , \quad \eta_K^* = - \frac{2}{3} \epsilon . \end{aligned} \tag{107}$$

In this case the asymptotic behaviour of the kinetic coefficients is given by:

$$K_{eff} \sim \lambda_{eff} \sim L_{eff} \sim \xi^{2\epsilon/3} . \tag{108}$$

Another interesting fact appears on a closer investigation of the flow equations: as we approach the λ line sufficiently close to the tricritical point, v first lies near its tricritical value, then crosses over to its critical value $\epsilon/5$. It may then happen that f is driven to an intermediate fixed point value before w_1 has the time to become significantly different from zero. If we look indeed at the stability of the fixed point $v^* = \epsilon/5$, $w_1^* = 0$, $f^* = 4\epsilon/5$ we obtain the transient exponents:

$$\omega_f = \frac{3}{2} f^* = \frac{6}{5}\epsilon , \tag{109}$$

$$\omega_{w_1} = -\frac{1}{2} f^* + v^* = -\frac{\epsilon}{5} , \tag{110}$$

and the smallness of $|\omega_{w_1}|$ is sufficient to warrant this possibility. We expect therefore the appearance of an intermediate region of purely dynamic nature, characterized by the behaviour:

$$\Gamma_{eff} \sim \xi^{3\epsilon/5} , \quad K_{eff} \sim \lambda_{eff} \sim L_{eff} \sim \xi^{2\epsilon/5} , \tag{111}$$

which lies between the "tricritical" and "critical" regions.

Which results of this analysis will hold good for real He^3-He^4 mixtures? Let us first note that the behaviour of v changes quantitatively in three dimensions, where the stable fixed point is at $v^* = 0$. Then we expect that $w_2^* < \infty$ on the λ line, what does not modify the asymptotic exponents but makes the analysis of correction terms much more complicated. Some sort of analysis like the one performed in [3] should become necessary. We may also note that if (98) is modified at higher orders in ϵ, the behaviour of the experimentally measured thermal conductivity may be qualitatively modified: since it is determined by a delicate interplay of diverging and vanishing quantities in which the leading corrections eventually dominate [23].

But the major problems lie in the divergence of w_2 at the tricritical point. A similar situation appears in model C of Halperin, Hohenberg and Ma [1,4] as well as in some models of bicritical behaviour [25]. As one computes second-order corrections, one formally obtains exponent functions which are singular (like $\ln w$) as $w \to \infty$. It appears as if the limits $w \to \infty$, $\epsilon \to 0$ could not be exchanged [5]. If one goes to $w \to \infty$ before taking $\epsilon \to 0$ one obtains frequency-dependent Z functions which prevent renormalization of the theory. On the other hand one may try to get out of the ϵ-expansion altogether by introducing values of w of $O(\exp C/\epsilon)$ [25,26] ; but in this case it is hard to believe that next order computations would not alter drastically the results.

Whatever the meaning of this difficulties are, they will be more apparent in tricritical He^3-He^4 mixtures than elsewhere, as emphasized by Nelson and Siggia [23] . The characteristic exponent for the divergence of w_2 is in fact $\alpha_t/\nu_t = \epsilon$, much larger than the corresponding one in model C. Probably a hint will come from experimental investigations or from real space analyses.

We have seen that however powerful the ϵ-expansion is, it cannot give all the answers, and that it is necessary to implement it with subtle extrapolation procedures if quantitative predictions are required. All available information is useful to allow for an interpretation of RG calculations. In particular a real space analysis of models with propagating modes would be welcome: for even if its interpretation problems are no smaller than those of the field-theoretic RG, they are not the same, and some hints may be obtained from the comparison.

* Mailing address: Istituto di Fisica "G.Marconi", Piazzale delle Scienze 5, 00185 Roma (Italy).

REFERENCES

[1] B.I.Halperin, P.C.Hohenberg, S.K.Ma, Phys.Rev.Lett. 29, 1548 (1972); Phys.Rev. B10, 139 (1974); Phys.Rev.B13, 4119 (1976).

[2] See the review article by P.C.Hohenberg and B.I.Halperin, Rev. Mod.Phys. 49, 435 (1977) and references therein.

[3] B.I.Halperin, P.C.Hohenberg, E.D.Siggia, Phys.Rev.lett.31, 1289 (1974); Phys.Rev. B13, 1299 (1976).

[4] E.Brézin, C.De Dominicis, Phys.Rev.B12, 4954 (1975).

[5] See e.g. P.C.Hohenberg and P.C.Martin, Ann.Phys.(N.Y.), 34, 291 (1965).

[6] B.I.Halperin, P.C.Hohenberg Phys.Rev.188, 898 (1969).

[7] R.A.Ferrell, N.Menyhàrd, H.Schmidt, F.Schwabl, P.Szépfalusy, Phys. Rev.Lett. 18, 891 (1967); Ann.Phys. (N.Y.) 47, 565 (1968).

[8] P.C.Martin, E.D.Siggia, H.A.Rose, Phys.Rev.A8 423 (1973);
C. De Dominicis, J. Phys. (Paris) C1, 247 (1976);
H.K. Janssen, Z. Phys. B23, 377 (1976).

[9] R.Bausch, H.K.Janssen, H.Wagner, Z.Physik B24, 113 (1976).

[10] C.De Dominicis, L.Peliti, Phys.Rev.Lett.38, 505 (1977); Phys.Rev. B18, 353 (1978).

[11] G.'t Hooft, M.Veltman, Nucl.Phys. B44, 189 (1972); C.G.Bollini, J.J. Giambiagi, Phys.Lett.B40, 566 (1970).

[12] G.Ahlers, in The Physics of Liquid and Solid Helium, K.H.Benneman and J.B.Petterson (Eds.) Part.I, Section 2. New York: Wiley (1976).

[13] G.Ahlers, Phys.Rev.Lett.21, 1159 (1968); in Proceedings of the Twelfth International Conference on Low Temperature Physics, E.Kanda (Ed.), p.21 Tokyo: Keigaku (1971).

[14] H.K.Janssen, Z.Physik B26, 187 (1977).

[15] V.Dohm, Z.Physik B33, 79 (1979).

[16] L.Sasvàri, F.Schwabl, P.Szépfalusy: Physica 81A, 108 (1975).

[17] B.I.Halperin, Phys.Rev.B11, 178 (1975).

[18] V.Dohm, Z.Physik B31, 327 (1978).

[19] R.A.Ferrell,V.Dohm, J.K.Bhattacharjee, Phys.Rev.Lett. 41, 1818 (1978). See also the communication by Ferrel to this Conference.

[20] G.Winterling, F.S.Holmes, T.J.Greytak: Phys.Rev.Lett.30, 427 (1973); G.Winterling, J.Miller, T.J.Greytak: Phys.Lett.48A, 343 (1974); M.F.Vinen, C.J.Palin, J.M.Lumley, D.L.Hurd, and J.M.Vaughan, in Low Temperature Physics, LT-14, M.Krusius and M.Vuorio (Eds.) Vol.I, p.191, Amsterdam: North-Holland (1975).

[21] P.C.Hohenberg, E.D.Siggia, B.I.Halperin, Phys.Rev.B14, 2865 (1876).

[22] E.D.Siggia, D.R.Nelson, Phys.Rev. B15, 1427 (1977).

[23] E.D.Siggia, Phys.Rev.B15, 2830 (1977).

[24] See e.g. the communication by H.Meyer, this Conference.

[25] V.Dohm, H.K.Janssen, Phys.Rev.39, 946 (1977); J.Appl.Phys.49, 1347 (1978) and to be published.

[26] K.K.Murata, Phys.Rev.B12, 4954 (1975).

INTRINSIC AND EXTRENSIC CENTRAL-PEAK PROPERTIES NEAR STRUCTURAL PHASE TRANSITIONS

K.A. Müller

IBM Zurich Research Laboratory, 8803 Rüschlikon, Switzerland*

* Present address : IBM T.J. Watson Research Center, Yorktown Heights, N.Y. 10598.

INTRINSIC AND EXTRINSIC CENTRAL-PEAK PROPERTIES NEAR STRUCTURAL PHASE TRANSITIONS

K.A. Müller
IBM Zurich Research Laboratory, 8803 Rüschlikon, Switzerland*

ABSTRACT

This review assesses the occurrence of slow collective dynamics *above* T_c in structural phase transitions, *distinct* from normal-, under-, or over-damped soft modes. This is done mainly from an experimental point of view but proper reference to important theoretical progress is given. After a reminiscence of the first observation of a central peak (c.p.) in $SrTiO_3$ and its possible origin, computer simulations and strongly anharmonic (soliton)-type theories related to this phenomenon will be discussed. Then, the results of theories in which slowly-relaxing defects are coupled to the order parameter are given. A comparison with existing experiments above T_c indicates that the width of c.p's have *extrinsic* origins, i.e., are impurity dominated. In contrast, electron paramagnetic resonance (EPR) measurements of the distribution of the order parameter in $SrTiO_3$ prove the existence of short-range order clusters in this displacive system at T_c. This short-range order accounts quantitatively for the non-vanishing soft mode, thus singling out an *intrinsic* property of the c.p. phenomenon. The vanishing c.p's in hydrogen-bonded ferroelectrics are put into perspective regarding their critical dimensionality. Examples of magnetic centers which relax slowly in these ferroelectrics, are discussed.

*Present address: IBM T.J. Watson Research Center, Yorktown Heights, New York 10598.

I. Introduction

Until 1971 structural and ferroelectric phase transitions showed, and were analyzed by, Landau-type mean-field behavior concerning the order parameter $\bar{\varphi}(T)$ and other static properties [1]. This included the "soft-phonon" mode $\omega_s(T)$ [2,3] present in these systems above T_c due to the finite mass of the atoms and their intertia, as opposed to magnetic systems. Accordingly, previous to that year both quantities, $\bar{\varphi}(T)$ and $\omega_s(T)$, appeared to vary proportionally to $|T-T_c|^{1/2}$. It was then found that the order parameter close to T_c would rather vary as $|T-T_c|^{1/3}$ in $SrTiO_3$ and $LaAlO_3$, a dependence common to short-range critical behavior [4]. This was not unexpected because the soft phonon for these crystals for $T > T_c$ is at the R-corner of the Brillouin zone and $\bar{\varphi}(T)$ has rotational character. Thus, a cancellation of long-range electron dipolar forces occurs as in antiferromagnets [5] resulting in short-range behavior.

The intensity $I(T)$ of the $(1/2, 1/2, 3/2)$ Bragg peak which is allowed below $T_c \simeq 105$ K on symmetry grounds, and is classically expected to behave as $I(T) \propto \bar{\varphi}(T)^2$, rather followed a $|T-T_c|^{2/3}$ dependence for $T < (T_c - 4\text{ K})$ but showed finite quasi-elastic scattering above T_c [6]. An inelastic neutron-scattering experiment by *Riste* et al. [6], then revealed in the van Hove function that the R-corner soft mode did not freeze-out. It remained finite, lost intensity on approaching T_c at the expense of a *central peak* (c.p.) in the structure factor $S(\vec{q},\omega)$, as shown in Fig. 1. Its width was beyond resolution. A subsequent study by *Shapiro* et al. [7] on $SrTiO_3$ and $KMnF_3$ confirmed the existence of this peak critically rising for $T \to T_c^+$ but its width could not be resolved either although the experimental resolution was improved over an order of magnitude to 0.02 meV (Fig. 2). *This c.p. feature distinct from the soft mode is the theme of the present review.* The soft mode itself may be underdamped as in $SrTiO_3$ or overdamped as in $KMnF_3$ or $LaAlO_3$ [8].

Due originally to *Hohenberg* [9], the form factor

$$S(\vec{q},\omega) = \frac{1}{2\pi} \int_{-\infty}^{+\infty} \langle \varphi_q(t)\, \varphi_q(0) \rangle\, e^{i\omega t}\, dt, \tag{1}$$

with $\varphi_q(t)$ the Fourier transform of the R-corner rotational order parameter $\varphi(\vec{r})$, has been cast in the following phenomenological form

$$S(\vec{q},\omega) \simeq \frac{n(\omega)+1}{\pi\omega}\, \mathrm{Im} \left\{ \frac{1}{\omega_{s0}^2(\vec{q},T) - \omega^2 - i\omega[\Gamma_0 + \Gamma(\omega,T)]} \right\}, \tag{2}$$

where $n(\omega)$ is the phonon occupation number and

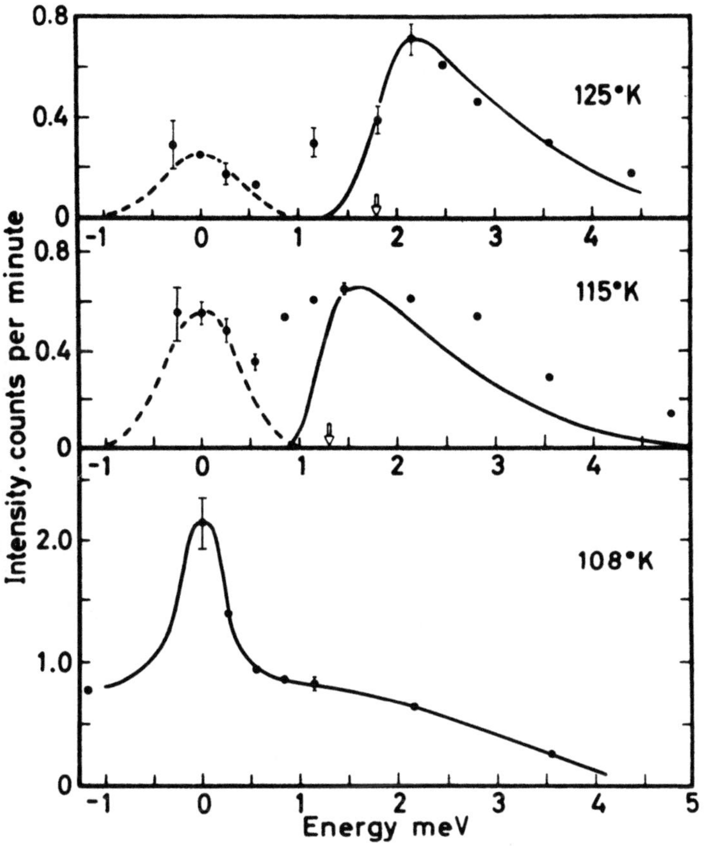

Fig. 1. Energy scans at the R-point for $SrTiO_3$ at three temperatures above the transition temperature T_c = 106 K. The dashed curves indicate the instrumental resolution as seen from the incoherent background. At 108 K the curve is only a guide to the eye. From *Riste* et al., Ref. [6]

$$\Gamma(\omega,T) = \delta^2(T)/(\gamma - i\omega). \tag{3}$$

For $\Gamma_0 \ll \delta^2/\gamma$ and $\omega \gg \gamma$, one gets the *soft mode* with $\omega_{s\infty}^2 = \omega_{s0}^2 + \delta^2(T)$ in the so-called collisionless regime [10],

$$S_{ph}(\vec{q},\omega) \simeq \frac{kT}{\pi} \quad \frac{\Gamma_0}{[\omega_{s\infty}^2(\vec{q},T) - \omega^2]^2 + \omega^2\Gamma_0^2} \quad , \tag{4}$$

and a central peak for small ω in the collision dominated regime

$$S_{cp}(\vec{q},\omega) = \frac{kT}{\pi} \quad \frac{\delta^2(T)}{\omega_{s\infty}^2 \quad \omega_{s0}^2} \left(\frac{\gamma'}{\omega^2+\gamma'^2}\right) \quad , \tag{5}$$

where $\gamma' = \gamma \quad \omega_{s0}^2/[\omega_{s0}^2 + \delta^2(T)]$ is the central-peak width and $\omega_{s0}^2(\vec{q},T)$ is the critical quantity proportional to $\chi^{-1}(\vec{q},T)$, χ being the static susceptibility of the system.

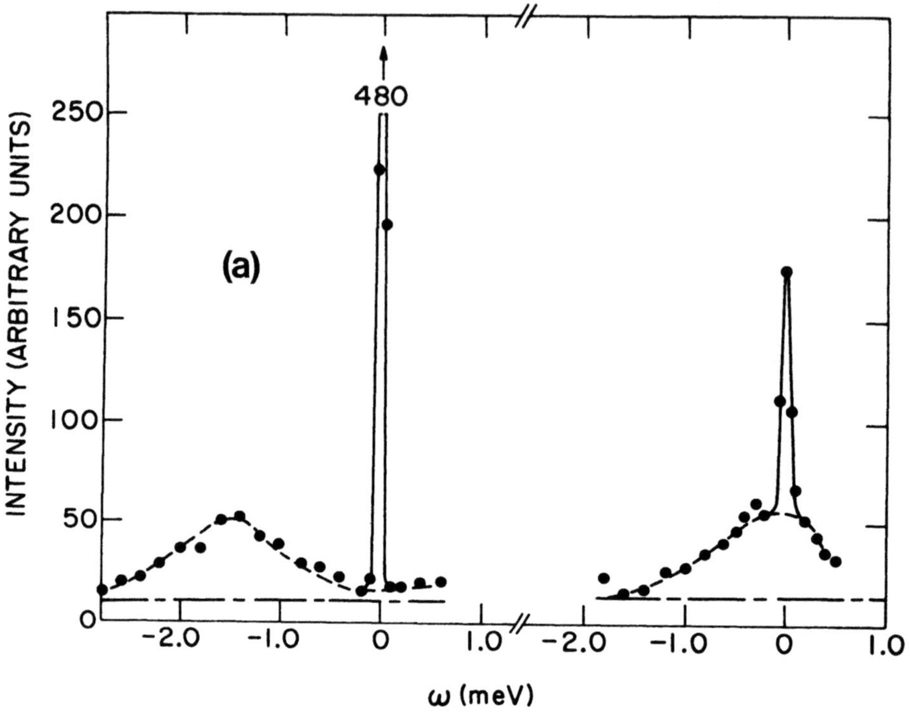

Fig. 2. High-resolution energy scans above T_c for $SrTiO_3$ and KMnF. The dashed line corresponds to the soft mode, and the solid line to the central component. From *Shapiro* et al., Ref. [7]. a) $SrTiO_3$, $T - T_c$ 15.2 K, b) $KMnF_3$, $T - T_c = 17.9$ K

The discovery of the c.p. in $SrTiO_3$ up to about 50 K above T_c, and the observation of EPR linewidth broadening over that temperature span [11] (see Fig. 3) started a large-scale research into the possible existence of a new fundamental excitation of solids. Theoretically and experimentally, attention was focussed on getting the two unknown quantities in Eq. (2): a) $\delta^2(T,\vec{q})$, responsible for the intensity of the c.p. as well as for the finite soft mode frequency $\omega_{s\infty}(T=T_c) = \delta(T=T_c)$, and b) $\gamma(\vec{q})$, responsible for the width of the c.p. Theoretically, in view of the form $\Gamma = \delta^2/(\gamma - i\omega)$, one looked for some *coupling of the soft phonon to slowly-relaxing intrinsic degrees of freedom* which, because of the symmetry of the $SrTiO_3$ system, had to be nonlinear. These efforts included coupling to acoustic modes, phonon-density fluctuations, thermal-diffusion modes, etc. None of them appeared satisfactory at the time and provoked criticism. For more detailed summaries, we refer the reader to the fine reviews of *Cowley* [10] and of *Feder* [12]. Both authors pointed out the

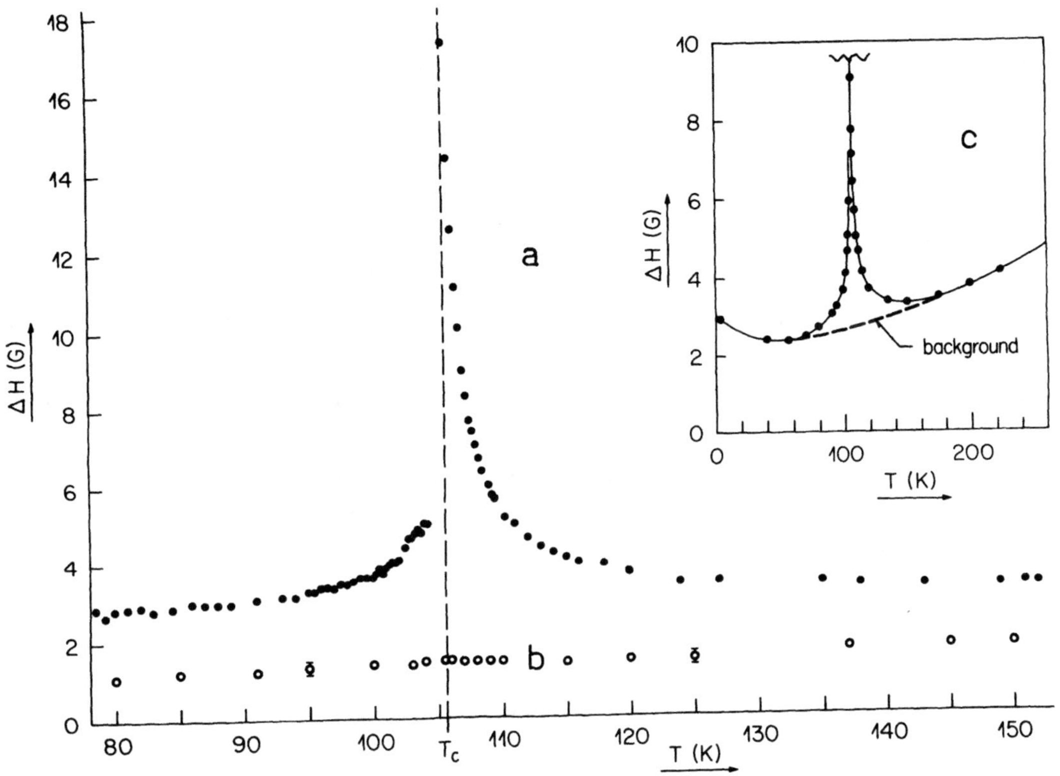

Fig. 3. The EPR linewidth of the $Fe^{3+}-V_0$ resonance in $SrTiO_3$ as a function of temperature, from *von Waldkirch* et al., Ref. [11]

severe difficulties encountered and *Feder* concluded: "As the phenomenon completely dominates the fluctuation spectrum near T_c, we are left with the nagging feeling that there is something essential we have not understood unless the central mode is satisfactorily explained". He expressed finally the view that *the central mode is intimately related with the critical fluctuations near the structural phase transitions,* barring the possibility of being due purely to impurity effects, as proposed by *Axe* [13]. This appears to be the case for the c.p. observed in Nb_3Sn [14].

In this paper we want to single out from the wealth of work done since these reviews were written, those dynamic properties which are intrinsic in nature and those extrinsic above T_c. We start by discussing results of computer simulations and non-linear theories in Sec. II. In this section the extrinsic theories of slowly-relaxing defects will also be presented as possible sources of the very narrow c.p's observed. We then review in Sec. III all published experiments on the c.p. linewidth γ' and show, by comparison with the computer and theoretical results of Sec. II, that they are impurity dominated. In Sec. IV, the question of which of the intrinsic properties are observable, is addressed. It turns out that it is the second quantity in Eq.(2), $\delta^2(T = T_c) = \omega_{s\infty}^2(T = T_c) \neq 0$, which is quantitatively related to the short-range order existing at T_c as detected by electron-paramagnetic resonance. The quantitative agreement is proof of a displacive to order-disorder crossover, on approaching T_c, in accordance with the expectations expressed above. The final section is devoted to central peaks observed in hydrogen-bonded KH_2PO_4-type ferroelectrics. In the rather lengthy Sec. V - which may be skipped by readers interested only in intrinsic properties - light-scattering and magnetic-resonance experiments on central peaks are summarized and shown to result almost totally from impurities either static or dynamic. A speculation involving the critical dimensionality d_c is offered at the end of this section to explain why pure uniaxial ferroelectrics only show classical mean-field static and dynamic properties without a central peak. Within the present text, a number of papers could not be directly referred to. In order to facilitate access to them they appear at the end of the reference list with full titles.

II. Computer Simulation, Anharmonic and Impurity Theory

a) Molecular-Dynamics Calculations and Strongly Anharmonic Theory

The difficulties with conventional anharmonic lattice perturbation theory led to the suspicion that the c.p. phenomenon might - close to T_c - result from strongly anharmonic effects, not tractable by perturbation expansions of the harmonic lattice. In order to see what additional excitations, if any, may be present, large-scale computer simulations were undertaken. The first paper of a series, pointing towards the heart of the problem, was that by *Schneider* and *Stoll* [15]. They simulated a two-dimensional $d = 2$ lattice with a model Hamiltonian of the form

$$\mathcal{H} = \sum_{\ell} \frac{M\dot{x}_{\ell}^2}{2} + \frac{\bar{A}}{2} \sum_{\ell} x_{\ell}^2 + \frac{B}{4} \sum_{\ell} x_{\ell}^4 + \frac{C}{2} \sum_{\langle \ell\ell' \rangle} (x_{\ell} - x_{\ell'})^2, \qquad (6)$$

where ℓ labels the particle with mass M in the ℓ-th unit cell. $M\dot{x}_{\ell}$ and x_{ℓ} are the momentum and displacement with respect to the square rigid lattice with constant a. $\bar{A}$,B and C are model parameters chosen such as to represent antiferrodistortive motions of the particles with short-range interactions only, as symbolized by $\langle \ell\ell' \rangle$. The second and third terms in Eq. (6) represent the harmonic and anharmonic single-particle potential, and the fourth their interaction.

At $T = 0$ the order parameter is given by $x = (-\bar{A}/B)^{1/2}$, thus for $\bar{A} < 0$ a finite T_c is guaranteed. The larger the $|\bar{A}|$ value the more order-disorder character the system will have, with a deep double-well single-particle potential. A convenient measure of the order-disorder versus displacive character is the ratio g of the depth of the single-particle well $E_w = \bar{A}^2/4B$ to kT_c, i.e., $g = \bar{A}^2/4B\ kT_c$. For $|\bar{A}| \to 0$ $g \to 0$ one approaches the classical displacive limit which has $T_c = 0$. *Schneider* and *Stoll* employed the molecular-dynamics technique, which solves the set of coupled Newton's equations associated with a model Hamiltonian, Eq.(6), according to a set of difference equations with a time increment. Using a lattice of 40×40 points, with periodic boundary conditions, they computed the displacement-displacement correlation function $\langle x_q(t)x_{-q}(0)\rangle$, and from it the corresponding dynamic structure factor $S(\vec{q},\omega)$. They simulated an order-disorder and a displacive system; the result for the latter with $g = 0.2$ is shown in Fig. 4 as a function of frequency at $(T - T_c)/T_c = 0.06$ for various wave vectors close to the reciprocal lattice point $\{3,1\}$ which becomes a Bragg spot below T_c. One sees *in addition* to the soft mode, a c.p. of comparable width. The soft mode stopped at a finite $\omega_{s\infty}^2$ and faded in intensity on approaching T_c. Simulations of the order-disorder system with $g > 1$ yielded only the well-known diffusive $\omega = 0$ centered excitation and no soft mode as expected.

In order to understand the physical meaning of their result, they displayed the dynamical-rotational motion of the particles graphically, as shown in Fig. 5. Therein, the dark signs represent local rotations having opposite sign to that expected at zero temperature. This figure reveals near T_c the existence of short-range clusters separated by walls where the local displacements change sign. The existence and dynamics of these clusters gives rise to the c.p. and inhibits the soft mode from freezing out because the symmetry is broken locally.

Similar results as those reviewed here were published later by *Aubry* [16] and others [17] for a linear array. In such a linear chain, no phase transition occurs at finite temperature but, classically, $T = 0$ can be viewed as a critical temperature. Consequently, the system exhibits all features of interest for $T \geq T_c$. *Krumhansl* and *Schrieffer* [18] then showed analytically, for such an array, the existence of an important class of non-linear solutions which indeed could not be represented by usual phonon perturbation expansions: the motion of "domain walls" or *solitons*. The particular species occurring in their linear chain model was the kink and anti-kink solitons. Recently physicists became aware of quite a number of other species such as breathers, etc. The reader is referred to a recent conference [19] held on the subject which shows the emphasis these highly-nonlinear mathematical solutions have attained.

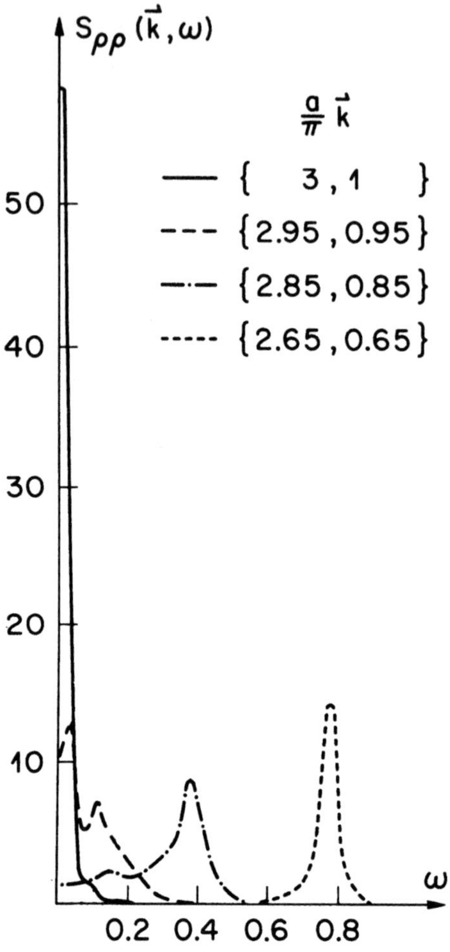

Fig. 4. The frequency- and wave-vector dependence of the spectral response of a d = 2 displacive model with g = 0.19 at $t \underset{\sim}{\sim} 0.06$, as determined by molecular dynamics from *Schneider* and *Stoll*, Ref. [15]

On the molecular-dynamics side *Schneider* and *Stoll* have considerably substantiated the picture by simulations for d = 3 and d = 4 dimensions [20]. They also considered a d = 2 lattice with anisotropy and a d = 3 lattice with two masses. Thus, the latter model also yields a phonon branch. They also computed the energy-energy correlations for $T \leq T_c$. To discuss all these results here would lead us too far, we therefore refer to a review of theirs [19]. The essential point is that for $T > T_c$ the picture we described for d = 2 is unaltered.

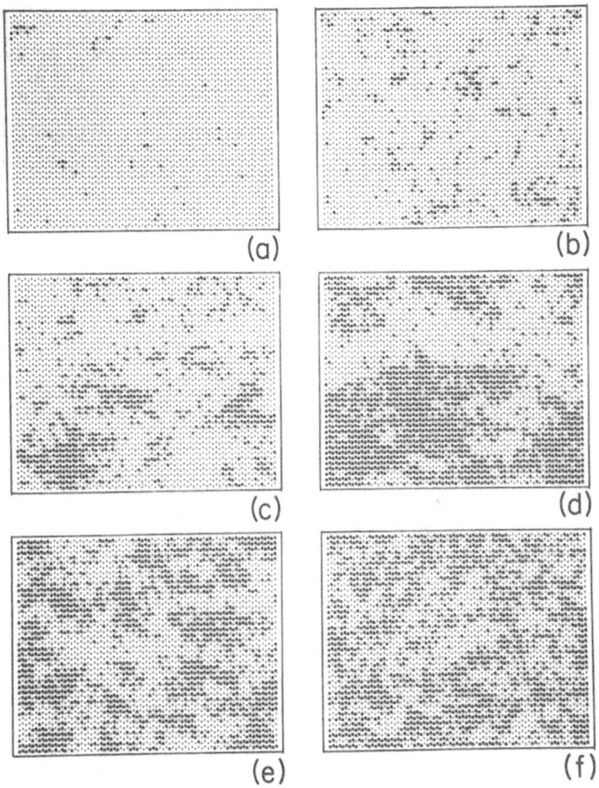

Fig. 5. Snapshots of instantaneous cluster configurations at $T = 0$. a) $k_BT = 0.1$; b) 0.22; c) 0.24; d) 0.26; e) 0.3; f) 0.8. $kT_c = 0.245$, from *Schneider* and *Stoll*, Ref. [15]

b) Coupling to Impurities

Due to the narrowness of the c.p's observed, scattering from static strain fields was discussed quite early by *Axe* et al. [13]. Such a simple mechanism would yield an intensity $I \propto \bar{\varphi}^{2}_{static} \propto 1/\omega_0^4$ and $\omega_0^2 = \omega_\infty^2$. But the latter formula was not obeyed experimentally and $I^2 \propto 1/(\omega_0^2\ \omega_\infty^2)$ with $\omega_0^2 \neq \omega_\infty^2$ was generally found. Furthermore, the temperature dependence of the EPR linewidth $\Delta H(T) \propto (T - T_c)^{-0.65}$ [11,21] could also not be accounted for by inhomogeneous broadening due to imperfections by *Folk* and *Schwabl* [22], who arrived at $\Delta H(T) = (T - T_c)^{-\nu(1/2 - n)} \approx (T - T_c)^{0.32}$ for $d = 3$.

The possibility that defects or impurities are relevant for c.p's near structural phase transitions (SPT) received a new and strong impetus from a paper by *Halperin* and *Varma* (HV) [23]. HV used mean-field theory for a displacive system containing random static or slowly-relaxing defect cells coupled linearly to the order parameter. The defects breaking the symmetry make transitions between two wells at $\pm x_d$ with a transition rate $\nu = \nu_0 \exp \Delta E/kT$, where ΔE is an activation energy of the order of 0.1 to 0.2 eV, $\nu_0 \approx 10^{-12}$ sec, their concentration is c (see Fig. 6). With these assumptions, they found a c.p. and calculated the two quantities characterizing its dynamic structure factor [Eq.(2)]

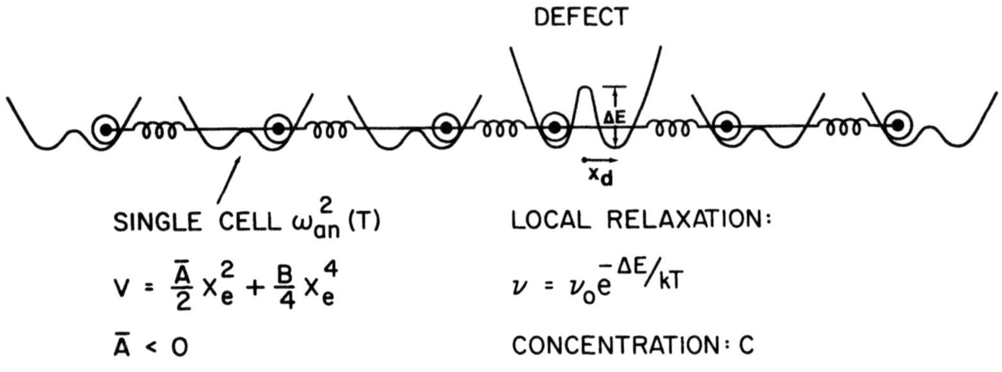

Fig. 6. A relaxational defect cell in an anharmonic lattice array

$$\delta^2(T) = c \left(\frac{m\omega_{an}^2(T)\, v_{q0}\, x_d^2}{kT} \right) \tag{7}$$

$$\gamma = \nu = \nu_0 \exp \Delta E/kT.$$

Here, m is the mass of the intrinsic atom, $\omega_{an}(T)$ the *uncoupled* intrinsic single-particle anharmonic frequency, and

$$v_{q0} = \sum_{\ell'} v_{\ell\ell'}\, e^{iq0(R_\ell - R_{\ell'})} \tag{8}$$

the Fourier transform of the bilinear interaction of particles in cells ℓ and ℓ'. The symbols $\omega_{an}(T)$ and v_{q0} used in Eq.(7) and characterizing the intrinsic lattice properties are those introduced by *Thomas* [24] previously in self-consistent mean-field (MF) theory. They are more often employed by experimentalists than those HV used. An important condition for a defect to be operative in producing a c.p. is that $dT_c/dc > 0$. From Eq.(7), one expects $\delta^2(T) \propto 1/T$. Close to T_c where the correlation length ξ becomes larger than the distance between particles, one does not expect Eq.(7) to hold due to cluster interactions.

Halperin and *Varma* compared their formula quantitatively to the known neutron and EPR data in $SrTiO_3$. They obtained agreement for a concentration of $c \approx 10^{-5}$ and a ratio of defect to intrinsic rotation of the octahedra of $x_d/x_i \approx 5$, i.e., very reasonable results, although the 1/T dependence was missing. Their findings prompted two further theoretical mean-field investigations.

The impurities considered by HV break the translational symmetry of the crystal and may give rise to local modes which can be investigated by resonance experiments. Due to the coupling to the soft mode, these local modes can become strongly temperature dependent. *Höck* and *Thomas* [25] calculated their behavior self-consistently with mean-field theory as well. They obtained a criterion for local-mode condensation at a temperature T_c^{loc} above T_c. If this is fulfilled, the local mode for displacive oscillatory defects is given by $\omega_{loc}^2 = K_1(T - T_c^{loc})$, and for a relaxing Ising-type impurity by $1/\tau_{loc} = \Gamma(T - T_c^{loc})$.

In a range $T_c \leq T \leq T_c^{loc}$, local order exists around the impurity site and the local order parameter at this site grows as $x_d \propto (T_c^{loc} - T)^{1/2}$ and decays exponentially as $(1/r)\, e^{-\xi/r}$, with increasing distance from the impurity. Here, ξ is the intrinsic correlation length of the host. Going beyond mean-field approximation no real freeze-out is expected but a qualitative change in local defect dynamics should be observed.

Quadratic coupling to the order parameter is another possibility [26]. Here, the impurity is at a symmetry conserving site in contrast to that which coupled linearly. If the impurity is softer than the rest of the lattice, a localized bond state ε_b below the soft-mode band exists, i.e., $\varepsilon_b - \bar{\varepsilon} = a > 0$ where $\bar{\varepsilon}$ is the bottom of the soft-mode band and $a = \alpha(T - T_0)$ with T_0 the intrinsic transition temperature. The local "transition temperature" is then given by $T_c^{loc} = T_0 + \bar{\varepsilon}/\alpha$. Below T_c^{loc} the high-temperature symmetry will be broken and the cluster couples linearly to the soft mode, giving a c.p. above the T_0 of the bulk.

A further step in understanding was achieved by a subsequent computer simulation by *Schneider* and *Stoll* [27]. They introduced in their $d = 2$ system, random slowly-relaxing defects coupled to the displacive system. The result was a c.p. considerably narrower than that found in their intrinsic simulations, plus the essentially unchanged broad soft phonon at finite frequency $\omega_\infty(T)$. The critical dynamics of the narrowed c.p. $\propto (T - T_c)^{-\gamma}$ remained unchanged. This simulation which avoids the shortcomings of the mean-field approximation, and includes impurity interactions near T_c, indicated where further progress could be expected experimentally in real $d = 3$ systems.

III. Experimental Central-Peak Widths in Displacive Structural Phase Transitions (SPT)

This section reviews the experimental width $\gamma' \equiv \Gamma'(\vec{q},T)$ [see Eq.(2)] in displacive systems as has been done in an unpublished report by the author [28]. The summary on c.p. in hydrogen-bonded ferroelectrics which was also contained therein, is deferred to Sec. V, amplified by most recent results and by a critical review of magnetic resonance work.

a) $SrTiO_3$

Since by far the largest effort has been devoted to this crystal, we dedicate the first subsection to it. As described in Sec. I, it has not been possible to resolve the energy width of the c.p. in the early inelastic neutron-scattering efforts [6,7]. The resolution has since been enormously improved from 0.02 meV to 0.08 μeV which corresponds to 20 MHz. This has been achieved by *Töpler* et al. with a refined back-scattering technique [29]. Even with this excellent resolution, no broadening $\propto \omega_{0s}^2(T)$ [Eq.(5)] could be observed up to $T_c + 12$ K, as seen in Fig. 7. The upper limit so found is in agreement with γ-ray scattering [30]. The most recent light-

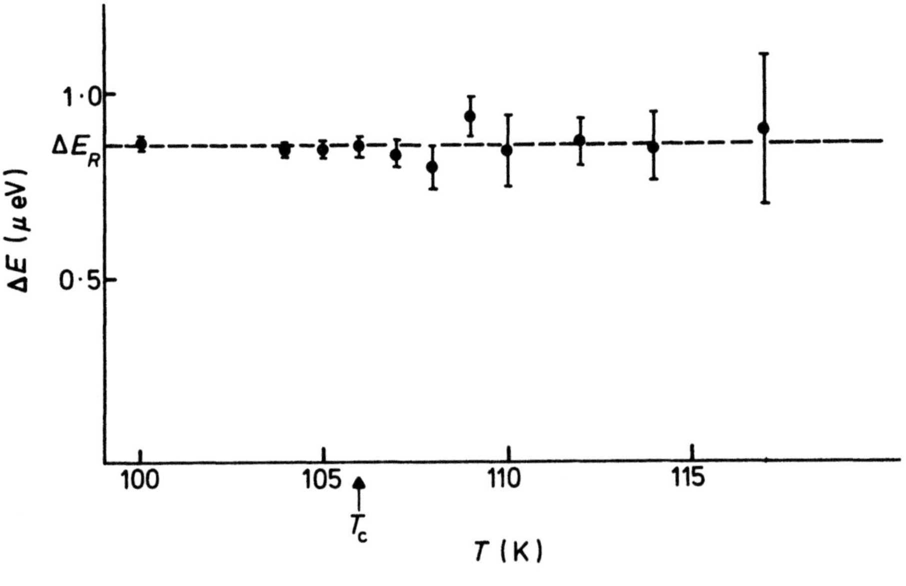

Fig. 7. Energy width of the $SrTiO_3$ c.p. back-scattering spectra as a function of temperature. The broken line indicates the width of the resolution function. After *Töpler* et al., Ref. [29]

scattering experiments of *Lyons* and *Fleury* [31] indicate two characteristic times of the relaxing self-energy τ_1 and τ_2: a shorter τ_1 with $\tau_1^{-1} \sim 15$ GHz is observable direct in the depolarized scattering for $T < T_c$. This feature is also found from an analysis of the Brillouin spectrum where, for $T < T_c$, the TA phonons couple linearly to the soft phonons. From the same analysis, a slower process τ_2 with $\tau_2^{-1} < 0.3$ GHz, not resolvable with the iodine-filter technique in depolarized light scattering, was inferred for $T \to T_c^-$. Its existence is what Courtens predicted from an analysis of neutron scattering, specific heat and birefringence data for $T > T_c$ [32].

The upper limit of τ_2^{-1} from the light scattering is in agreement with the other work discussed, and the result of the EPR linewidth and shape measurements. The latter yielded a local relaxation rate of $\Gamma_\ell = 60$ MHz at $T = T_c + 2$ K, from which, under certain assumptions, one calculates a collective central-peak width at T_c of $\Gamma' \sim 1/10\ \Gamma_\ell$ [33]. A most recent extended analysis of the EPR-linewidth data has further narrowed down Γ_ℓ to less than 6 MHz between T_c and $T_c + 11$ K [34] for radial local fluctuations of the oxygen. Thus $\Gamma' < 0.6$ MHz which represents the narrowest resolution attained so far. A summary of the present findings is given in Table 1.

Thus, it appears that above T_c the central-peak dynamics is very slow and most probably results from a linear coupling of the soft mode to slowly-relaxing defects as follows from *Halperin* and *Varma's* [23] analysis as well as from the later one by *Schneider* and *Stoll* [27].

The above-mentioned more or less experimental agreement of Γ_c for the central-peak width in $SrTiO_3$ is not a necessity, as the various techniques used probe different properties of the crystals. The detailed analysis of the depth-dependence of critical γ- and X-ray scattering in $SrTiO_3$ by *Darlington* et al. [30] has shown that the penetration depth at 14.4 keV is of the order of 1 to 10 µm. *Aso* finds from birefringence data [35] that surface strain extends up to 1 mm into the bulk crystal. This means that the thickness one explores with γ-rays is about two orders of magnitude thinner than the region of surface strain, i.e., highly perturbed. There will be a sufficient number of static defects of the proper symmetry to couple linearly with the soft mode - as considered by *Halperin* and *Varma* - to produce a narrow static central peak.

In neutron scattering, the quasi-elastic peak one observes is the sum of all effects in the bulk, i.e., coupling to point defects, dislocations or volume strains. That, away from T_c, the central peak must come from these crystal imperfections is also borne out by a recent inelastic neutron study by *Currat* et al. [36]. From Fig. 8 ,

Table 1 Central-peak width and other relevant data for $SrTiO_3$, unresolved width: given, resolved: underlined

Technique	Minimum width & Resolved width	Date	Authors	Reference
Inel. neutrons	< 0.3 meV	1971	*Riste* et al.	[6]
$\delta(T=T_c) = 0.13$ THz	< 0.02 meV	1972	*Shapiro* et al.	[7]
γ-rays (1 μ surface only)	< 0.01 μeV = 2.5 MHz	1975	*Darlington* et al.	[30]
Light scattering				
Raman	$T < T_c$, 15 GHz, < 0.3 GHz	1977	*Lyons & Fleury*	[31]
Brillouin		1979	*Su*	[31]
Inel. neutrons $\delta(T=T_c) = 0.13$ THz	< 0.02 meV	1978	*Currat* et al.	[36]
Inel. neutrons back-scattering (T_c + 12 K)	< 0.08 μeV = 20 MHz	1977	*Töpler* et al.	[29]
EPR $\nu_L < 6$ MHz (T_c + 11 K)	$\lesssim 1/10\nu_L = 0.6$ MHz	1979	*Reiter* et al.	[34]

one can see that away from T_c, the parameter $\delta^2(T)$, which characterizes the central-peak strength, differs for the crystal used in Grenoble and the one of *Shapiro* et al. [7] in Brookhaven. This points to the fact that the central peak in the bulk is extrinsic in neutron scattering away from T_c. *However, on approaching T_c, the magnitude of these quantities merged for the crystals used. Actually, the intensity at T_c is the same as the one obtained by Lyons and Fleury* [31] *from light scattering approaching T_c from below.* Most recently, *Hastings* et al. [37] observed a systematic enhancement of the central-peak intensity upon reduction of $SrTiO_3$ with hydrogen. The results provide direct experimental evidence for the involvement of a defect mechanism for the central-peak formation in $SrTiO_3$. Increasing the amount of Ti^{3+} upon reduction from 6×10^{17} to 20^{20} cm^{-3} they found an enhancement of a few times the least reduced sample. One possible defect involved is Ti^{3+} on a Sr^{2+} site which EPR studies had shown earlier to couple linearly to the order parameter of the transition [38]. It resulted from the EPR study that the center relaxes with times shorter than $\tau_{Ti} = 1/\Delta\nu = 3 \times 10^{-9}$ sec, and thus is an H-V center for times *shorter*

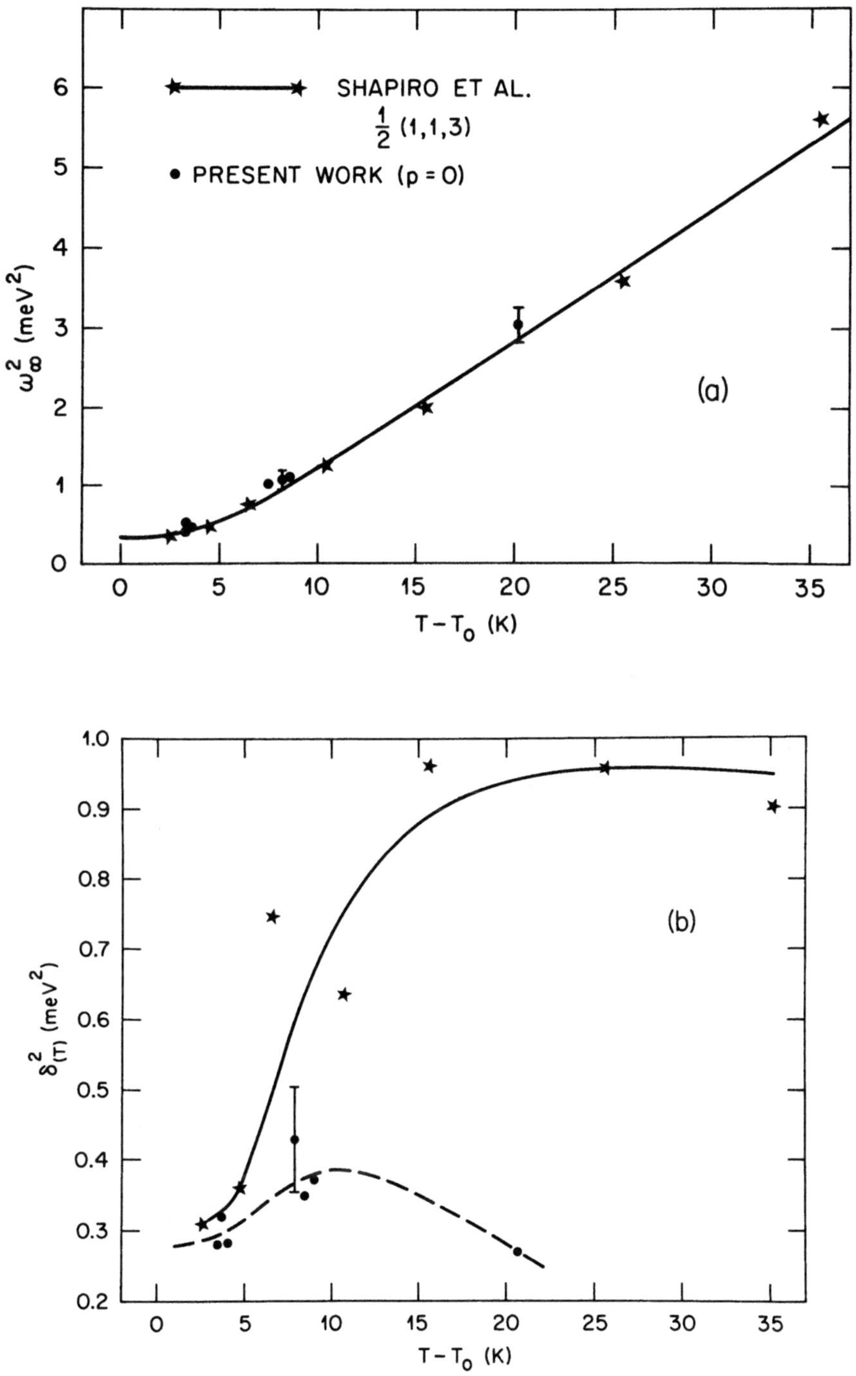

Fig. 8. R_{25} soft-mode in $SrTiO_3$. a) Comparison between two determinations of the quasiharmonic frequency $\omega_\infty^2(T)$. b) Same comparison in terms of $\delta^2(T)$, the parameter which characterizes the central-peak strength. After *Currat* et al., Ref. [36]

than 3×10^{-9} sec. The EPR studies on non-conducting $SrTiO_3$ using the $Fe^{3+} - V_0$ center as a probe mentioned earlier [11,33], yields local relaxation times longer than 3×10^{-9} sec, namely, $\tau_\ell > 1.3 \; 10^{-8}$ sec. Thus, in non-reduced crystals, the Ti^{3+} is not the defect center which induces the central peak.

In nuclear magnetic resonance (NMR) and electron paramagnetic resonance (EPR), the lines observed arise from magnetic moments which come from regions of the crystal which are only slightly disturbed. Too strong disturbances cause the lines to shift so much that they do not contribute to the line intensity. Resonance lines are, of course, observed if the crystal is strained homogeneously, as for instance, in monodomain $SrTiO_3$ samples. In other words, the magnetic-resonance experiment acts as a discriminator against regions in the crystal which are much too perturbed.

The three kinds of experiments discussed above, namely: a) X- and γ-ray scattering; b) inelastic neutron-, Raman-, Brillouin- and Rayleigh scattering, and c) NMR and EPR are different probes; furthermore, ultrasound absorption may be yet another category. Experiments a) and c) are disjoint in probing the crystal, whereas experiments of type b) integrate over the whole crystal in the case of neutron scattering, and over selected parts in light scattering. If the crystal is relatively defect-rich, the quasi-elastic scattering will then, in the bulk, result mainly from coupling to those defects and not from the almost undisturbed microscopic regions of the crystal.

b) Observations in Antiferrodistortive Crystals

We shall first discuss two other crystals with zone-boundary soft modes: $KMnF_3$ and $NaNbO_3$ in which techniques of groups a), b) and c), reviewed above, definitely yield different relaxation rates. Cubic $KMnF_3$ also undergoes the same kind of structural phase transition as $SrTiO_3$ near second-order at 185 K. From critical ultrasound absorption, *Hatta* et al. [39] arrived at a critical Γ' width of 20 MHz at $T = T_c + 1$ K and 170 MHz at $T = T_c + 4$ K. On the other hand, using *Mössbauer* γ-ray scattering *Hanisch* and *Drosg* [40] found an upper limit of the central peak width of 10^{-8} eV or 2.5 MHz up to $T_c + 3$ K. Despite the difference of the results in this material, even the faster collective relaxation rate Γ' reported is very slow near $T_c + 1$ K. In contrast to $SrTiO_3$, which shows a well-resolved underdamped phonon, the latter is overdamped in $KMnF_3$ and the central peak sits on top of it [41]. The soft-phonon dispersion is more anisotropic than in $SrTiO_3$. This is even more so in the case of cubic $NaNbO_3$, where the behavior is near two-dimensional (2-d) above T_c + 10 K with $T_c = 640^\circ$C as observed in X-ray and inelastic neutron scattering by *Denoyer* et al. [42]. The soft mode condenses at the M point of the Brillouin zone.

In the high-temperature region above $T_c + 10$ K the soft phonon is overdamped and narrows considerably on cooling below $T_c + 10$ K. This has been taken earlier as evidence for a setting-in of three-dimensional correlations on approaching T_c. However, more recently the existence of an unresolved central peak extending 350 K above T_c was found and its domination over the overdamped mode for $T \geq T_c$ accounts for the narrowing of the excitation observed near $\omega \simeq 0$ [42].

As for $KMnF_3$, γ-ray experiments yield an unmeasurable narrow width in $NaNbO_3$ [30]. Earlier, *Rigamonti's* [43] group accounted for their NMR ^{23}Na relaxation measurements in $NaNbO_3$, shown in Fig. 9, by using a constant and large cubic-phonon anisotropy of $\Delta = 1/50$ down to $T_c + 4$ K and a local relaxation rate of $\Gamma_\ell = 300$ MHz at that temperature. *Rigamonti's* experiments yield a dynamic width which is larger than the upper limit of the γ-ray experiments outside the experimental error. As discussed earlier, this means that the two experiments probe different parts of the crystal: the NMR studies the least-perturbed of the bulk whereas the γ-ray probes the highly perturbed surface. Owing to the very flat soft-phonon branch between the cubic R and M points in $NaNbO_3$, this results in near two-dimensional correlated sheets above $\sim T_c + 10$ K [42]. This anisotropy is in approximate agreement with the NMR analysis. Therefore, this material should be the closest to the two-dimensional computer simulations of *Schneider* and *Stoll* [15] which shows an intrinsic broad central peak of width comparable to the soft phonon. *Rigamonti's* local relaxation

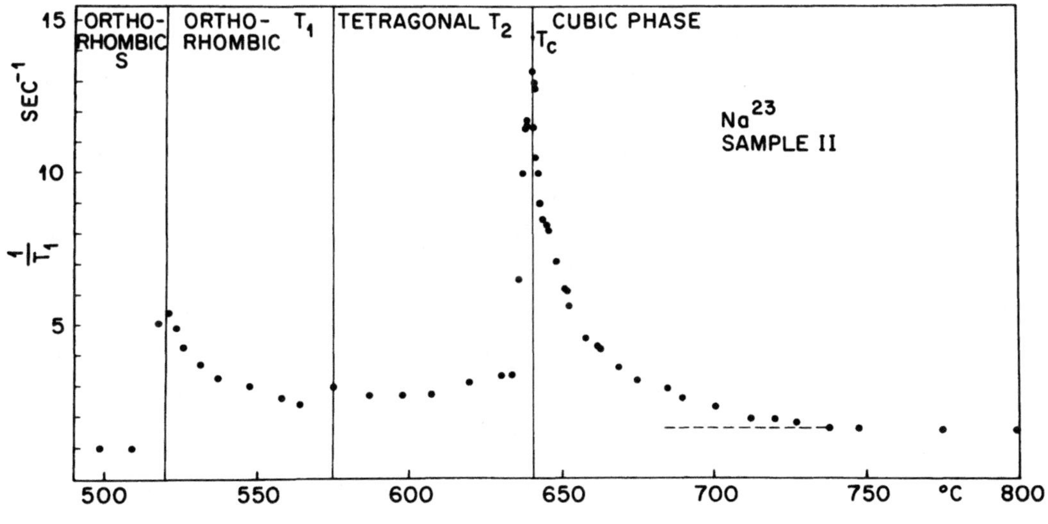

Fig. 9. Temperature behavior of the Na^{23} NMR relaxation rate in $NaNbO_3$. After *Avogadro* et al., Ref. [43]

rate of 300 MHz at $T_c + 4$ K is much narrower than the overdamped soft phonon of $\Gamma_c \sim 50$ GHz [42]. Thus, again, such a narrow width can only result from coupling to slowly-relaxing defects. As the local relaxation rate of 300 MHz at $T_c + 4$ K is larger than in $SrTiO_3$, the defects are dynamic. Thus, they have to follow the dynamics as simulated in a recent 2-d model calculation with relaxing impurities by *Schneider* and *Stoll* [27].

At 193 K, $RbCaF_3$ undergoes the same kind of phase transition as $SrTiO_3$ as well. Recently, in this fluoride crystal using inelastic neutron scattering *Almairac* et al. [44] observed a central peak whose width could not be resolved. However, they found that the parameter $\delta^2(T)$ decreases when $(T - T_c)$ increases as required by the *Halperin-Varma* mean-field theory [23]. Such a behavior was also found in $SrTiO_3$ by *Currat* et al. [36] and very recently in $NaNbO_3$ [42], however near T_c, $\delta^2(T)$ decreases faster than $1/T$ as the mean-field theory predicts. Further away $\delta^2(T)$ is mean-field like in $NaNbO_3$ [42]. In $LaAlO_3$ where an unresolved c.p. was observed, $\delta^2(T)$ is independent of temperature [8]. It has been argued [44] that the increase in $\delta^2(T)$ observed by *Shapiro* et al. could also result from ignoring the term $\lambda_3 q_\alpha q_\beta$ in the dynamical matrix. On the other hand, this term was also neglected in the paper by *Currat* et al. [36], and $\delta^2(T)$ nevertheless decreases for that $SrTiO_3$ sample.

c) Ferroelectric Displacive Crystals

In lead germanate $Pb_5Ge_3O_{11}$, a quasistatic central peak has been observed by neutron scattering above and below T_c by *Cowley* et al. [45]. From accurate Brillouin data, the main part is narrower than 80 MHz [46]. Its intensity diverges as $(T - T_c)^{-1}$ for $T \to T_c^-$. Its origin may be mainly static as bright spots are visually observed [47]. For $T < T_c$, the intensity can be accounted for by local density variations, but for $T > T_c$ the only possibility allowed by symmetry is at present the frozen defect cell mechanism of *Halperin* and *Varma* [23]. Furthermore, *Fleury* and *Lyons* [46], using the iodine-filter technique in Brillouin scattering found, in addition, a dynamic feature of smaller intensity whose width is 0.13 cm^{-1}. This feature cannot be accounted for by classical coupling theory. Therefore, these authors suggest that this intensity could result from quasi-mobile defects.

Cubic $KTaO_3$ doped with several percent of Li, Na and Nb was investigated by *Yacoby* [48]. From the first-order Raman spectra he concludes that the Li^+ and Na^+ ions on K^+ sites or Nb^{5+} on Ta^{5+} sites lack inversion symmetry on the time scale of the Raman experiment. From these and birefringence data he concludes that the centers are dynamic and relax with a time constant slower than 10^{-10} sec. A dynamic central peak should therefore exist in this doped material. However, as the doping was quite high (several percent), it is not clear whether the slowly-relaxing effects observed are due to the single Li^+ or Na^+ ion "rattling" slowly in the too large K^+ cell, or whether it is due to the cooperative interaction among these impurities. Most recently, he measured in a $KTaO_3$ sample containing about 6% $KNbO_3$, the first-order TO_2 Raman intensity $I_N(T)$ above and below $T_c = 52$ K [49]. $I_N(T)$ increases critically on approaching T_c^+ from above as $I_N(T)^{-1}\,T \propto (T - T_c)$ (see Fig. 10). This critical increase is shown to result from odd symmetry distortions with correlation length exceeding several unit cells and induced by the Nb ions on Ta sites. The analytic form of the divergence in $I_N(T)$ is compatible with the *Halperin-Varma* model. Assuming a continuous phase transition to be present does not account for the data. It thus appears to be the first scattering investigation confirming the H-V model for a particular *known* impurity.

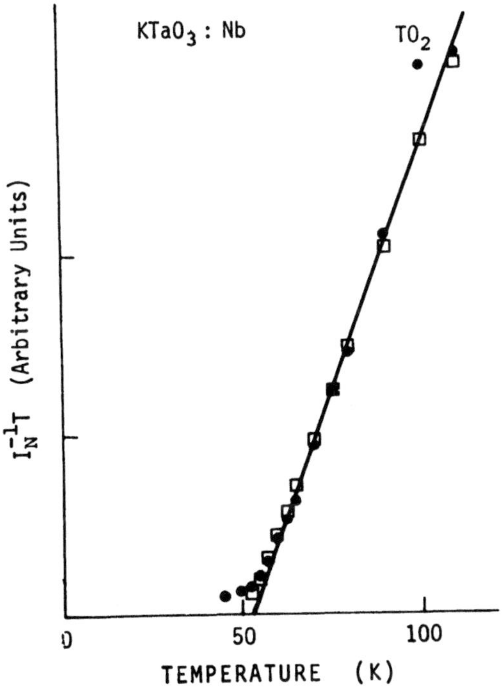

Fig. 10. Temperature dependence of $I_N^{-1}T$, of TO_2 Raman mode. Experimental values are designated by points. The straight line is the asymptotic dependence and it extrapolates to $T_c = 52$ K. After *Yacoby*, Ref. [49]

Höchli et al. investigated the Li:$KTaO_3$ system by dielectric constant measurements as a function of frequency [50]. They found that the Li^+ motion becomes slower than 10^{-1} Hz near 4 K, thus one might say that it freezes out. However, the soft-mode energy is enhanced on doping with Li^+. Thus, would $KTaO_3$ be a ferrolectric with finite T_c then $dT_c/dc < 0$, and Li^+ is not a H-V center. Thus the Li (in contrast to the Na and Nb) doping with $dT_c/dc > 0$ of $KTaO_3$, is in this sense already outside the scope of our present review. We mentioned it to complete the picture, as we also do for the ferroelectric SbSI in which *Steigmeier* et al. [51] observed a broad c.p. of 1 THz width below T_c with Raman scattering.

We can summarize by saying that the narrow width of critical central peaks observed at commensurate structural and ferroelectric transitions for $T > T_c$, are all defect-induced, be they static or dynamic. However, their nature has not yet been determined. The work reviewed in subsections b) and c) is condensed in Table 2.

Table 2 Central-peak width and other relevant data for displacive SPT including ferroelectric ones

Crystal Technique	Min. width & Resolved width	Date	Authors	Reference
$KMnF_3$ (overdamped soft phonon)				
Inel. neutrons	< 0.02 meV	1972	*Shapiro* et al.	[7]
Ultrasound (T_c + 1 K):	20 MHz			
(T_c + 4 K):	170 MHz	1974	*Hatta* et al.	[39]
γ-rays (1 μ surface) (T_c + 3 K):	< 2.5 MHz	1976	*Hanisch & Drosg*	[40]
$LaAlO_3$ (overdamped phonon)				
Inel. neutrons	< 0.02 MHz	1973	*Kiems* et al.	[8]
$NaNbO_3$ (overdamped phonon)				
Inel. neutrons	< 0.8 meV	1977	*Denoyer* et al.	[42]
NMR (T_c + 4 K)	300 MHz	1974	*Avogadro*	[43]
$Pb_5Ge_3O_{11}$				
Inel. neutrons	< 0.02 meV	1976	*Cowley* et al.	[45]
Brillouin scattering	< 80 MHz	1976	*Fleury & Lyons*	[46]
SbSI				
Raman scattering ($T < T_c$)	1 THz	1971	*Steigmeier* et al.	[51]

IV. Order-Disorder Behavior at Displacive SPT

a) The Displacive → Order-Disorder Crossover Concept

The wealth of experimental evidence for an extrinsic origin of the observed width of the critical c.p's for $T \geq T_c$, as presented in the preceding section, led to considerable concern as to whether the phenomenon may not be entirely extrinsic in nature. However, the finite frequency of the soft mode $\omega_{s\infty}(T = T_c)$ in $SrTiO_3$, independent of the sample at T_c [36] pointed to a possible intrinsic observable quantity, i.e., $\delta^2(T = T_c)$. The findings of computer simulations underlined this possibility, as $\omega_{s\infty}(T = T_c)$ in these studies was independent of the presence, or absence, of slowly-relaxing defects, the latter only influencing the width of the c.p. [28].

We remind the reader that we are only interested in experimental critical c.p's resulting from *novel* excitations of the solid as distinct from known ones. Coupling of the soft mode to the latter also gives rise to c.p's, e.g., heat diffusion modes or multiphonon states [32], etc. These are in themselves of interest but do not represent phenomena which fall outside the accepted picture of the slightly anharmonic lattice undergoing an SPT. As the above-mentioned c.p's mainly occur below T_c, it is experimentally better to confine oneself to $T \geq T_c$ as has mainly been done throughout this overview.

Within the possible context of a novel excitation of the solid near SPT's, the situation has been reviewed very recently by *Bruce* [52]. Going beyond the known theoretical and experimental results he showed the direction in which further experimental progress had to be expected. We summarize in the following the view he expressed. Starting with a comparison of the two extreme idealized cases of SPT's, the deep-well order-disorder system on the one hand, and the displacive system on the other, he pointed out the shortcomings of both approaches. In the former, only the dynamical hopping across the deep local double well is taken into account, leading to the existence of cluster walls and their diffusion (as in magnetic systems), but neglecting the phonons in the quasi-statically disordered lattice. In the other extreme is the classical soft-phonon picture. There, using the independent-site approximation, i.e., replacing $x_\ell x_{\ell'}$ by $x_\ell \langle x_{\ell'} \rangle$, the short-range correlations between particles are neglected. In this process, the particles move in an effective single-minimum potential well, each one following an independent Gaussian random fluctuation. The response of the system has a single pole at the soft-phonon frequency. In another approach called the independent mode approximation, x_ℓ^4 in Eq. (6), is replaced by $3\, x_\ell^2 \langle x_\ell^2 \rangle$. This approach includes correlations. It yields an optimum harmonic representation and a soft mode as well. However, the system undergoes characteristically a first-order transition to the ordered phase. This is a consequence of taking only harmonic modes into account whereas the behavior is inherently anharmonic near T_c.

From the two extreme cases of order-disorder and near-harmonic displacive models in which only domain wall motion or soft modes result, respectively, *Bruce* summarized the contemporary theoretical view as follows: "...... With the onset of criticality, the growth in short-range order promotes a crossover from a regime in which collective behavior has the classical displacive form, to a regime in which the collective behavior displays features better described in the language traditionally reserved for order-disorder systems". This picture is evidently in agreement with the computer simulations in $d = 1$ to 4 dimensions [16-18,21] as well as with the analytical analysis in $d = 1$ dimensions [19,20] which we summarized in Sec. II a). It, furthermore, constitutes an appealing graphic expression of universality, in that it offers a vivid description of the underlying character of the spectrum of excitations and of why deviations from classical static exponents have been observed in SPT. In other words, it correlates the occurrence of non-classical static behavior with the formation of order-disorder precursor clusters. *Bruce* amplified his statements by considering the local potential V_s^* of cluster coordinates after application of renormalization-group (RG) transformations: $V_b = |R_b V$. In $d = 3$, $g^* = V_s^*/kT_c = 0.55$ is of the order of unity. Referring to our discussion in Sec. II a), this suggests that a near-displacive system with $g \underset{\sim}{\sim} 1/10$ before renormalization becomes more order-disorder-like after, or equivalently, when the correlation length increases on approaching T_c. For $d = 2$ and 1, this behavior is more pronounced. For $d > d^*$ when the Gaussian fixed point is stable, the harmonic approximation should hold and $g^* \underset{\sim}{\sim} 0$. It thus appears that for $d = 1$ and 2, clusters are "mutually exclusive", i.e., at any one time a given region of the crystal can only be occupied by a single cluster. In $d = 4$ [53], clusters of ordering and anti-ordering can interpenetrate and the idea of a cluster wall seems less important.

The quite appealing picture reviewed above however, lacked experimental verification. Since the local potential V_s is of importance, it was natural to look for experiments which can give information, if not on the V_s of a cluster, then on a closely-related property: the local probability distribution $P(\varphi)$ of the order parameter φ. In the following it is shown that EPR achieved this goal.

b) EPR as a Probe for the Local Probability Distribution

The accurate values of $\langle\varphi(T)\rangle$ which led to the discovery of static critical phenomena in $SrTiO_3$ were obtained by measuring Fe^{3+} EPR line-positions. The EPR of Fe^{3+} ions substitutional on B sites reflects the local orientation of BO_6 octahedra [4]. For a fixed direction θ of the external magnetic field $\vec{H}$ in a (001) crystallographic plane, the secular resonance field H_r is given by [54]

$$H_r = H_0(\theta,\nu) + A(\theta,\nu)\varphi^{[001]} + O(\varphi^2) \tag{9}$$

where $\varphi^{[001]}$ is the octahedral rotation around the [001] axis and H_0 is the resonance field in the absence of rotation. The sensitivity parameter $A(\theta) = \partial H_r/\partial\theta$ measures the shift of the resonance field upon rotation by θ . H_0 and A both depend also on the applied microwave frequency. It could be shown that the second-order term, proportional to φ^2, becomes negligibly small for certain values of θ. The constant A is largest for the $Fe^{3+}-V_0$ pair center due to the large anisotropy of H_0 along the pair axis and the rotation $\varphi^{[001]}$ it measures is proportional to that measured by the Fe^{3+} [4]. The $Fe^{3+}-V_0$ center consists of a trivalent iron impurity on a B site with a nearest-neighbor oxygen vacancy [55]. Its resonance pattern has been analyzed above and below T_c [54]. For $H \parallel [110]$ and K-band, A = 26 Gauss/degree for an [001] octahedral rotation. The local random fluctuations φ_ℓ are of the order of a degree, whereas the background linewidth is about 3 Gauss or 0.10 degrees. Thus, owing to the high sensitivity of this center, the stochastic variation $\delta\varphi_\ell(t) = \varphi_\ell(t) - \varphi$ could be observed, as shown in Fig. 3. The possibly time-dependent departure $\delta H(t)$ from $H_0 + A\langle\varphi\rangle$ is a random function proportional to $\delta\varphi_\ell(t)$ [from Eq.(9)].

In magnetic resonance, one can distinguish two extreme cases with respect to the fluctuation time τ of the order parameter. If τ is much faster than the characteristic measuring time of the experiment τ_m one calls it a "fast-motion regime", if τ is much slower than τ_m, a "slow-motion regime". τ_m is essentially given by the inverse homogeneous EPR linewidth $\tau_m = 1/\Delta\nu = \hbar/g\beta\Delta H$ [56]. Whereas the fast-motion regime had been observed previously by NMR, it was the above-cited $SrTiO_3$ experiment [4] where, for the first time, a slow-motion regime close to T_c was reported. In this case, the main critical part of the fluctuation time τ of $\varphi(t)$ becomes quasi-static compared to the EPR measuring time τ_m. This means that at each site ℓ of an $Fe^{3+}-V_0$ center, the local rotation $\varphi(\vec{R}_\ell,t)$ is seen at rest by the EPR experiment. Thus, the shape of the EPR line $L(H) = L(\varphi)$ [from Eq.(9)] is proportional to the probability distribution $P(\varphi)$ of the ensemble. In EPR, the derivatives $dL(H)/dH$ of the absorption lines $L(H)$ are recorded using Zeeman modulation. For those not familiar with the technique, dL/dH and $L(H)$ are shown in Fig. 11 for $\langle\varphi\rangle = 0$ above T_c. Thus for slow motion, the curves are proportional to $P(\varphi)$ and $dP/d\varphi_\ell$, respectively. From this proportionality, it also follows from Eq.(9) that the full

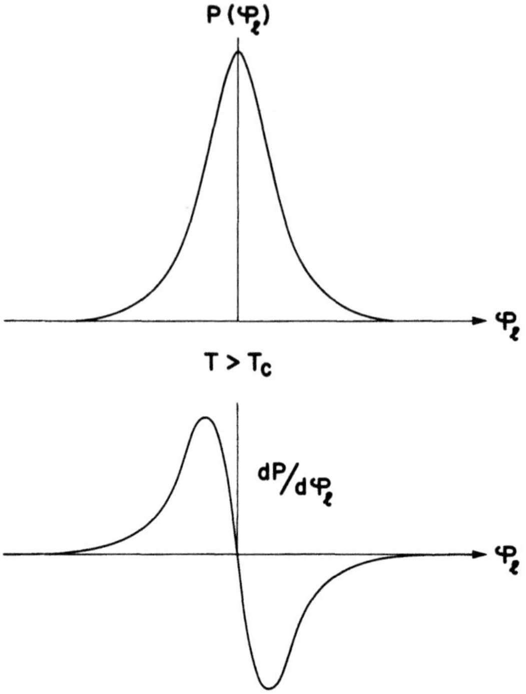

Fig. 11. Probability distribution $P(\varphi_\ell)$ and the derivative $dP/d\varphi_\ell$ which in the slow-motion regime are proportional to the line shape $L(H)$ and its derivative dL/dH, respectively

EPR linewidth $\Delta H(T)$ in the slow-motion regime is proportional to the mean-square fluctuation amplitude [57]

$$\Delta H_s(T) = 2\,A \langle \delta\,\varphi^2 \rangle^{1/2} . \tag{10}$$

Thus, at T_c the linewidth is always finite as is the local fluctuation amplitude.

On increasing the temperature above T_c the extra EPR linewidth $\Delta H_s(T)$ narrows down to the background width ΔH_b (see Fig. 3). An analysis of this narrowing with a theory of *Schwabl* [58] suggested a crossover to the fast-motion regime [57] near T_c+ 4 K. His theory was based on the phenomenological expression for the c.p. [Eqs. (2) and (3)], but contained various other parameters. This crossover seemed further substantiated by a change of the lineshape near T_c+ 3 K to a Lorentz form [34] which is usually the hallmark for a fast autocorrelation time. However, in a very detailed recent study, the sensitivity parameter $A(\theta)$ was varied at a set of fixed temperatures. It showed unambiguously that $\Delta H_s(T)$ is strictly proportional to $A(\theta)$ [34] (see Fig. 12). In the fast-motion regime, it would be proportional to $A(\theta)^2\, H(\theta)$. Thus, one always sees static local fluctuations throughout the critical regime for the secular part of the line, at least up to $T_c + 11$ K = 116 K. The static Lorentzian shape away from T_c is possibly caused by impurity-induced strain fields, the theory of which was developed in Ref. [34] as well.

The background width $\Delta H_b(\theta)$ varies also with θ but in an entirely different manner. Therefore, $\Delta H_s(\theta)$ and $\Delta H_b(\theta)$ could be analytically separated, since for both variations analytic expressions could be computed. $\Delta H_b(T)$ is determined by the spectral density

$$J(\omega) = \sum_{\vec{q}} S(\vec{q},\omega)$$

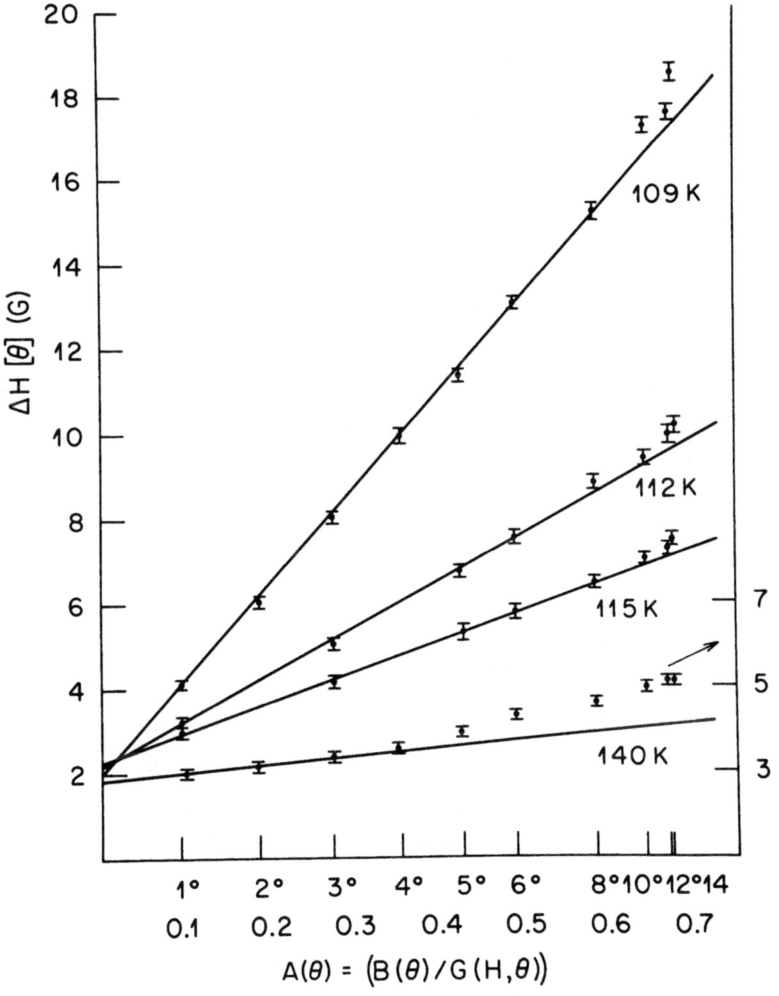

Fig. 12. Variation of the EPR linewidth $\Delta H(\theta)$ as a function of the sensitivity parameter $A(\theta)$ for various temperatures between 109 and 140 K. After *Reiter* et al., Ref. [34]

at $\omega_1 = 5.2\ cm^{-1}$ and $\omega_2 = 0.4\ cm^{-1}$ of the quasi-harmonic fluctuations of the octahedra. It is essentially a Raman nonsecular process between the $\pm 1/2$ ground state of the $Fe^{3+}(S = 5/2)$ ion and its two excited Kramers doublets $\pm 3/2$ and $\pm 5/2$ [55] located at energies $\hbar\omega_{1,2}$. Because the fluctuations are much faster than the reciprocal EPR linewidth, this background contribution is in the fast-motion regime which explains its narrow width of 2-3 Gauss. The latter is the limit of resolution of the local relaxation rate and corresponds to 6 MHz, as given in Table 1.

From the EPR linewidth data in a monodomain sample [59], using (9) the mean spread at T_c was

$$(\langle \delta \varphi_c^2 \rangle)^{1/2} = 0.4^o \quad \text{and} \quad (\langle \delta \varphi_a^2 \rangle)^{1/2} = 0.15$$

where c is the direction parallel to the monodomain axis and a is perpendicular to it [60]. *Courtens* [61] confirmed these values by an independent birefringence experiment in which he observed a cusp near T_c. This optical experiment was certainly fast as compared to the fluctuations involved. Thus, one can be confident that the EPR lineshapes reflect quantitatively $P(\varphi)$ in the critical region, i.e., as long as $\Delta H_s(\varphi) >> \Delta H_b(\varphi)$.

From the analysis of the lineshapes of the $Fe^{3+} - V_0$ center in the slow-motion regime, it was found that the probability $P(\varphi_\ell)$ to observe φ_ℓ deviated from the expected Gaussian, above and below T_c. Above T_c the shoulders of the lines fell more rapidly to zero than a Gaussian and $P(\varphi_\ell)$ was better described by adding a term $e^{-\varphi_\ell^4}$. Close to T_c, up to 20% of a normalized function of this form had to be used. This deviation from a Gaussian will be the basis for the new analysis described in the next subsection.

For $T < T_c$ where $\langle \varphi \rangle \neq 0$, and the symmetry is broken, an asymmetry in $P(\varphi_\ell)$ was observed [62]. The asymmetry a_s appeared to be a critical quantity diverging for $T \to T_c^-$ as $a_s \propto [(T - T_c)/T_c]^{-1}$. This prompted molecular dynamics simulations and analytic research. The first outcome was [63] that a_s diverges at some temperature $T^* \neq T_c$ because $P(\varphi)$ exhibits a double-peak structure disappearing at $T = T^*$. The ratio T^*/T_c was dependent on the model parameter chosen and, in the displacive limit, even $T^*/T_c \lesssim 1$ could occur, whereas for order-disorder systems always $T^* > T_c$. In a subsequent effort [64] employing R-G techniques $T^* > T_c$, i.e., a double-peaked $P(\varphi)$ for T_c was derived for d = 1 but for d = 3 the answer remained open. However, in the latter paper the question of a possible displacive order-disorder crossover was posed, the notion of which we gave in Sec. IVa), and experimental proof is reviewed in the next subsection.

c) Evidence for Short-Range Order at T_c in $SrTiO_3$

Motivated by the cluster picture, it has been supposed most recently that the time-dependent local scalar coordinate $\varphi(t)$, whose ensemble average is the order parameter for the displacive SPT, can be written in the form [65]

$$\varphi(t) = \sigma(t) + y(t). \tag{11}$$

Here, the coordinate $y(t)$ is taken to be a Gaussian random variable, with correlation time τ_y, and mean-square amplitude $(\langle y^2\rangle)^{1/2}$: this variable describes the quasi-harmonic fluctuations about the instantaneous quasi-equilibrium position set by the value of the coordinate $\sigma(t)$, which reflects the influence of the clusters. y may also include the distribution resulting from impurities and strains which are always present in a real crystal. The simplest variant (sv) of the cluster picture suggests that the coordinate σ be taken to undergo Markovian hopping between *two* values, $\pm\sigma_0$, with a transition probability per unit time $1/\tau_\sigma$. This is clearly an oversimplification: the large thickness of cluster walls in displacive systems, demands a more refined variant (rv) of the cluster picture, allowing for a continuous *distribution*, $P(\sigma)$.

From a comparison of the temperature dependences of the order parameter $\langle\varphi(t)\rangle$ and that of the soft mode $\omega_s(T)$, one can estimate σ_0 as follows: for $T < T_c$, the long-range order $\langle\varphi\rangle_{\ell r}$ in monodomain single crystals as measured by EPR, was found by *Steigmeier* and *Auderset* [66] - and confirmed by *Yacoby* [67] - to be proportional to

$$\omega_s(T) = 0.69 \quad \langle\varphi(T)\rangle_{\ell r}, \tag{12}$$

where ω_s is measured in THz and φ in angular degrees [68]. At T_c the soft mode is not frozen-out due to the existence of clusters which prefigurate the low-temperature phase. $\omega_{s\infty}(T = T_c) = 0.13$ THz [6,36] due to the existence of the short-range order $\langle\varphi(T)\rangle_{sr}$. One estimates from Eq.(12)

$$\langle\varphi(T)\rangle_{sr} = 0.19^\circ. \tag{13}$$

High-resolution structural X-ray studies attempted to measure this quantity [69]. However, X-rays and neutron scattering probe the crystal much faster than the harmonic thermal fluctuations of the lattice. The latter have amplitudes for the rotational order parameter of $(\langle y^2\rangle)^{1/2} = 2.1^\circ$ [69]. Thus, $\langle y^2\rangle / \langle\varphi(T)\rangle^2_{sr} = 122$ and X-ray and neutron-scattering structural investigations which determine the probability distribution function through its Fourier transform (the Debye-Waller factor) cannot, in general, be expected to detect the cluster-induced nonlinear local behavior. The quasi-harmonic fluctuations y mask the nonlinear σ_0 character.

Consider now EPR as a probe to determine $P(\varphi)$. We have seen in the preceding subsection, that the quasi-harmonic fluctuations with spectral density $J(\omega)$ are only contributing to the non-critical background EPR linewidth $\Delta H_b(\varphi) \approx 2$ Gauss, called non-secular fast in Ref. [34]. In other words, for the discrimination against the harmonic fluctuations it is sufficient that $P[\sigma(t)]$ can be probed in the present case.

There are, however, limitations to this undertaking for the following reasons: a) The ratio $r = (\sigma_s^2 \tau_\sigma / \langle y^2 \rangle \tau_y) >> 1$ has to be fulfilled. This is easily the case because the cluster fluctuations τ_σ have been slowed down by impurities to a frequency slower than $\Delta H_b(\varphi)\, g\beta/h \sim 6$ MHz as compared to τ_y which should be, at most, 10^{-11}sec from the soft-phonon width [6]. b) The distribution of $P(y)$ also contains contributions from static strains y_{st} thus $\langle y_{st}^2 \rangle \lesssim \sigma_0^2$. As we shall see, this limit is more stringent than another limitation discussed in Ref. [65]. c) If the sensitivity parameter $B = \partial\omega/\partial\varphi = A\, H_r\, \partial g/\partial\varphi$ gets smaller than $1/\sigma_0\tau_\sigma$, i.e., B (or A) is too small and τ_σ too fast, the double-peaked form of $P(\varphi)$ is erased. Now $B = 1.6 \times 10^8$sec/deg [34] and, expecting $\sigma_0 = 0.2^o$ with $\tau_\sigma > 10^{-7}$sec, $1/(\sigma_0 \cdot 10^{-7}) < 5 \cdot 10^7$sec/deg $< B$ is fulfilled as well.

We now return to discuss the distribution function $P(\sigma)$ itself in EPR, as recently published. A comparison of the *derivative* of this function at T_c with that of a Gaussian of the same peak height, clearly reveals the non-Gaussian nature of $P(\sigma)$ (Fig. 13c). The extent of its anharmonic character is characterized by its representation as a superposition of two symmetrically displaced Gaussians for y (Fig. 13a), whose separation provides one adjustable parameter σ_0. The Ising variable σ_0 is chosen so as to optimize the corresponding representation of the derivative spectrum (Fig. 13b). Note that $\langle y_{st}^2 \rangle \approx \sigma_0^2$ and condition b), the strain-induced broadening, limit the resolution of the experiment.

These results established, unambiguously and for the first time, the existence of local precursor order persisting for times long in compairson with 10^{-7}sec. More explicitly, Fig. 13 suggests that, near but above $T_c = 105.3$ K, the oxygen octahedra in $SrTiO_3$ oscillate typically about quasi-equilibrium positions displaced by $\sigma_0 \approx 0.22^o$ from the high-symmetry position. The striking accord with the value in Eq. (13), derived from $\omega_{s\infty}$ constitutes strong evidence of the overall coherence for the displacive order-disorder crossover as an intrinsic property of displacive phase transitions on approaching T_c.

Starting from the Hamiltonian Eq. (6) *Bunde* and *Lam* [70] calculated rigorous bounds for the ratio $\langle \varphi(T=0) \rangle^2 / \langle \delta\varphi(T=T_c)^2 \rangle \leq 3$ where the symbols $\langle\ \rangle$ represent classical thermal averages. In comparing their result with the measured $\langle\varphi(T = 0)\rangle = 2.1^{o}$ and $\langle\delta\varphi(T = T_c)^2\rangle$ data from EPR, they claimed a violation of their lower bound by a factor of 8. They therefore concluded that the Hamiltonian (6) is not applicable to SPT studies. What these authors did not recognize is the motional

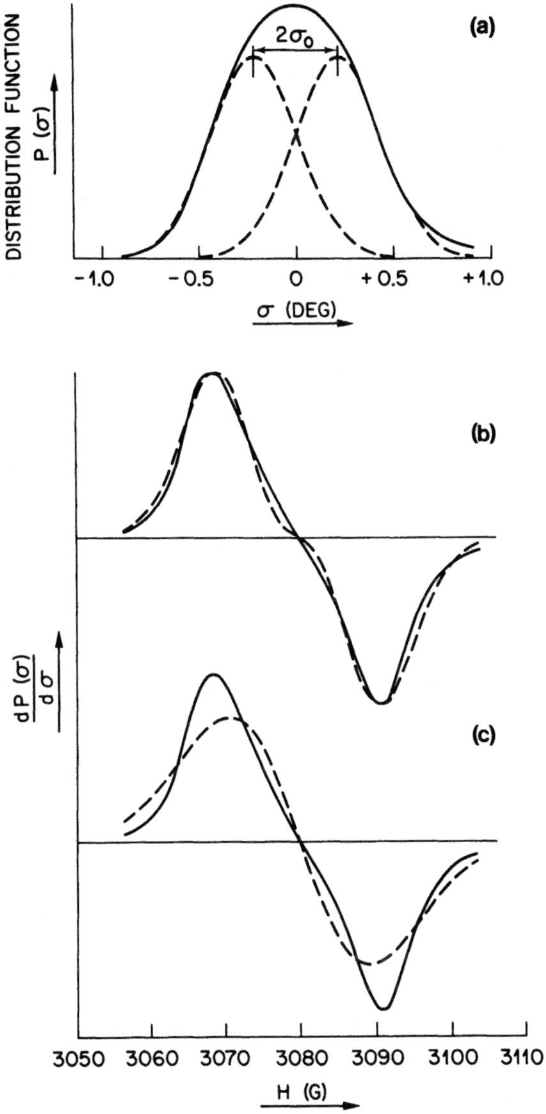

Fig. 13. a) The experimental distribution function $P(\sigma)$ at $T = 105.5$ K (solid line) together with two displaced Gaussians with whose superposition (not shown) the spectrum is modeled. b) The derivative of the experimental $P(\sigma)$ (solid line), and of its double-Gaussian representation. c) The derivative of $P(\sigma)$ and its single-Gaussian representation, Ref. [65]

narrowing of the y(t) components of $\varphi(t)$ occurring in EPR. If the proper X-ray value of $\langle \delta\varphi(T = T_c)^2 \rangle^{1/2} = 2.2^o$ is used, which yields the *total fluctuation* entering the theory, their upper bound applies and their criticism is pointless.

V. Hydrogen Bonded Ferroelectrics

a) The Disappearing Central Peak in Light Scattering

Lagakos and *Cummins* [71] were the first to report a critical central peak in a hydrogen-bonded ferroelectric, the well-known KH_2PO_4. In their Brillouin experiments, a quasi-elastic component was observed to increase dramatically near the 122 K phase transition. The peak was narrower than their experimental resolution width of 10^{-9} s. This was much narrower than predicted by the paraelectric coupling theory of *Cowley* et al. [72]. More recently, *Durvasula* and *Gammon* [73] reported a speckled interference pattern in the scattering column of their Brillouin experiments in the same crystal. This implies that the scattering centers have time-independent amplitudes and positions, i.e., the central peak is of static origin.

Courtens [74] then assumed that the deuterium, present in natural abundance, may be a *Halperin-Varma* center in hydrogen-bonded ferroelectrics; the deuterium impurity relaxing much slower than the intrinsic tunneling hydrogens. He compared the central-peak intensity, observed by *Lagakos* and *Cummins* [71], and *Durvasula* and *Gammon* [73], to a mean-field theory which takes into account the deuterium impurity. However, most recent experiments show that the deuterium does not freeze-out either [75] and is not the cause of the central peak in KH_2PO_4. The quasi-elastic central peak in the paraelectric phase could be suppressed to 2% of its initial value by annealing for 18 h at 140°C. The soft-branch Brillouin spectra could then be fitted with a classical coupled-mode expression without any deviation larger than 1% that would suggest additional dynamic contributions, and with parameters in good agreement with available data.

The very recent findings of *Courtens* in KH_2PO_4 are in line with the previous ones in $KH_3(SeO_3)_2$ of *Yagi* et al. [76]. In this material, a critical central peak was also observed in the Brillouin spectrum of as-grown crystals. However, after repeated thermal cycling through T_c this central peak almost disappeared (see Fig. 14). Thus, the origin is definitely of extrinsic nature. Owing to this disappearance, strains coupling linearly to the order parameter have been invoked. They are released when the sample remains below T_c for sufficiently long. This thermal cycling process is ineffective in KH_2PO_4 to reduce the c.p. [75] and the thermal anneal used for

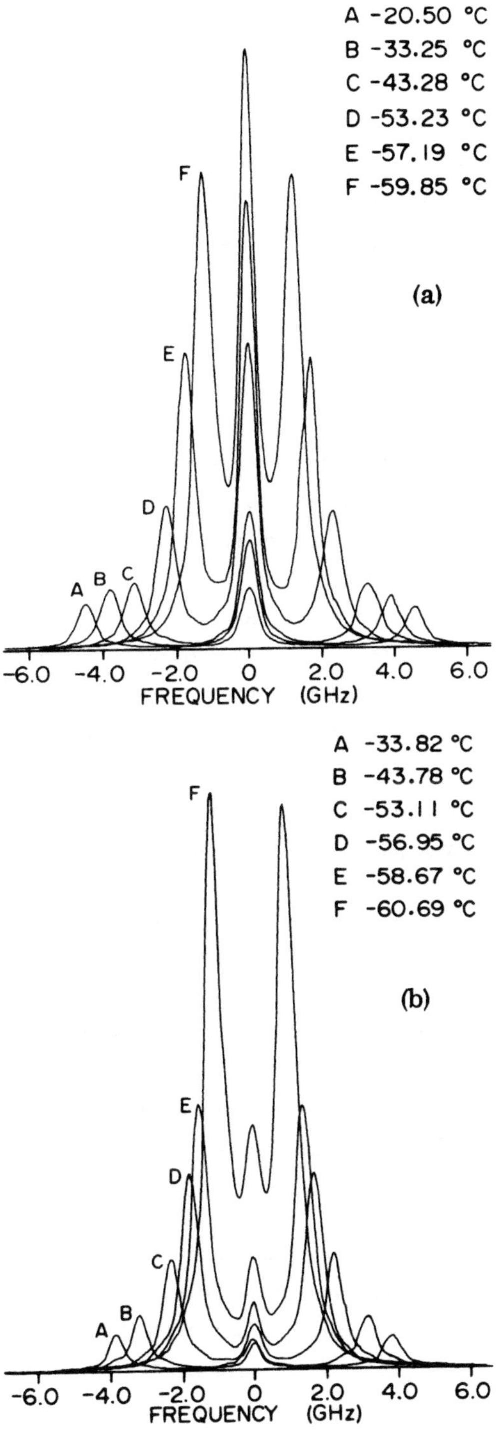

Fig. 14. The central peak of $KH_3(SeO_3)_2$. a) The Brillouin spectra associated with the soft acoustic mode observed with the as-grown sample. b) The spectra observed after keeping the sample below $-30^{\circ}C$ for about 24 h, Ref. [76]

KH_2PO_4 suggests local cancellation of intrinsic defects.

In squaric acid ($C_4O_4H_2$) high-resolution NMR techniques have revealed static low-temperature clusters above the phase transition [77]. Their local reorientation time was found to be longer than 0.1 sec. Those near-static clusters were found to coexist with dynamic clusters reorienting faster than 10^{-5} sec, and it was therefore assumed that impurities act as pinning points. This result thus falls into the same picture as arrived at by light scattering.

b) EPR of Halperin-Varma Centers

A crossing over from high- to low-temperature EPR spectra was first observed for the AsO_4^{-4} radical by *Blinc* et al. in KH_2AsO_4 [78]. Crossover was detected in the Zeeman-splitting $\tilde{g}$ tensor as in the hyperfine interaction $\tilde{A}$. It was ascribed by this group [79] and especially by *Dalal* et al. [80], until recently [81], as resulting from *intrinsic* slow polar fluctuations. However, *Lamotte* et al. [82] had already pointed out several years ago that the high-temperature constants of the axial $\tilde{g}$ and $\tilde{A}$ tensors observed at low temperatures. If they reflected intrinsic fluctuations, they should differ. owing to the existence of the four lateral configurations. *Lamotte* et al. concluded, quite correctly, that the fluctuations observed were not intrinsic in nature and that a perturbed zone around the AsO_4^{-4} existed with different local dynamics in the paraelectric phase and with different local order in $NH_4H_2AsO_4$.

More recently, *Adriaenssens* [83] amplified the view of *Lamotte* et al. on the AsO_4^{-4} center. He based this on applied electric-field experiments. Upon reversal of the field $\vec{E}$, the two local AsO_4^{-4} configurations can only be switched below T_c^F but not above T_c^F. Thus, he concluded that the two polar Slater configurations only reverse through the intermediary of the surrounding domains existing below T_c^F. This means that the AsO_4^{-4} center is coupled mainly to the long-range dipolar field in the lattice and less via the short-range proton coupling.

Adriaenssens described the local polarization dynamics of the AsO_4^{-4} for $T > T_c$ in a passive manner as clusters. We replace this by the self-consistent approach of *Höck* and *Thomas* [25] where the center not only reacts but also acts on its surroundings and can possibly freeze-out locally. In this sense, it is a *Halperin-Varma* center.

The well-behaved motionally-narrowing EPR data on the AsO_4^{-4} radical could be accounted for by a single thermally-activated relaxation time $\tau(T) = \tau_0 \exp E/kT$ [82]. This may be taken as an indication that no local freeze-out occurs, the impurity being coupled sufficiently strongly to other temperature-independent modes. It thus behaves as foreseen by *Halperin* and *Varma* [23]. Therefore, for $T \to T_C^F$ a critical *collective* slowing-down motion should occur due to the soft-mode coupling of *Halperin-Varma* character [23], and an extrinsic collective central peak is present. Its intensity depends on the concentration of the defects, however this does not apply for the *local* relaxation rate of the particular defect, as long as the correlation length ξ is smaller than the distance between defects. The latter property has been verified for the AsO_4^{-4}-polar centers in 15% $NH_4H_2AsO_4$ - 85% $NH_4H_2PO_4$ crystals, for concentrations of $10^{17} - 10^{18} cm^{-3}$ from 100 to 180 K above T_c (i.e., very small ξ)[84]. *The important distinction between the local EPR relaxation rate observed at the defect and collective central-peak width* Γ *was not made in the above work nor in a recent paper by Blinc et al.* [85]. *Thus, the conclusions reached in these papers are not correct.* In the paper by *Blinc* et al., a dependence of the local relaxation rate on T_c in mixed crystals of $NH_4H_2AsO_4$ - $NH_4H_2PO_4$ and KH_2AsO_4 - KH_2PO_4, has been observed. This is expected on the basis of the self-consistent *Höck-Thomas* local-defect theory. A full-sized isotope effect had been previously reported for the CrO_4^{-5} center, to be discussed next.

EPR experiments of paramagnetic Cr^{5+}, substitutional for As^{5+} in KH_2AsO_4 and for P^{5+} in KH_2PO_4, were reported by *Müller* et al. [86]. They showed the existence of slow reorientation times of 10^{-8} s below a dynamic reorientation temperature T^{*AF}= 60 K and 100 K above the Curie temperature T_c^F, shifting proportional to T_c^F upon deuteration. T^{*AF} is defined as the temperature at which the orientation time τ of the center becomes comparable to the time-scale of the EPR experiment $\Delta\nu^{-1}$ = 10^{-8}sec. Figure 15 shows schematically the isotope shift observed on the ferroelectric Curie temperatures T_c^F and T^{*AF}. Upon deuteration, T_c^F and T_c^{*AF} are shifted by similar amounts. From the g values, the ground-state wave-function d^1 of $d_{x^2-y^2}$ character was deduced [87]. Then, *Gaillard* et al. [88] showed conclusively from ENDOR that the $d_{x^2-y^2}$ function couples with two near protons in *lateral* configurations indicating a *Halperin-Varma*-type center. A lateral configuration determines an antiferroelectric character locally. This is expressed by the notation T^{*AF}. In such a lateral configuration, each of the two near protons H^+ can get maximum electrostatic interaction with one negative lobe of the tilted $x^2 - y^2$ wave functions. These x^2 and y^2 lobes are oriented at 90° from each other. Owing to the essential interactions of the $d_{x^2-y^2}$ wavefunction with two lateral protons at a time, a coupling to the rest of the lattice occurs via these protons to the next nearest-neighbor ones by their reorientation along O - - H - O bridges, i.e., clearly by short-range interaction J_{sr}. The latter is also mainly responsible for the shift in the bulk ferroelectric T_c^F and antiferroelectric T_c^{AF} [89]. Therefore,

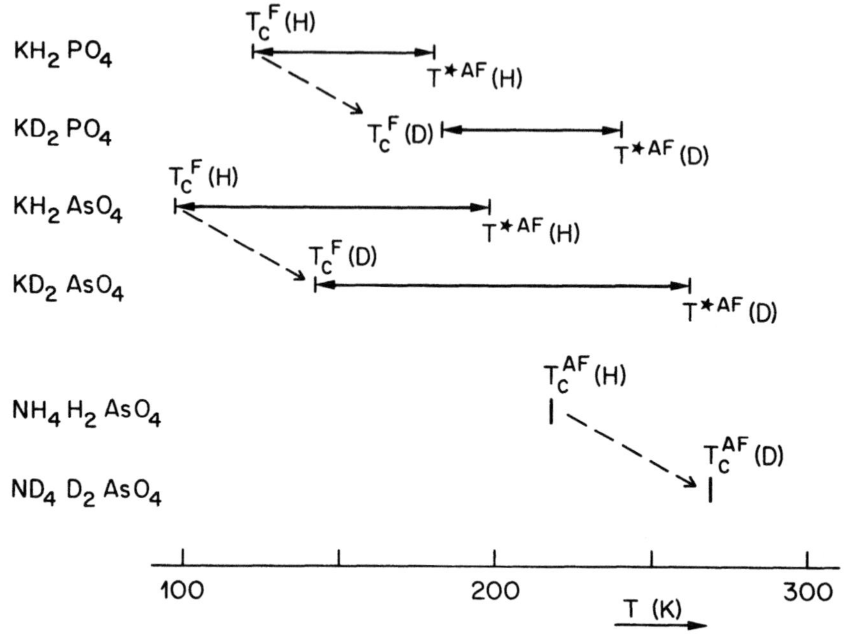

Fig. 15. Comparison of the ferroelectric Curie temperatures T_c^F in KH_2PO_4 and KH_2AsO_4 and their partially-deuterated isomorphs, with the antiferroelectric Cr^{5+} EPR crossover temperatures T^{*AF}. The antiferroelectric Curie temperatures T^{AF} in normal and partially-deuterated $NH_4H_2AsO_4$ are included as well. The distances $T^{*AF} - T_c^F > 0$ are marked with arrows. After *Müller* and *Berlinger*, Ref. [89]

the remarkable shift upon deuteration in local-dynamic slowing-down of the $(CrO_4)H_2$ center in approximate proportion to the ferro- and antiferro-electric bulk T_c^F's results from the isotope effect of the short-range interaction J_{sr} for bulk and impurity.

The observed activation energies of the local lateral reorientation times are of the order of 0.2 eV in pure and mixed crystals of KD_2AsO_4 [89] and KD_2PO_4 [86]. This is similar to what was found for the AsO_4^{-4} center, and is of the order of magnitude required by the *Halperin* and *Varma* theory and by T_c^F's of 200 K. The AsO_4^{-4} couples mainly to $\vec{q} = 0$ ferroelectric excitations and the CrO_4^{-3} to $\vec{q} = \pi/a$ antiferroelectric ones. The AsO_4^{-4} radical interacts via long-range dipolar and unipolar fields [83] due to its extra negative charge. For the CrO_4^{-3} it is the $d_{x^2-y^2}$ wave function coupled with nearest protons and the short-range interaction with those in neighboring cells.

It should be noted that in contrast to NMR, the EPR *always* sees the conformation of the next-neighbor ions to 80%. This is also true for Fe^{3+}, Cr^{3+} and Cu^{2+} [90,91]. The absence of any evidence for slow motion in their spectra in hydrogen-bonded crystals simply means that they are not Halperin-Varma centers, and no intrinsic slow motions of time constants of 10^{-8} sec, 100 K above T_c, are present [85].

c) Nuclear Magnetic Resonance (NMR) in KH_2AsO_4

A decade ago *Blinc* and *Bjorkstam* [92] reported in the paraelectric phase of KH_2AsO_4 and RbH_2AsO_4, an ^{75}As NMR spectrum which has the symmetry properties of the ferroelectric phase but differs from that in the latter phase. We call this spectrum hereafter B, due to its discoverers, and the one in the ferroelectric phase spectrum F. As there are two polarities, we have actually to distinguish between $B^{\pm}$ and $F^{\pm}$. The spectrum observed in the paraelectric phase, well above T_c, is termed P and is axial along the c-axis. It reflects the intrinsic paraelectric phase and results from the appropriate time average over all possible rapidly-fluctuating polar and nonpolar protonic Slater configurations of the AsO_4 tetrahedron and its four neighboring protons.

A first interpretation of spectrum B was attempted by *Bjorkstam* [93] in terms of the Zeeman energy $\gamma\omega$ perturbed by stochastic quadrupolar interaction due to ferroelectric polar configurations. This theory did not yield quantitative agreement with the observed anisotropy. In contrast, more recently, *Adriaenssens* [94] could calculate the anisotropy quantitatively by using a model in which he assumed that the ^{75}As nuclei see (during the lifetime of their spin state of 10^{-3} sec) with a 90% probability, the six-possibly fast-fluctuating protonic Slater configurations of the $(AsO_4)H_4$ group, yielding for p = 100% spectrum P, and for 1 - p = 10% of their lifetime one *specific up-* (or down-) polarized configuration $F^{\pm}$. The averaging is such that a particular nucleus does not see for a polarized configuration $F^{\pm}$, *longer than* 10^{-5} *sec at a time.*

From this quantitative agreement *Adriaenssens* assigned spectrum B as resulting from intrinsic fully-polarized clusters moving very rapidly in a "sea of paraelectric matter". The spin lifetimes involved in average spectra P and $F^{\pm}$, as compared to spectrum $B^{\pm}$, are such that the lifetime for the ferroelectric clusters $F^{\pm}$ is 10^{-2} to 10^{-3} sec. Such lifetimes have also been found from dielectric measurements in the same crystals. Thus, his interpretation has been taken over to a certain extent by *Blinc* [95], as proof for the existence of *intrinsic* fully-polarized ferroelectric clusters in KH_2AsO_4. More recently, *Blinc* and his collaborators [95] found a dip in the spectral density $J(\omega)$ for $\omega < 10^3$ proving the existence of slow ferroelectric fluctuations in agreement with the observation of spectrum B. However, in deuterated KD_2AsO_4 no dip in $J(\omega)$ nor spectrum B was detected.

If the spectrum B is indeed intrinsic and results from the anharmonicity of the lattice, it is not easy to see why it is not observed in a deuterated crystal. Of greater importance is that such slow intrinsic dynamics have not been reported by either NMR or dielectric studies in other well-annealed and clean hydrogen-bonded ferroelectrics.

In proposing the rapidly-diffusing intrinsic ferroelectric clusters above T_c *Adriaenssens* invoked the recent molecular-dynamics studies and the analytical investigations of *Krumhansl* and *Schrieffer* [18]. However, the picture which emerged from these studies is that of up- and down-polarized clusters separated by soliton-like walls. This is at variance with fully-polarized ferroelectric clusters moving rapidly in a sea of paraelectric matter. Moreover, the speed of the clusters deduced is orders of magnitude faster than what intrinsic three-dimensional molecular-dynamics studies yield. Furthermore, no quantitative measurement of the ratio of spectrum B to P as a function of temperature has been published for $T > T_c$. For intrinsic clusters one also expects, according to the theory of *Binder* [96], spectrum F to be present for $T > T_c$. Crossing over to spectrum F is expected at $T^* > T_c$ for $\Delta\nu T \simeq 1$, where $(T^* - T_c)^\rho \propto \Delta\nu$ and $\rho = 1$ for a uniaxial ferroelectric. $\Delta\nu$ in the present case is the quadrupole splitting and depends on the direction of the external magnetic field to the crystal axis. For instance, in the case of the *Halperin-Varma* CrO_4^{-3} center, $\rho \simeq 2.2$ has been found thus definitely proving its extrinsic character [89].

These considerations indicate that the observed ferroelectric clusters of 10^{-2} sec lifetime, are specific to the KH_2AsO_4 and RbH_2AsO_4 crystals. Furthermore, upon application of a polarizing electric field E^+ in the paraelectric-phase, spectrum B^+ was enhanced over spectrum B^-, *but upon switching E^+ off, the disproportionality of B^+ versus B^- only disappeared after about a month!* [97]. This observation is reminiscent of the pre-anneal area in KH_2PO_4 light-scattering studies.

d) A Speculation

The experimental facts on light-scattering [Sec. V a)], magnetic resonance [Sec. V b) and c)] as well as dielectric dispersion measurements - reviewed by *Peterson* [98] - all indicate the extrinsic character of the central-peak phenomenon in the restricted sense used throughout the present review. One is then led to the question of why no intrinsic c.p. intensity in hydrogen-bonded uniaxial $d = 3$ ferroelectrics is observed above T_c, whereas in $SrTiO_3$ and related compounds, one finds the finite soft phonon at T_c, $\omega_{s\infty}^2(T = T_c) = \delta^2(T_c)$.

We speculate - amplifying a suggestion by *Courtens* [75]- that this may be due to the critical dimensionality d_c of the systems. In $SrTiO_3$ and related compounds $d_c = 4$ [99]. These antiferrodistortive systems are clearly short range in character as opposed to the dipolar ferroelectric systems. Thus, for hydrogen-bonded uniaxial $n = 1$, KH_2PO_4 ferroelectrics, $d_c = 3$ if one neglects the piezoelectric coupling to acoustic modes. However, this coupling is so strong that one cannot safely do so; including it d_c is pushed down to 2.5 [100]. Therefore, in real crystals with $d = 3$, for the $SrTiO_3$-category, the inequality $d < d_c$ holds and for the KH_2PO_4-category $d > d_c$.

From R-G theory we know that for the case $d < d_c$, *correlated* fluctuations near T_c are relevant and determine the behavior of the system [99]. In the case of antiferrodistortive transitions these fluctuations result in short-range correlated regions and prevent the freezing-out of the soft phonons. Thus, they cause an intrinsic central-peak phenomenon [52]. For KH_2PO_4 on the other hand, with $d > d_c$, the Gaussian fixed point is stable. Thus, the fluctuations of the order parameter are of Gaussian *uncorrelated* character [101]. This is what mean-field theory always assumes. The soft phonon or the collective relaxation of the system should be described by the latter, as observed. If the system's dimensionality is marginal, i.e., $d = d_c$ as for TGS, logarithmic corrections to the Gaussian fixed-point behavior are predicted [102] and observed [103]; thus we also expect an intrinsic c.p. to exist. In computer simulations for a short-range system with $d_c = 4$, this was found [20]. As was emphasized in Sec. IV a), the lattice dimensionality parameter in RG $\varepsilon = d_c - d$, tells one whether or not sharply-defined clusters occur on a local scale. For ε positive and of order one near T_c, the cluster picture is dominant whereas for $\varepsilon < 0$ it is not relevant. It may however be that the above considerations are valid rather in an asymptotic sense, i.e., the c.p. phenomenon is observed for $\varepsilon < 0$ but disappears only for $-\varepsilon \gtrsim 1$, according to Schneider [104].

At incommensurate SPT within Landau theory, c.p's are expected [52], and most recent experiments indicate their presence [105]. They result from the phason mode becoming soft. Its origin being phase slips (or phase solitons) between commensurate domains of the lattice.

Acknowledgement

The author would like to thank C.P. Enz for the invitation to give this review and his help in editing it, as well as A. Bruce, E. Courtens, T. Schneider and W. Berlinger for reading and suggesting improvements on the manuscript; and to P. Jones for substantial help in typing the text.

References

1. L. Landau, E. Lifshitz: Physique Statistique (MIR editions, Moscow 1967)
2. W. Cochran: Advances in Physics 9, 387 (1960)
3. E.J. Samuelsen, E. Andersen, J. Feder (eds.): *Structural Phase Transitions and Soft Modes*, (Oslo, 1971); J.F. Scott: Rev. Mod. Phys. 46, 83 (1974)
4. K.A. Müller, W. Berlinger: Phys. Rev. Lett. 26, 13 (1971)
5. M.E. Fisher: Rev. Mod. Phys. 46, 597 (1974)
6. T. Riste, E.J. Samuelsen, K. Otnes, J. Feder: Solid State Commun. 9, 1455 (1971)
7. S.M. Shapiro, J.D. Axe, G. Shirane, T. Riste: Phys. Rev. B 6, 4332 (1972)
8. J.K. Kjems, G. Shirane, K.A. Müller, H.J. Scheel: Phys. Rev. B 8, 1119 (1973)
9. P. Hohenberg: (1971) unpublished
10. R.A. Cowley: Ferroelectrics 6, 163 (1974)
11. Th. von Waldkirch, K.A. Müller, W. Berlinger, H. Thomas: Phys. Rev. Lett. 28, 503 (1972)
12. J. Feder: In *Local Properties at Structural Phase Transitions*, ed. by K.A. Müller and A. Rigamonti (North-Holland, Amsterdam 1976) p.312; a more recent paper by R. Bausch and B.I. Halperin: Phys. Rev. B 18, 190 (1978), vindicates earlier nonlinear coupling theories of the thermal diffusion mode
13. J.D. Axe, S.M. Shapiro, G. Shirane, T. Riste: In *Anharmonic Lattices, Structural Transitions and Melting*, ed. by T. Riste (Noordhoff-Leiden, 1974) p.23
14. G. Shirane, J.D. Axe: Phys. Rev. Lett. 27, 1803 (1971)
15. T. Schneider, E. Stoll: Phys. Rev. Lett. 31, 1254 (1973); Phys. Rev. B 13, 1216 (1976)
16. S. Aubry: J. Chem. Phys. 62, 3217 (1975); 64, 3392 (1976)
17. T.R. Koehler, A.R. Bishop, J.A. Krumhansl, J.R. Schrieffer: Solid State Commun. 17, 1515 (1975)
18. J.A. Krumhansl, J.R. Schrieffer: Phys. Rev. B 11, 3535 (1975)
19. A.R. Bishop, T. Schneider (eds.): *Solitons and Condensed Matter Physics*, Springer Series in Solid State Sciences (Springer, Berlin, Heidelberg, New York 1978)
20. T. Schneider, E. Stoll: Phys. Rev. Lett. 36, 1501 (1976); Phys. Rev. B 17, 1302 (1978); Phys. Rev. Lett. 41, 964 (1978)
21. Th. von Waldkirch, K.A. Müller, W. Berlinger: Phys. Rev. B, 7, 1052 (1973)
22. R. Folk, F. Schwabl: Solid State Commun. 15, 937 (1974)
23. B.I. Halperin, C.M. Varma: Phys. Rev. B 14, 4030 (1976)
24. H. Thomas in Ref. 3, p.15
25. K.H. Höck, H. Thomas: Z. Physik B 27, 267 (1977)
26. H. Schmidt, F. Schwabl: Phys. Lett. A 61, 476 (1977)
27. T. Schneider, E. Stoll: Phys. Rev. B 16, 2220 (1977)
28. K.A. Müller: IBM Internal Report RZ982 (1978)
29. J. Töpler, B. Alefeld, A. Heidemann: J. Phys. C 10, 635 (1977)
30. C.N.W. Darlington, W.J. Fitzgerald, D.A. O'Connor: Phys. Lett. 54A, 35 (1975); C.N.W. Darlington, D.A. O'Connor: J. Phys. C 9, 3561 (1976) and *Lattice Dynamics*, Proc. Int. Conf., ed. by H. Balkanski (Flammarion, Paris 1977) p.750

31. K.B. Lyons, P.A. Fleury: Solid State Commun. 23, 477 (1977). The 15 GHz central peak was confirmed by C.B. Su, with Brillouin scattering (Thesis, Brandeis University 1979) unpublished
32. E. Courtens: Phys. Rev. Lett. 37, 1584 (1976)
33. K.A. Müller, W. Berlinger, C.H. West, P. Heller: Phys. Rev. Lett. 32, 160 (1974)
34. G.F. Reiter, W. Berlinger, K.A. Müller, P. Heller: (to be submitted for publication)
35. K. Aso: J. Appl. Phys. Jpn., 15, 1243 (1976)
36. R. Currat, K.A. Müller, W. Berlinger, F. Denoyer: Phys. Rev. B 17, 2937 (1978)
37. J.B. Hastings, S.M. Shapiro, B.C. Frazer: Phys. Rev. Lett. 40, 237 (1978)
38. O.F. Schirmer, K.A. Müller: Phys. Rev. B 7, 2986 (1973)
39. I. Hatta, M. Matsubo, S. Sawada: J. Phys. C 7, L 299 (1974)
40. K. Hanisch, M. Drosg: Phys. Lett. 58A, 415 (1976)
41. K. Gesi, J.D. Axe, G. Shirane, A. Linz: Phys. Rev. B 5, 1933 (1972)
42. F. Denoyer, M. Lambert, A. Comes, R. Currat: Solid State Commun. 18, 441 (1976); F. Denoyer, R. Currat: *Neutron Inelastic Scattering*, Proc. Symp. IAEA (Vienna 1977) p.273
43. A. Avogadro, G. Bonera, F. Borsa, A. Rigamonti: Phys. Rev. B 9, 3905 (1974)
44. R. Almairac, M. Rousseau, J.Y. Gesland, J. Nouet, B. Hennion: J. de Physique 38, 1429 (1977)
45. R.A. Cowley, J.D. Axe, M. Iizumi: Phys. Rev. Lett. 36, 806 (1976)
46. P.A. Fleury, K.B. Lyons: Phys. Rev. Lett. 37, 1088 (1976)
47. D.J. Lockwood, J.W. Arthur, W. Taylor, T.J. Hosea: Solid State Commun. 20, 703 (1976)
48. Y. Yacoby: *Lattice Dynamics*, Proc. Int. Conf., ed by M. Balkanski (Flammarion, Paris 1977) p.453; Y. Yacoby, S. Just: Solid State Commun. 15, 715 (1974)
49. Y. Yacoby, Z. Physik B 31, 275 (1978)
50. U.T. Höchli, H.E. Weibel, L.A. Boatner: Phys. Rev. Lett. 41, 1440 (1978)
51. E.F. Steigmeier, G. Harbeke, R.K. Wehner: *Light Scattering in Solids*, ed. by M. Balkanski (Flammarion, Paris 1971) p.396
52. A.D. Bruce: In *Solitons and Condensed-Matter Physics*, ed. by A.R. Bishop, T. Schneider (Springer, Berlin, Heidelberg, New York 1978) p.116
53. E. Stoll: Phys. Lett. 58A, 121 (1976)
54. Th. von. Waldkirch, K.A. Müller, W. Berlinger: Phys. Rev. B 7, 1052 (1973)
55. E.S. Kirkpatrick, K.A. Müller, R.S. Rubins: Phys. Rev., A86, 135 (1964)
56. A. Abragam: *The Principles of Nuclear Magnetism* (Clarendon, Oxford 1961) Chap.10
57. Th. von Waldkirch, K.A. Müller, W. Berlinger: Phys. Rev. B 5, 4324 (1972)
58. F. Schwabl: Phys. Rev. Lett. 28, 500 (1972); Z. Phys. 254, 57 (1972)
59. K.A. Müller, W. Berlinger, M. Capizzi, H. Gränicher: Solid State Commun. 8, 549 (1970)
60. K.A. Müller: In *Anharmonic Lattices, Structural Transitions and Melting*, ed. by T. Riste (Noordhoff-Leiden, 1974)
61. E. Courtens: Phys. Rev. Lett. 29, 1380 (1972)
62. K.A. Müller, W. Berlinger: Phys. Rev. Lett. 29, 715 (1972)
63. T. Schneider, E. Stoll: Phys. Rev. B 10, 2004 (1974)
64. A.D. Bruce, T. Schneider: Phys. Rev. B 16, 3991 (1977)
65. A.D. Bruce, K.A. Müller, W. Berlinger: Phys. Rev. Lett. 42, 185 (1979)
66. E.F. Steigmeier, H. Auderset: Solid State Commun. 12, 565 (1973)
67. Y. Yacoby, S. Yust, W.W. Kruhler: Ferroelectrics 12, 117 (1976)
68. A theoretical rationale for Eq.(12) to hold has been given in the paper by F. Szépfalusy at this conference
69. G.M. Meyer, R.J. Nelmes, I. Hutton: *Lattice Dynamics*, Proc. Int. Conf., ed. by H. Balkanski (Flammarion, Paris 1977) p.652
70. A. Bunde, L. Lam: Z. Phys. B 28, 225 (1977)
71. N. Lagakos, H.Z. Cummins: Phys. Rev. B 10, 1063 (1974)
72. R.A. Cowley, G.J. Coombs, R.S. Katiyar, J.F. Ryan, J.F. Scott: J. Phys. C 4, L 203 (1971)
73. L.N. Durvasula, R.W. Gammon: Phys. Rev. Lett. 38, 1081 (1977)

74. E. Courtens: Phys. Rev. Lett. 39, 561 (1977)
75. E. Courtens: Phys. Rev. Lett. 41, 1171 (1978)
76. T. Yagi, H. Tanaka, I. Tatsuzaki: Phys. Rev. Lett. 38, 609 (1977)
77. M. Mehring, D. Suwelack: Phys. Rev. Lett. 42, 317 (1979)
78. R. Blinc, P. Cevc, M. Schara: Phys. Rev. 159, 411 (1967)
79. R. Blinc, B. Zeks: In *Soft Modes in Ferroelectrics and Antiferroelectrics*, ed. by E.P. Wohlfarth (North-Holland, Amsterdam 1974)
80. N.S. Dalal, C.A. McDowell, R. Srinivasan: Mol. Phys. 24, 1052 (1972); J. Chem. Phys. 60, 3787 (1974)
81. N.S. Dalal, J.A. Hebden, D.E. Kennedy, C.A. McDowell: J. Chem. Phys. 66, 4425 (1977)
82. B. Lamotte, J. Gaillard, O. Constantinescu: J. Chem. Phys. 57, 3319 (1972)
83. G.J. Adriaenssens: J. Mag. Res. 25, 511 (1977)
84. L.V. Gonzaga, A.S. Chaves, R. Blinc, G.M. Ribeiro, R. Gazzinelli, G. Rius: Solid State Commun. 27, 495 (1978)
85. R. Blinc, P. Cevc, G. Cevc: J. Chem. Phys. 70, 153 (1979)
86. K.A. Müller, N.S. Dalal, W. Berlinger: Phys. Rev. Lett. 36, 1504 (1976)
87. K.A. Müller, W. Berlinger: Phys. Rev. Lett. 37, 916 (1976)
88. J. Gaillard, P. Gloux, K.A. Müller: Phys. Rev. Lett. 38, 1216 (1977)
89. K.A. Müller, W. Berlinger: Z. Phys. B 31, 151 (1978)
90. D.J. Newman, W. Urban,: Adv. Phys. 24, 793 (1975)
91. E. Siegel, K.A. Müller: Bull. Am. Phys. Soc. 24, 509 (1979)
92. R. Blinc, J.L. Bjorkstam: Phys. Rev. Lett. 23, 788 (1969)
93. J.L. Bjorkstam: Adv. Mag. Res. 7, 1 (1974)
94. G.J. Adriaenssens: Phys. Rev. B 12, 5116 (1975)
95. R. Blinc: Ferroelectrics 20, 121 (1978); R. Blinc, J. Slak, F.C. Sa Barreto, A.S.T. Pires: Phys. Rev. Lett. 42, 1000 (1979)
96. R. Binder: Solid State Commun. 24, 4011 (1977)
97. G.J. Adriaenssens, J.L. Bjorkstam: Phys. Status Solidi A 18, 129 (1973), and private communication
98. J. Peterson: to be published
99. M.E. Fisher: Rev. Mod. Phys. 46, 597 (1975)
100. R. Folk, H. Iro, F. Schwabl: Z. Phys. B 25, 69 (1976)
101. S.K. Ma: *Modern Theory of Critical Phenomena*, ed. by W. A Benjamin Inc. (London, Amsterdam)
102. A.I. Larkin, D.E. Khmelnitskii: JETP 56, 2087 (1969)
103. B.A. Strukov: Ferrolectrics 12, 97 (1976)
104. T. Schneider (private communication)
105. D.W. Bechtle, J.F. Scott, D.J. Lockwood: Phys. Rev. B 18, 6213 (1978)

Additional Literature

- R.L. Armstrong, C.A. Martin: Phys. Rev. Lett. 35, 294 (1975); R.L. Armstrong, D. Mintz, M. D'Iorio: Phys. Rev. B 15, 2840 (1977)
  New Evidence for the Formation of Dynamic Clusters at Temperatures in the Vicinity of a Structural Phase Transition
- C.P. Enz: Helv. Phys. Acta 47, 749 (1974)
  Theory of Coupled Hydrodynamic Modes Applied Above Structural Phase Transitions
- I. Ohnari, S. Takada: Progr. Theor. Phys. Jpn. 61, 11 (1979)
  Central Peak and Thermal Diffusion Mode near a Structural Phase Transition
- G.F. Reiter, N. Tzoar: Phys. Rev. B 14, 208 (1976)
  Critical Fluctuations in the Weakly Anharmonic Crystal Model
- E. Šimánek, R. Ward: Solid State Commun. 18, 841 (1976)
  On the Hydrodynamic Theory of a Central Peak in Displacive Structural Phase Transitions

SYSTEMS WITH QUENCHED RANDOM IMPURITIES,

AN OVERVIEW OF DYNAMICS, REPLICAS AND FRUSTRATION APPROACHES

C. De Dominicis

Service de Physique Théorique, C.E.A. Saclay
BP no 2, 91190 Gif-sur-Yvette, France

SYSTEMS WITH QUENCHED RANDOM IMPURITIES, AN OVERVIEW OF DYNAMICS, REPLICAS AND FRUSTRATION APPROACHES.

C. DE DOMINICIS

Service de Physique Théorique, C.E.A. Saclay,
BP n°2, 91190 Gif-sur-Yvette, France

ABSTRACT

Systems with quenched random impurities are briefly discussed. Landau-Ginzburg systems with (*i*) random magnetic field, (*ii*) random temperature, (*iii*) random temperature and no coupling (localization model). Spin systems with random bonds (spin glass model). Dynamics is used to derive an equation of state and time evolution of a hypothetical frozen-in phase. Results are compared with those obtained with replicas. Mean field and fluctuations results for the Edwards-Anderson model are reviewed, pointing to an unsatisfactory situation. The low temperature description that uses the frustration concept of Toulouse-Anderson for spin glasses is also described and discussed.

1. INTRODUCTION

Critical properties of pure systems, both static and dynamic, are by now fairly well understood. Likewise for systems with impurities that are mobile on the scale of usual observation times. These "annealed" impurities act as ordinary degrees of freedom and their effect on critical properties has been elucidated by M. Fisher[1]. For example, critical exponents are unchanged if the specific heat exponent (of the host system) α is negative. They undergo a simple renormalization if α is positive.

The situation is much more complex for frozen or quenched impurities. Despite abundant work in the last years, we do not yet understand, in particular for spin glasses, what is really going on. For the interested reader, this is a lucky period that has seen, in the last few months, the flourishing of a beautiful set of notes by Anderson[2], Joffrin[3], Kirkpatrick[4], Lubensky[5], Thouless[6] (1978 Les Houches Session), and reviews by Blandin[7] and by Binder and Stauffer[8].

The present contribution begins with a brief overview of quenched systems (§ 2) with random magnetic field, random temperature or potential (and its limiting case of pure random potentials as occurs in the localization problem), and finally random bonds as an idealization of spin glasses. In dealing with quenched random systems, two techniques are available : replicas and dynamics (§ 3) . We present how the dynamic approach allows to write an equation of state and its time evolution for hypothetical frozen-in phase in Landau-Ginzburg systems (random field or temperature) (§ 4) . Its relation to results obtained via replicas is displayed. The random bond (Edwards – Anderson) model is then treated with replica techniques and results are reviewed for mean field, fluctuations and renormalization group thereof (§ 5) . In view of the confused situation existing for the presumed ordered phase, the low temperature approach based on frustration is reviewed (§ 6) with some emphasis on the role played by defects (§ 7). We end up by pointing to a few well defined puzzles that would deserve further work (§ 8) .

2. QUENCHED RANDOM SYSTEMS : AN OVERVIEW

We start from the pure system as given by

$$\mathcal{H} = \sum_{j\ell} J_{j\ell}\, \sigma_j \cdot \sigma_\ell - h_j \cdot \sigma_j \tag{2.1}$$

the nearest neighbor classical spin interaction, or from the Landau-Ginzburg-Wilson Hamiltonian that is known to be equivalent for most critical properties of ordinary phase transitions

$$\mathcal{H} = \int d^d x \left\{ \frac{1}{2}\, \varphi(-\nabla^2 + r_o)\varphi + \frac{u_o}{4!}(\varphi.\varphi)^2 - h.\varphi \right\} \quad . \tag{2.2}$$

Adding random impurities into (2.1,2) will manifest itself[9] by the fact that the parameters a ($a = h_j$, $J_{j\ell}$ or h, r_o, u_o,...) acquire a space (or site) dependent random part $a \to a + a_R(x)$ that is we have,

$$\mathcal{H} \to \mathcal{H} + \mathcal{H}_R \quad .$$

These new random degrees of freedom are then characterized by a probability law $\mathcal{P}(a_R(x))$. Experiments are done on one given system , that is with one given set of values $\{a_R\}$, values determined by the way in which we prepare the system e.g. fast cooling. However an argument due to Brout[10] shows that the thermodynamic average of an *observable* A for a large system, $\langle A\rangle_{a_R}$ (i.e. for a given set of $\{a_R\}$) is identical with its probability-weighted average*

$$\langle A\rangle_{a_R} \simeq \int \mathcal{D}a_R \, \mathcal{P}(a_R) \, \langle A\rangle_{a_R} \equiv \overline{\langle A\rangle} \quad . \tag{2.3}$$

This is fortunate because $\overline{\langle A\rangle}$ is usually easier to deal with than $\langle A\rangle_{a_R}$.

i) <u>*Random magnetic field*</u> : This is the simplest type of randomness. We have

$$\mathcal{H}_R = \sum_j h_{j_R} \cdot \sigma_j \tag{2.4}$$

or with continuous fields

$$\mathcal{H}_R = \int d^dx \; h_R(x) \cdot \varphi(x) \quad , \tag{2.5}$$

and we may take a probability law which, in its simplest form, is given by a Gaussian distribution

$$\langle h_R\rangle = 0 \tag{2.6}$$

$$\langle h_R^a(x) \; h_R^b(x')\rangle = \Delta_h \, \delta_{ab} \, \delta(x-x') \quad . \tag{2.7}$$

a) Lower critical dimension : If the field is strong, clearly the spin field will align along the magnetic field destroying the order (e.g. ferromagnetic) that exists below T_c. The order is also destroyed for infinitesimal h_R if the systems reaches a lower free energy by breaking up into small domains. A heuristic argument, due to Harris[11] considers two neighboring cells (of linear dimension L) where $\langle h_R\rangle_{cell}$ is up and down respectively and balances the energy gained by aligning the spins along $\langle h_R\rangle_{cell}$ inside each cell, i.e.

$$(\Delta_h \; L^d)^{1/2}$$

against the energy cost of creating a wall between the cells. Namely within a factor proportional to the spin coupling, L^{d-1} for an Ising spin, L^{d-2} for a spin with $n > 1$. The order is thus unstable to the presence of a small random magnetic field at

---

* Strictly speaking the argument fails as the correlation length becomes infinite. It is generally assumed to work in practice. See section 1.1 in T. Lubensky, Ref.[5].

dimension $d < 4$ (for $n \geq 2$) and $d < 2$ (for $n = 1$) this last boundary being modified by a more elaborate examination.

b) *Time long-range order* : Whatever the dimension, at low T, the spins are then, either all aligned (ferromagnetism) or following the random magnetic field. Being quenched, the field has no time dependence and as a result one always has *time long-range order*, i.e. in the Landau Ginzburg case

$$\langle \varphi(xt)\ \varphi(x't')\rangle_{t-t'\to\infty} \neq 0 \tag{2.8}$$

ii) Random bonds : This is, as is well known, the most interesting case. Indeed magnetic impurities (typically Fe, Mn) in normal metals (Cu,Au) interact via conduction electrons, resulting, for large separations $|r_1-r_2|$, in the Ruderman-Kittel-Kasuya-Yosida interaction

$$\frac{J\ \cos 2k_F|r_1-r_2|}{(k_F|r_1-r_2|)}\ \vec{\sigma}_1.\vec{\sigma}_2 \tag{2.9}$$

k_F is the Fermi vector of the conduction electrons. At concentrations large enough, impurities are close by, the interactions remain ferromagnetic. At low concentration (a few percent) impurities are on the average at distances where the sign of (2.9) fluctuates. This is a system where interactions are randomly ferro-and antiferromagnetic, the celebrated spin-glass. This situation is idealized by Edwards and Anderson[12] as a system with spins on a lattice with a nearest neighbor random coupling $J^R_{j\ell}$. The probability law $\mathcal{P}(J^R)$ that governs the coupling is taken either to be Gaussian as above or in the so-called *ideal spin glass* as $J^R = \pm 1$ with equal probability. The emphasis is on *competition*. As the temperature is lowered, spins receive contradictory orders to align along this or that direction, the result is a freezing-in at a critical temperature T_g . The freezing-in at T_g shows no space long-range order due to the randomness in the distribution of ferro-and antiferromagnetic bonds. Edwards and Anderson characterize the new phase below T_g by time long-range order, that is by the "order parameter"

$$\langle \sigma_j(t)\ \sigma_j(t')\rangle_{t-t'\to\infty} = Q_j \quad . \tag{2.10}$$

New features that appear both in experiments[13,3,4] and Monte Carlo computations[8] include

- a sharp cusp in the zero field magnetic susceptibility at T_g , that is rounded off by magnetic field
- no cusp or singularity in the specific heat
- below T_g hysteresis with non exponential relaxation

iii) Random temperature (or potential) : Transformation of the spin-variable Hamiltonian (2.1) into a continuous field-variable Hamiltonian like (2.2) introduces a nonlinear dependence into $J_{j\ell}$. One may think that we may obtain an equivalent physical description by taking all the couplings $(r_o, u_o, \ldots)$ as random variables * i.e.

$$\mathcal{H}_R = \int d^d x \left\{ \frac{1}{2} \left[k_R(x) \; (\nabla\varphi)(\nabla\varphi) + V_R(x) \; \varphi.\varphi \right] + \frac{u_R(x)}{4!} (\varphi.\varphi)^2 + \ldots \right\} \qquad (2.11)$$

It is immediately recognized that if one wants to investigate the effect on the ferromagnetic critical region only the random temperature (or potential) term $V_R(x)\ \varphi.\varphi$ is relevant. Let us again take $\mathcal{P}(V_R)$ as Gaussian

$$\langle V_R \rangle = 0 \qquad (2.12)$$

$$\langle V_R(x) \; V_R(x') \rangle = \Delta_v \; \delta(x-x') \qquad (2.13)$$

In random pockets of space the local temperature falls below T_c and one has local formation of *spontaneous* random microferromagnets, that is if the temperature is low enough, and the density of impurities ($\sim\Delta v$) is not too small. These micromagnets act in turn as random sources for the rest of the system, inducing thus random but frozen alignement as when one has a quenched random magnetic field. If this is true, for concentrated enough impurities, below a freezing temperature T_g, one may expect time long-range order

$$\langle \varphi(xt) \; \varphi(x't') \rangle_{t-t' \to \infty} = Q(x-x') \qquad (2.14)$$

and at the same time, no space long-range order

$$\langle \varphi(xt) \rangle = 0 \quad . \qquad (2.15)$$

A slightly more elaborate form of the above Harris argument** shows however that when $\alpha < 0$ the random temperature impurities are not effective (weakly random systems). When $\alpha > 0$, for small enough values of the concentration a new fixed point arises that governs critical properties, giving rise to new exponents (Fig.1).

If the concentration is increased the flow lines lead the system in a *run-away* region where hopefully it is then characterized by time long-range order (2.14). We shall see however that, in the ordered phase, the system is plagued by instabilities.

---

*With appropriate stabilizing higher-order terms to keep (2.11) meaningful.

**See e.g. section 1.2 in D.J. Thouless, ref.[15].

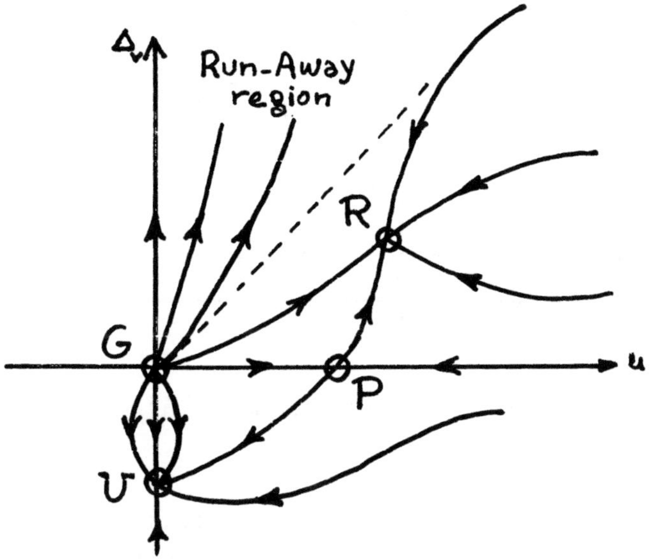

Fig. 1 - Flow lines in Δ_V , u space ($\alpha > 0$). Fixed points : R : random, P : pure, G : Gaussian, U : unphysical after Aharony [14].

iv) Pure random potential : A limiting case of the above situation is obtained when the (repulsive) coupling in φ^4 is set equal to zero. One has a non-interacting system in the presence of a random potential $V_R(x)$, which is the limit of the tight binding Anderson model[16] as the lattice spacing tends to zero.

If E_λ , $f_\lambda(x)$ are eigenvalues and eigenfunctions of $-\nabla^2 + V_R(x)$, then one has

$$\langle\varphi(x)\ \varphi(x')\rangle = \overline{\sum_\lambda \frac{f_\lambda(x)\ f_\lambda(x')}{E - E_\lambda + io}} \tag{2.16}$$

where the bar stands for average over the $V_R(x)$ distribution and we write $-E$ instead of r_o , E being the energy. It follows from (2.16) that the level density $\rho(E)$ takes the form

$$\rho(E) = \pi^{-1}\ \mathrm{Im}\ \langle\varphi(x)\ \varphi(x')\rangle = \overline{\sum_\lambda \delta(E-E_\lambda)} \tag{2.17}$$

The level density of the free system is $\propto \sqrt{E}$ (in dimension $d = 3$). The presence of the random potential allows for bound states in the $E < 0$ region. As a result $\rho(E)$ develops a tail into $E < 0$ that extends to infinity if the probability law $\mathcal{P}(V_R)$ is long tailed, e.g. for the Gaussian white noise potential as in (2.13). If $\mathcal{P}(V_R)$ is short tailed, $\rho(E)$ vanishes below E_o , the bottom of the modified band.

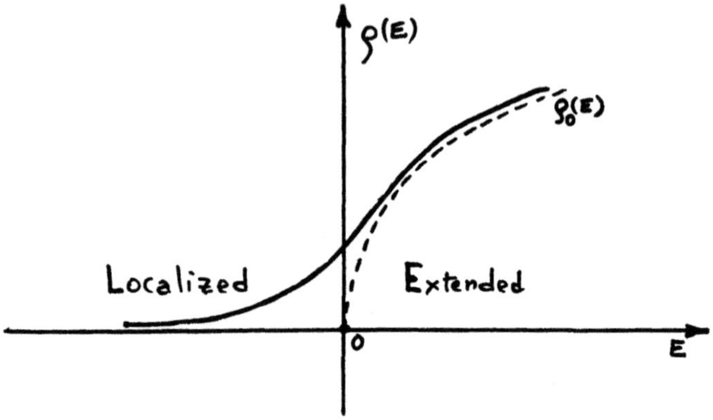

Fig.2 - Side of the band modification by random potential V_R .

The $E \lesssim 0$ region (due to bound states) is the region of localized states. For $E \gtrsim 0$ we have extended states. Somewhere in between lies the "mobility edge" E_c that separates between localized and delocalized systems. We see below how the behavior in the tail is obtained ($|E|^{2-d/2} >> \Delta_v$) as a mean field result and fluctuations thereof. The critical behavior around E_c is a much more delicate problem.

v) *Outlook* : After this overview from far away we want now to take the usual steps involved in studying critical phenomena : mean field, fluctuations, renormalization group. The task is here much more difficult for several reasons that are both physical and technical. (*i*) Physical, because we are not completely sure that the time long-range order parameter is sufficient (or may be even relevant) to describe what happens at low temperature (below T_g). For this reason we also describe the view given by low temperature approaches together with the type of "order parameters" they may suggest. (*ii*) Technical, this is because we deal with quenched systems and averages have to be taken on observables as in (2.3) (not on the partition function). More of this now.

3. REPLICAS VS DYNAMICS

For quenched averages as in (2.3), if A is the free energy, we need to compute

$$- \frac{1}{T} \overline{\ln Z} = - \frac{1}{T} \int \mathcal{D}a_R \; \mathcal{P}(a_R) \; \ln \operatorname{tr} e^{-\mathcal{H}_R\{\sigma, a_R\}} \tag{3.1}$$

Edwards and Anderson[16] circumvented the difficulty of having to take the log before averaging, by introducing n identical systems (replicas) and using the identity

$$\ln Z = \frac{\partial}{\partial n} (Z^n) \Big|_{n=0} \quad . \tag{3.2}$$

At the same time they had to replace the time long-range order parameter (2.14) which is presumed to describe the frozen-in phase

$$< \sigma_j(t) \, \sigma_j(t') >_{t-t' \to \infty} = Q_j \tag{3.3}$$

by

$$< \sigma_j^{\alpha} \, \sigma_j^{\beta} >_{\alpha \neq \beta} = Q_j^{\alpha\beta} \quad . \tag{3.4}$$

We shall see below that the (average) order parameters (3.3) and (3.4) are equivalent in the $n=0$ limit. Of course one is always reluctant to discard a bona fide physical quantity like (3.3) for a quantity in replica space. Besides, replica results seemed to be marred by pathologies in the zero temperature limit (Sherrington, Kirkpatrick [17], Aharony[18]). This was enough to motivate a search to work directly with (3.3). Among others, Grinstein, Ma and Mazenko[19] have indeed used perturbation expansion of time-dependent Landau-Ginzburg models to study dilute random systems, Ma and Rudnick[20] to study, through resummations, the phase below the freezing-in temperature T_g. We use here a Lagrangian systematic expansion[21], which is basically an extension of the Martin, Siggia and Rose[22] approach adapted to random systems[23]. This allows us to discuss general properties of the ordered phase and reveals that frozen-in Landau Ginzburg systems are unstable, unless constrained in a proper way.

The beauty of working with dynamics is, not only that it keeps closer to physics but it makes unnecessary any trick to get across the log of (3.1). Basically the reason why one can get rid of replicas by looking from a dynamic point of view is that propagators *retarded in time* give a vanishing contribution for closed loops, just like replicas give a vanishing weight ($n=0$) for closed loops. In fact it could work with any dynamic equations provided one insures that at $t=\infty$ the starting (static) system is recovered.

Consider first a Landau-Ginzburg Hamiltonian $\mathcal{H}\{\varphi\}$ and a Langevin equation that describes the relaxation towards equilibrium[24]

$$\frac{\partial \varphi(xt)}{\partial t} = -\Gamma_o \frac{\delta \mathcal{H}\{\varphi\}}{\delta \varphi(x,t)} + \zeta(xt) \tag{3.5}$$

Γ_o is a bare kinetic coefficient, $\zeta(xt)$ a dynamic random noise, i.e. governed by a probability law. To simplify we take again a white noise law

$$<\zeta(xt)> = 0 \tag{3.6}$$

$$<\zeta(xt) \, \zeta(x't')> = C \, \delta(x-x') \, \delta(t-t') \tag{3.7}$$

which is justified if the noise takes care of the rapidly varying degrees of freedom. If we choose $C = 2\Gamma_o$, then we are insured (via a Fokker-Planck description) that the

$t = \infty$ limit is governed by the density matrix $\exp - \mathcal{H}$ that describes equilibrium (statics). From (3.4) one derives the generating functional[25]

$$\hat{Z}_\zeta(\ell) = \int \mathcal{D}\varphi \exp\left\{\int d^dx\, dt\, \ell(xt)\, \varphi(xt)\right\} \prod_{xt} \delta\left(\frac{\partial\varphi}{\Gamma_o \partial t} + \frac{\delta\mathcal{H}}{\delta\varphi} - \frac{\zeta}{\Gamma_o}\right) J \quad (3.8)$$

$$= \int \mathcal{D}\varphi\, \mathcal{D}\hat{\varphi} \exp\left\{\int d^dx\, dt \left(i\hat{\varphi}\left[\frac{\partial\varphi}{\Gamma_o\partial t} + \frac{\delta\mathcal{H}}{\delta\varphi} - \frac{\zeta}{\Gamma_o}\right] + \ell\varphi\right)\right\} J \quad (3.9)$$

where the source $\ell(xt)$ couples to the field φ and the magnetic field to $i\hat{\varphi}$. The Jacobian J insures that the δ functions are properly normalized, that is $\hat{Z}_\zeta(\ell=0) = 1$. As a result $\hat{Z}_\zeta(\ell)$ is directly an observable. By averaging over the dynamic noise one has

$$\hat{Z}(\ell) = \int \mathcal{D}\varphi\, \mathcal{D}\hat{\varphi} \exp\left\{\int d^dx\, dt\, \ell(xt)\, \varphi(xt) + \mathcal{L}(\varphi,\hat{\varphi})\right\} J \quad (3.10)$$

$$\mathcal{L}\{\varphi,\hat{\varphi}\} = \int d^dx\, dt \left[i\hat{\varphi}\left(\Gamma_o^{-1}\frac{\partial\varphi}{\partial t} + \frac{\delta\mathcal{H}}{\delta\varphi}\right) + \Gamma_o^{-1}\, i\hat{\varphi}\, i\hat{\varphi}\right] \quad (3.11)$$

where $\mathcal{L}$ generates the Martin-Siggia-Rose perturbation expansion with bare propagators

$$\langle i\hat{\varphi}\varphi\rangle_o \equiv R_o(k,\omega) = \left[-\frac{i\omega}{\Gamma_o} + r_o + k^2\right]^{-1} \quad (3.12)$$

$$\langle\varphi\varphi\rangle_o \equiv C_o(k,\omega) = 2\Gamma_o^{-1}\left|-\frac{i\omega}{\Gamma_o} + r_o + k^2\right|^{-2} \quad (3.13)$$

respectively linear response, and correlation function. Closed loops built with the (retarded) linear response, identically vanish. The generating functional for responses and correlations (3.10,11) may now be used for systems with random impurities, since we do not have yet specified what $\mathcal{H}$ is.

4. RANDOM LANDAU-GINZBURG SYSTEMS[23]

i) Random magnetic field :

If we add to $\mathcal{H}$ the random magnetic term (2.5)

$$\mathcal{H}_R = \int d^dx\, h_R(x)\, \varphi(x)$$

then $\mathcal{L}$ in (3.10) becomes $\mathcal{L} + \mathcal{L}_R$ with

$$\mathcal{L}_R = \int d^dx\, h_R(x)\, \left[\int dt\, i\hat{\varphi}(xt)\right] \quad (4.1)$$

where $h_R(x)$ remains time independent, because the random degrees of freedom are *quenched*. Again we are allowed to average $\hat{Z}(\ell)$ over the probability law $\mathcal{P}(h_R)$, which, for the Gaussian characterized by (2.6,7), replaces (4.1) by

$$\mathcal{L}_{eff} = \Delta_h \int d^dx \left[\int dt\, i\hat{\varphi}(xt)\right]^2 \quad . \quad (4.2)$$

It is instructive to look at the system with no coupling ($u_o = 0$). In $\mathcal{L} + \mathcal{L}_{eff}$ one has a local source (Γ_o^{-1}) and a completely delocalized (Δ_h) source. The response (3.12) is unchanged but the correlation has two terms corresponding to the two sources

$$<\varphi\,\varphi>_o \;=\; \frac{2\Gamma_o^{-1}}{\left|-\frac{i\omega}{\Gamma_o} + r_o + k^2\right|^2} \;+\; \frac{\Delta h\;\delta(\omega)}{\left|-\frac{i\omega}{\Gamma_o} + r_o + k^2\right|^2} \tag{4.3}$$

that is in Fourier transform

$$<\varphi(t)\;\varphi(t')>_o \;=\; \frac{1}{r_o+k^2}\;\exp\left\{-\,|t-t'|\;[r_o+k^2]\right\} + \frac{\Delta_h}{|r_o+k^2|^2} \tag{4.4}$$

displaying a *decaying term* and a *time persistent term*. This last feature is preserved when the coupling is turned on, and the correlation function keeps a time persistent part (proportional to $\delta(\omega)$).

ii) *Random temperature* :

Now we have (2.11) for the random part

$$\mathcal{H}_R \;=\; \int d^dx\; V_R(x)\;\varphi^2(x)$$

and then for the Lagrangian

$$\mathcal{L}_R \;=\; \int d^dx\; V_R(x)\left[\int dt\;i\hat{\varphi}(xt)\;\varphi(xt)\right] \quad . \tag{4.5}$$

Averaging with cumulants (2.12,13) we obtain

$$\mathcal{L}_{eff} \;=\; \Delta_v \int d^dx\left[\int dt\;i\hat{\varphi}(xt)\;\varphi(xt)\right]^2 \tag{4.6}$$

again completely delocalized in time. The Lagrangian $\mathcal{L} + \mathcal{L}_{eff}$ governs the Landau-Ginzburg system with random temperature. It is handily used with renormalization theory to study critical behavior of dilute systems in 4-ε dimensions, e.g. to compute the phase diagram of Fig.1. Of greater interest is to establish the *equation of state* for the presumed (averaged) order parameter Q, below T_g .

Equation of state : The equation of state is obtained by writing the Dyson equation for the matrix propagator

$$\mathbb{G} = \begin{pmatrix} C & R \\ R^* & \hat{C} \end{pmatrix} \tag{4.7}$$

where the components are the correlation and response and $\hat{C} = <i\hat{\varphi}\;i\hat{\varphi}>$ a vanishing correlation in the absence of sources for the field φ . Namely one has

$$\mathbb{G}^{-1} = \mathbb{G}_o^{-1} - \frac{\delta}{\delta\mathbb{G}}\,\mathcal{K}\{\mathbb{G}\} \tag{4.8}$$

with

$$\mathbb{G}_o^{-1} = \begin{pmatrix} 0 & R_o^{-1} \\ (R_o^*)^{-1} & \Delta_h \delta(\omega) + 2\Gamma_o^{-1} \end{pmatrix} \tag{4.9}$$

and $\mathcal{H}\{\mathbb{G}\}$ the 1-irreducible functional (no self energy insertions) expressed in terms of the full propagator ($\mathbb{G}$). Taking the C component of (4.8) we get

$$\frac{C(k\omega)}{|R(k\omega)|^2} = 2\Gamma_o^{-1} + \Delta_h\ \delta(\omega) - \frac{\delta\mathcal{H}\{C,R,R^*,\hat{C}\}}{\delta\hat{C}}\Bigg|_{\hat{C}=0} \quad , \tag{4.10}$$

a corresponding equation exists for R, related to (4.10) by fluctuation-dissipation. In (4.10) the last term is the mass operator associated with the correlation function

$$\frac{\delta\mathcal{H}}{\delta\hat{C}} \equiv -\Sigma_{\hat{\varphi}\hat{\varphi}} = \tag{4.11}$$

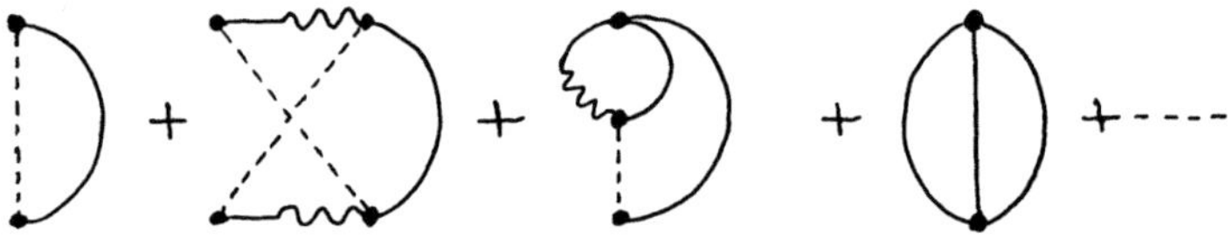

where plain lines stand for C, mixed lines for R. The local vertex $u_o\, i\hat{\varphi}(t)\ \varphi^3(t)$ and the delocalized vertex $\Delta_v(i\hat{\varphi}(t)\ \varphi(t))(i\hat{\varphi}(t')\ \varphi(t'))$ have obvious graphic representations. The Δ_v interaction preserves frequency along the solid lines (complete delocalization), this is why (4.11) can sustain a persistent component for $C(k,\omega)$. If we write

$$C(k,\omega) = \tilde{C}(k,\omega) + Q(k)\ \delta(\omega) \tag{4.12}$$

separating a decaying and a persistent part, we now have the option that below T_g (4.11) splits into *two equations*, a function component for $\tilde{C}(k,\omega)$, and a distribution component for $Q(k)\ \delta(\omega)$ the latter being the *equation of state*. Namely we have

$$\Delta_h = \frac{Q(k)}{R(k)^2} + \Sigma^d_{\hat{\varphi}\hat{\varphi}}\{Q,\tilde{C},R\} \tag{4.13}$$

$$-\Sigma^d_{\hat{\varphi}\hat{\varphi}} = \tag{4.14}$$

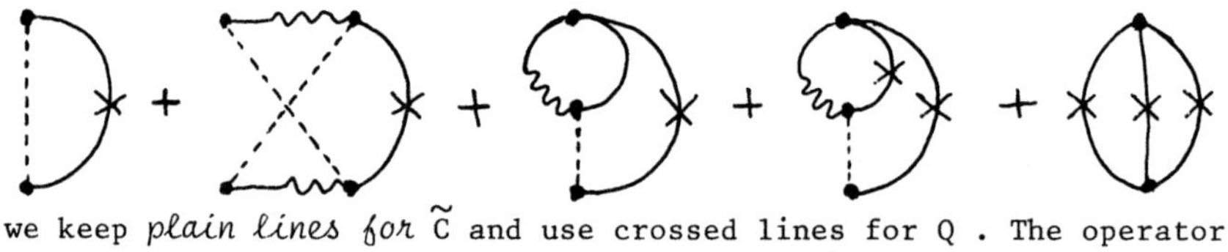

where now we keep *plain lines for* $\tilde{C}$ and use crossed lines for Q. The operator Σ^d is the same as the one defined by Ma-Rudnick if one keeps all 1-irreducible graphs that fall apart into two pieces by cutting all Δ_v (dotted) lines and all Q (crossed) lines. Eqs.(4.13,14) express the field Δ_h in terms of the (observable) conjugate variable Q which appears explicitly and also via R and $\tilde{C}$. If $\Delta_h \neq 0$, there always is a non-zero Q solution to (4.13,14). In zero field, $\Delta_h = 0$, a non-zero solution may exist below T_g defined by the existence of a zero eigenvalue for the matrix

$$\frac{\delta\Delta_h(k)}{\delta Q(\ell)} \quad . \tag{4.15}$$

To lowest order we get the well known equation

$$1 = \Delta_v \int \frac{d^d k}{(r_{og} + k^2)^2} \tag{4.16}$$

for the freezing temperature T_g .

- *Asymptotic time dependence* : The companion equation generated by (4.10) yields

$$\Gamma_o = \frac{C(k,\omega)}{|R(k,\omega)|^2} + \Sigma^c_{\hat{\varphi}\hat{\varphi}} \tag{4.17}$$

$$-\Sigma^c_{\hat{\varphi}\hat{\varphi}} = -\Sigma_{\hat{\varphi}\hat{\varphi}} + \Sigma^d_{\hat{\varphi}\hat{\varphi}} =$$

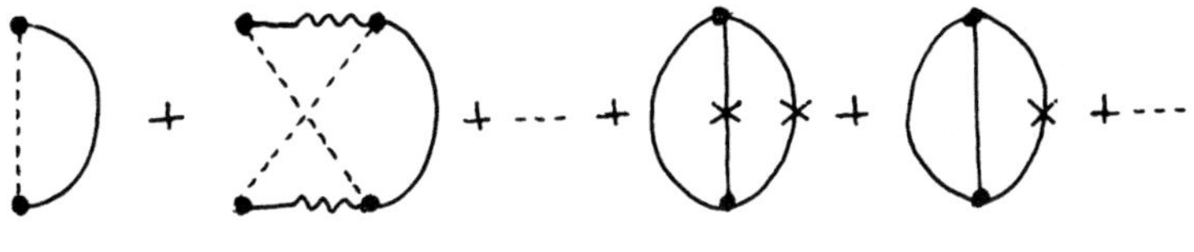

determining $\tilde{C}(k,\omega)$, the decaying part, also related by fluctuation-dissipation to $R(k,\omega)$. Suppose that one removes all the Δ_v (dotted) lines from (4.18), the external vertices at t and t' remain connected by at least one $\tilde{C}$ (and possibly some Q's). In the u_o^2 term, e.g. there may exist two or three such $\tilde{C}$'s . Ma and Rudnick have shown that if one keeps consistently only terms where $\Sigma^c_{\hat{\varphi}\hat{\varphi}}$ is prevented from being persistent by one single $\tilde{C}$ connection, then $\tilde{C}(\omega) \sim \omega^{1/2}$ for small frequencies, i.e. $\tilde{C}(t-t') \simeq (t-t')^{-1/2}$. In this approximation u_o^2 terms should be absent and indeed, near T_g they do not contribute to (4.14). However below T_g one may expect that Q increases to the extent that u_o^2 terms will become significant in (4.14), these terms are *attractive* and will tend to destabilize the ordered state after a time of the order $\sim |T_f/(T-T_f)|^4$. Notice that the destabilizing Q^3 terms of the equation of state correspond to Q^4 terms in the free energy (or in a Hamiltonian $\mathcal{H}\{Q\}$). However there is *no* free energy functional[25'] of $C, R, \hat{C}$ that would generate (4.17), or its C and R components, by stationary variations of $\mathcal{G}$ (or C and R, see further).

iii) Role of probability tails :

It is instructive to look at the very small repulsion limit. We only have to suppress graphs containing u_o vertices. The equation for $\Sigma_{\hat{\varphi}\hat{\varphi}}$ is now linear in $C(k,\omega)$. One may again define a T_g where one has a zero eigenvalue for (4.15) and where fluctuations of Q become infinite. However, if one considers the connected part of $\langle \varphi(t_1)\, \varphi(t_2)\, \varphi(t_3)\, \varphi(t_4) \rangle$ for all time intervals becoming infinite, it turns out to have[26] a time persistent part that appears (whose fluctuations become infinite) *before* $\langle \varphi(t_1)\, \varphi(t_2) \rangle_{t_1 - t_2 \to \infty}$, as the temperature is lowered i.e. associated with a $T_g^{(2)} > T_g$. And so on for higher correlation functions in time.

In terms of the localization problem ($u_o \equiv 0$) this signals that a singularity appears at a negative energy $E_o < \ldots < E_g^{(2)} < E_g < 0$ which in the case of long tail (Gaussian) probability law for V_R is rejected at $-\infty$. If we take a short tailed law (e.g. double peaked at $\pm\Delta_v^{1/2}$) then E_o (the bottom of the band) is at a finite distance $-\Delta_v$. The quartic term $+\frac{\Delta_v}{2}\int d^dx\, A^2(x)$ in $\mathcal{L}_R$ (4.6)

$$A(x) = \int dt\, i\hat{\varphi}(xt)\, \varphi(xt) \tag{4.19}$$

is replaced by

$$+ \int d^dx \;\ln \mathrm{ch}\, \Delta_v^{1/2}\, A(x) \tag{4.20}$$

which for small values of the arguments gives back (4.6).

In fact both negative cumulants and the repulsive coupling u_o play against the appearance of higher time persistent correlations at a temperature $T > T_g$ (i.e. $|E| > |E_g|$), but the result of that delicate balance is difficult to foretell (see also further via replicas).

iv) <u>*Replica approach*</u> : We now work with replicas indexed by $\alpha = 1,2,\ldots,n$, we have

$$\mathcal{H} = \int d^dx \;\frac{1}{2}\vec{\varphi}(r_o-\nabla^2)\vec{\varphi} + \frac{u_o}{4!}\sum_\alpha (\varphi^\alpha)^4 + h\sum_\alpha \varphi^\alpha - \sum_{\alpha\neq\beta} H^{\alpha\beta}\,\varphi^\alpha\,\varphi^\alpha \tag{4.21}$$

$$\mathcal{H}_R = \int d^dx \;\frac{1}{2}\, V_R(x)\; \vec{\varphi}.\vec{\varphi} \tag{4.22}$$

which leads to (Gaussian law for V_R) $\mathcal{H} + \mathcal{H}_{eff}$ with

$$\mathcal{H}_{eff} = -\frac{\Delta_v}{2}\int d^dx \left(\frac{1}{2}\,\vec{\varphi}.\vec{\varphi}\right)^2 \quad . \tag{4.23}$$

- *Free energy and equation of state* : Consider the correlation matrix $\mathcal{G}$ whose components are

$$G^{\alpha\beta} \equiv Q^{\alpha\beta} \qquad \alpha \neq \beta \tag{4.24}$$

$$G^{\alpha\alpha} = \langle\varphi^\alpha\, \varphi^\alpha\rangle \quad . \tag{4.25}$$

In this case the *free energy* (Legendre transform of $\ln \overline{Z^n}$) is given by the stationary functional *

---

* This in contradistinction with the dynamic approach where there is no *free energy* functional whose stationarity equations (upon varying R and C) are Eq.(4.10) (and the corresponding equation for R). One way to construct a (non stationary) free energy vs C and R is quoted in Ref.[23]. It would be redundant and wrong to use the MSR generating functional *and* replicas as in Tanaka [27]. If one insists upon obtaining a free energy functional stationary for the dynamic equations of motion the following can be done.

i) Use the Lagrangian approach described in Ref. 28 that generates response functions. For example $R(\omega_\nu)$, where ω_ν is discrete and has to be analytically continued in the proper way. This approach only works for purely relaxational systems (no mode coupling).

ii) Use replicas as above then get

$$G^{\alpha\beta}(\omega_\nu) = Q\,\delta_{\omega_\nu;0}$$

$$G^{\alpha\alpha}(\omega_\nu) = R(\omega_\nu) + Q\,\delta_{\omega_\nu;0}$$

$$R^{-1}(\omega_\nu) = -\frac{i\omega}{\Gamma_o} + r_o + k^2 + K(\omega_\nu) - L\,\delta_{\omega_\nu;0}$$

instead of (4.27-29) and (4.32). Eq.(4.31) is unchanged. The only change in (4.30) is that $G^{\alpha\alpha}$ contains a frequency dependent piece (R).

$$-\Gamma = \frac{\partial}{\partial n}\bigg|_{n=0} \mathrm{Tr}\,\ln \mathbb{G} - [\mathbb{G}^{o}]^{-1}\mathbb{G} + \mathcal{K}\{\mathbb{G}\} \tag{4.26}$$

and its associated Dyson equation (4.8), with $\mathbb{G}^{o} = \delta_{\alpha\beta}(r_o+k^2)^{-1}$. If we look for solutions with symmetry preserved in replica space, letting

$$Q^{\alpha\beta} = Q(k) \tag{4.27}$$

$$G^{\alpha\alpha} = G(k) \tag{4.28}$$

$$R(k) = G(k) - Q(k) \tag{4.29}$$

we obtain the free energy

$$-\Gamma = \sum_k \left\{\ln R(k) + \frac{Q(k)}{R(k)} - (r_o+k^2)\,G\right\} + \frac{\partial}{\partial n}\mathcal{K}\{G,Q\}\bigg|_{n=0} \tag{4.30}$$

For Dyson equations that give the *equation of state*, we have

$$Q(k) = R^2(k)\,L(k) \tag{4.31}$$

$$R^{-1}(k) = (r_o+k^2) + L(k) - K(k) \tag{4.32}$$

where

$$L \equiv \frac{\delta\mathcal{K}}{\delta Q^{\alpha\beta}} =$$ 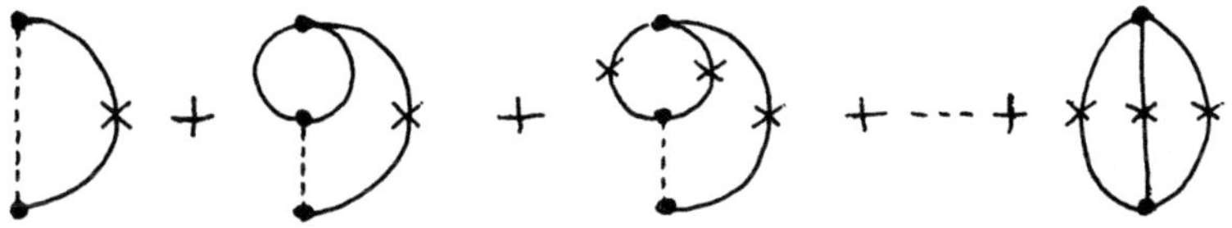 (4.33)

$$K \equiv \frac{\delta\mathcal{K}}{\delta G^{\alpha\alpha}} =$$ 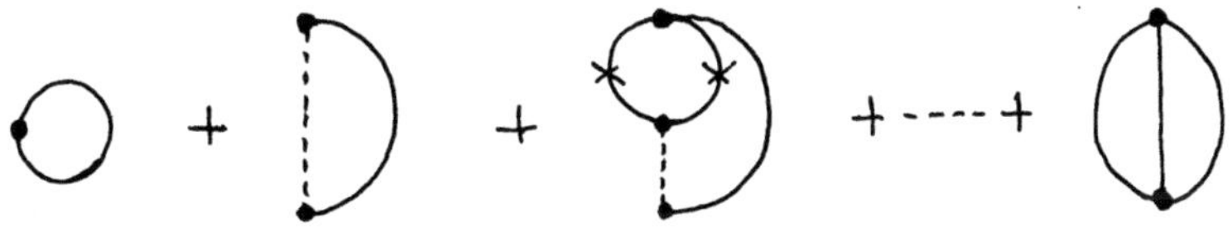 (4.34)

Here crossed lines again stand for Q and plain lines for $G \equiv R+Q$. R(k) is the zero frequency response function, the equal time (purely decaying) correlation $\tilde{C}(t=t')$. Although (4.31,33,34) look different from (4.13,14) one easily checks their identity to few lowest orders. They probably are identical (though we have no direct proof). Note that G is even in Q, R is odd.

- *Stability* : If we rewrite (4.21,23) as

$$\mathcal{H}+\mathcal{H}_{eff} = \int d^dx \left\{\frac{1}{2}\,\vec{\varphi}(r_o-\nabla^2)\vec{\varphi} - (3\Delta-u_o)\,\frac{1}{4!}\sum_\alpha (\varphi^\alpha)^4 - \frac{\Delta_v}{8}\sum_{\alpha\neq\beta}(\varphi^\alpha\varphi^\beta)^2 - \sum_{\alpha\neq\beta} H^{\alpha\beta}\varphi^\alpha\varphi^\beta\right\} \tag{4.35}$$

where we have a field $H^{\alpha\beta}$ coupling to the presumed order parameter, we may now integrate out the φ variables and only keep the variables $Q^{\alpha\beta}$ trivially related to the order parameter

$$\mathcal{H}\{Q^{\alpha\beta}\} = \Delta_v^{-1} \int d^d x \sum_{\alpha\beta} (Q^{\alpha\beta})^2 + S\{Q^{\alpha\beta}\} \tag{4.36}$$

with

$$\exp - S\{Q^{\alpha\beta}\} = \int \mathcal{D}\varphi \exp \int d^d x \left\{ \sum_{\alpha\neq\beta} (H^{\alpha\beta}+Q^{\alpha\beta})\varphi^\alpha\varphi^\beta - \frac{1}{2}\vec{\varphi}(r_o-\nabla^2)\vec{\varphi} + (3\Delta-u_o)\frac{1}{4!}\sum_\alpha (\varphi^\alpha)^4 \right\} \tag{4.37}$$

On this form it is now clear that the system turns unstable in the run-away region of the coupling parameters.

- *Classical localized solutions* : In the localization problem ($u_o = 0$) we are precisely interested in the lifetime of various correlations or their imaginary part, that arise on account of their unstable Hamiltonian. Far in the tail $|E|^{(4-d)/2} \gg \Delta_v$ mean field should be a good approximation, as given by saddle point approximation. This has been done by J. Cardy[29] using the techniques developed for instanton calculations, with the result

$$\rho(E) \sim E^{\frac{d}{4}(5-d)} \exp\left\{-C\,\frac{E^{(4-d)/2}}{\Delta_v}\right\} . \tag{4.38}$$

Although the prefactor (coming from fluctuations) looks different from most results in the literature, it can be reconciled if one modifies Edwards[30] and Lifschitz[31] computations for unaccounted collective modes, and Zittartz-Langer[32] who were the first to treat the collective modes, for a trivial approximation. In fact (4.38) gives the correct result even for $d = 0$ and $d = 1$. Fluctuations are computed around the classical solution, $O(n)$ symmetric, of the equations of motion

$$(-\nabla^2 + |E|)\,\vec{\varphi}_{cl} = \frac{\Delta_v}{2}\,(\vec{\varphi}_{cl}\cdot\vec{\varphi}_{cl})\,\vec{\varphi}_{cl} . \tag{4.39}$$

The dynamic approach provides another set up to look for classical solutions $\varphi_{cl}(x,t)$ that would be easier to handle when there is no more $O(n)$ symmetry, e.g. in the presence of φ^4 repulsion.

To describe the "ordered" phase (region with extended φ_{cl} solutions) $E > E_c$, Imry and Aharony[32'] have proposed, in close analogy to the spin glass, an "order parameter" $\varphi_+^\alpha(x)\ \varphi_-^\beta(x)$ where $\varphi_\pm$ is governed by the random Hamiltonian $E \pm io - \nabla^2 + V_R(x)$. The other operators $\varphi_+\varphi_+$, $\varphi_-\varphi_-$ are non-critical (they give regular contributions to the four-point function or the current correlation), hence the effective Hamiltonian is quartic (not cubic as in (4.37)) and $d_c = 4$. Recently Abrahams, Anderson, Licciardello and Ramakrishnan[33] have used instead the energy shift under changing boundary conditions (related to the conductivity), for which they are able to give renormalization group equations and determine the behavior near E_c .

5. RANDOM BOND HAMILTONIAN

If we turn now to the more interesting random bond Hamiltonian, we may use a Langevin-Glauber equation instead of (3.5) to describe relaxation to equilibrium. The previous dynamic approach follows. However the constraint ($\vec{\sigma}^2 = 1$) renders the Langevin-Glauber equation non-linear in the random coupling ($J^R_{k\ell}$), or conversely if we want to keep it linear, we have to use two coupled equations (for the spin field and its conjugate variable). Here we follow the literature and use replicas.

i) Effective spin glass Hamiltonian :

We use the spin glass Hamiltonian of Edwards and Anderson[16], i.e. after averaging over a Gaussian probability law of variance Δ_J, we have

$$\mathcal{H}_{eff} = \sum_{jk}\sum_{\alpha\neq\beta} \frac{1}{2}(\Delta_J)_{jk}\ (\sigma^\alpha_j\sigma^\beta_j)(\sigma^\alpha_k\sigma^\alpha_k) - \sum h_j\sigma^\alpha_j - \sum H^{\alpha\beta}_j\ \sigma^\alpha_j\sigma^\beta_j \tag{5.1}$$

with $\Delta_J \equiv \Delta/T^2$. Tracing out the spin variables we obtain

$$\mathcal{H}\{Q^{\alpha\beta}_j\} = \sum_{\substack{jk\\ \alpha\neq\beta}} (\Delta_J)^{-1}_{jk}\ Q^{\alpha\beta}_j\ Q^{\alpha\beta}_k + \sum_j S_j\{Q^{\alpha\beta}\} \tag{5.2}$$

$$\exp - S_j\{Q^{\alpha\beta}\} = \mathop{\mathrm{Tr}}_{\sigma^\alpha}\ \exp \sum_{\alpha\neq\beta} (H^{\alpha\beta}_j + Q^{\alpha\beta}_j)\ \sigma^\alpha_j\sigma^\beta_j \quad . \tag{5.3}$$

On (5.2,3) we already see that the stability troubles that were plaguing the Landau-Ginzburg model for large values of $Q^{\alpha\beta}$, do not exist here.

The usual successive steps to study the transition region in the ordered phase (mean field, fluctuations, renormalization group) must now be taken on (5.2,3), where one has forced in the presumed order parameter

$$\text{limit} < \sigma^\alpha_j\ \sigma^\beta_j >_{\alpha\neq\beta} \Big|_{n=0} = \delta_{jk}\ q \tag{5.4}$$

and q is trivially proportional to the average value of Q .

ii) Mean field :

Edwards and Anderson derive from (5.2,3) an exact mean field equation when all averaged $Q^{\alpha\beta}$ are taken identical (no symmetry breaking in replica space). One gets ($Q = q\Delta_J$, $\Delta_J \equiv \Delta/T^2$),

$$-\frac{F}{NT} = \frac{\Delta_J}{4}\left(\frac{Q}{\Delta_J} - 1\right)^2 + \int_{-\infty}^{+\infty} \frac{dx}{\sqrt{2\pi}}\ e^{-x^2/2}\ \ln 2\mathrm{ch}\ Q^{1/2}x \tag{5.5}$$

and for the free energy per site and through stationarity,

$$\frac{Q}{\Delta_J} = \int_{-\infty}^{+\infty} \frac{dx}{\sqrt{2\pi}}\ e^{-x^2/2}\ \mathrm{th}^2\ Q^{1/2}x \tag{5.6}$$

which nicely gives a second order transition for Q with $T_g = \Delta^{1/2}$ and a linear behavior in $T_g - T / T_g$ near T_g . It also gives a cusp in the susceptibility χ (χ is linear in Q). Unfortunately the specific heat also has a cusp. Mean field is believed to become exact in the limit of an infinite number of interacting neighbours. The model of Sherrington - Kirkpatrick[17] has that feature and was solved exactly giving the same result as the mean field result of Anderson - Edwards. It also showed a negative entropy at $T = 0$ due to inversion of the order of limits ($n = 0$ and thermodynamic limit). This led Thouless, Anderson and Palmer[34] to derive the mean field result without replicas, showing that below T_g , the above solution is on the "wrong branch" On the "correct" branch the free energy is at a saddle point $F \sim (Q-Q_o)^3$. This saddle structure is conjectured to persist at all $T < T_g$. It also has a vanishing entropy at $T = 0$. The TAP technique which involves resumming infinite subseries of diagrams, is difficult to use in a systematic way. Blandin[7,35] has derived an equivalent result near T_g , to Q^3 order, using replicas and breaking the symmetry in a "physical way". Defining

$$Q = \langle \sigma S \rangle = \lim_{H_j \to 0} \frac{\partial}{\partial H_j} \ln \operatorname{Tr} \exp\left(\mathcal{H}\{\sigma\} + \mathcal{H}\{S\}\right) + \sum_j H_j \sigma_j S_j \tag{5.7}$$

where $\mathcal{H}\{\sigma\}$ and $\mathcal{H}\{S\}$ may be thought as two identical systems (or the same system at a long time difference). To compute the log , replicas are used for each system ($n = 2m$) and mean field is sought with variables ($\alpha = 1,2,...m$)

$$\begin{aligned} Q &= \langle \sigma^\alpha S^\alpha \rangle \\ P &= \langle S^\alpha S^\beta \rangle = \langle \sigma^\alpha \sigma^\beta \rangle & \alpha \neq \beta \\ R &= \langle S^\alpha \sigma^\beta \rangle & \alpha \neq \beta \end{aligned} \tag{5.8}$$

Since the field H couples to $\sigma^\alpha S^\alpha$, Q is preferentially aligned and $Q \geq P,R$.

The free energy for the "physical" order parameter Q is obtained by eliminating P,R (as functions of Q defined by their mean field value)

$$\frac{\partial F}{\partial P} = \frac{\partial F}{\partial R} = 0$$

and substituting in $F\{Q,P,R\}$. The result is surprising because it gives the saddle point behavior of TAP for $T < T_g$ with little cost. Blandin symmetry breaking when carried out to all orders gives ($P = R$)

$$-\frac{F}{NT} = \frac{\Delta_J}{4} \left[2\left(\frac{R}{\Delta_J} - 1\right)^2 - \left(\frac{Q}{\Delta_J} - 1\right)^2\right] + \int_{-\infty}^{+\infty} \frac{dx}{\sqrt{2\pi}} e^{-x^2/2} \frac{1}{2} \ln\left\{2\operatorname{ch} 2x R^{1/2} + 2e^{-2(Q-R)}\right\} \tag{5.9}$$

instead of (5.6) and the stationarity equations (for variation of R, Q) instead of (5.5). According to Parisi[36] the entropy resulting from (5.9) although better than the one from (5.6) is still negative.

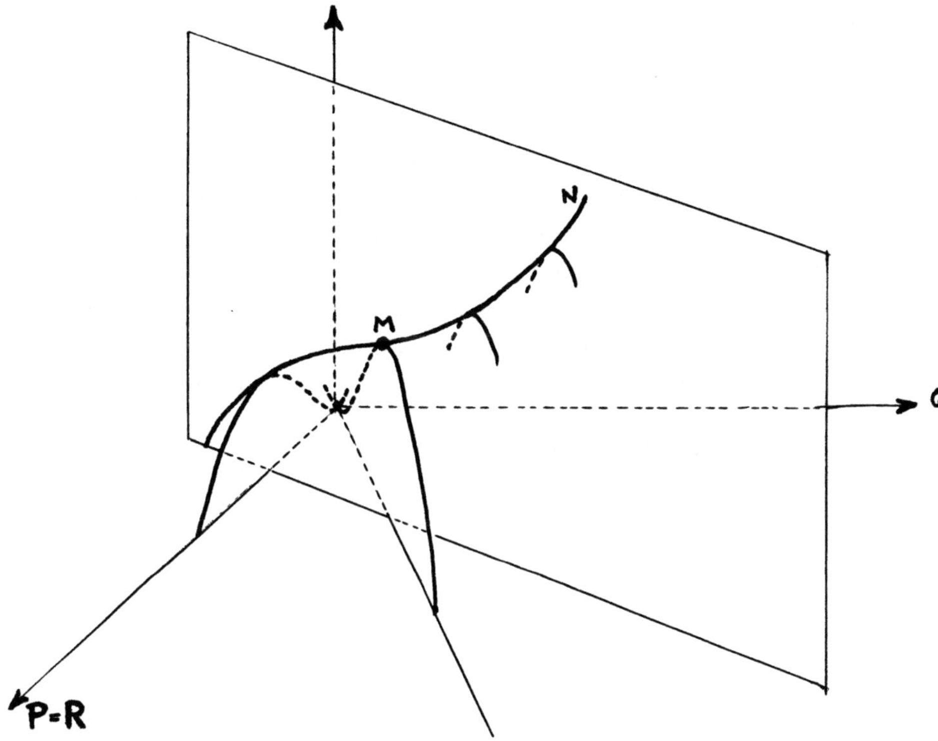

Fig.3 - Blandin free energy surface as function of Q,R . The TAP curve is M,N .

iii) Fluctuations :

They are used to test the stability of mean field solutions. From (5.2,3) one obtains in the continuum limit (and after rescaling)

$$\mathcal{H}\{Q^{\alpha\beta}\} = \int d^dx \left\{ \frac{1}{4} \sum r_o (Q^{\alpha\beta})^2 + (\nabla Q^{\alpha\beta})^2 - \frac{W}{6} \sum Q^{\alpha\beta} Q^{\beta\gamma} Q^{\gamma\alpha} \right\} + \dots \tag{5.9}$$

with

$$r_o = z^2 \left(\frac{T^2}{T_g^2} - 1 \right) , \qquad W = z^3 , \tag{5.10}$$

z is the coordination number.

The loop expansion rules tell us to write

$$Q^{\alpha\beta} = \langle Q^{\alpha\beta} \rangle + q^{\alpha\beta} \tag{5.11}$$

where $\langle Q^{\alpha\beta}\rangle = Q$ as given by mean field, then to keep terms quadratic in $q^{\alpha\beta}$. The spectrum studied by Bray and Moore[37,39] and Pytte and Rudnick[38] is threefold (there are three types of propagators $G_a = \langle q^{\alpha\beta} q^{\alpha\beta}\rangle$, $G_b = \langle q^{\alpha\beta} q^{\alpha\gamma}\rangle$, $G_c = \langle q^{\alpha\beta} q^{\gamma\delta}\rangle$) two are massive and degenerate in the $n = 0$ limit (mass $\equiv |r_o|$) one is massless and corresponds to the combination[38]

$$G_a - 2G_b + G_c \simeq \overline{(\langle\sigma_i\sigma_j\rangle - \langle\sigma_i\rangle \langle\sigma_j\rangle)^2} \equiv \chi_R(k) \tag{5.12}$$

called the replicon mode *. The above combination is easily shown to be the susceptibility with respect to the external coupling H (it is most obvious on Blandin's definition (5.7)).A simple but unfortunately, we believe, wrong argument of Bray and Moore[40] shows that χ_R is always infinite in the ordered phase when $H = 0$ in analogy with the transverse mode of ferromagnets. Such a property,in agreement with the fact that (5.12) has a massless mode, would be of course welcome.

Trouble seems to creep in when (5.9) is pushed to higher orders[38,39]

$$\int d^dx \left\{ -\frac{u}{8} \sum Q^{\alpha\beta}Q^{\beta\gamma}Q^{\gamma\delta}Q^{\delta\alpha} + \frac{x}{4} \sum (Q^{\alpha\beta})^2 (Q^{\alpha\gamma})^2 - \frac{y}{8} \sum (Q^{\alpha\beta})^4 + \dots \right\} \tag{5.13}$$

with
$$u = x = \frac{3y}{2} = z^4 \quad , \tag{5.14}$$

where the destabilizing $(Q^{\alpha\beta})^4$ terms are the attractive Q^3 terms of the equation of state (4.14). Indeed they destabilize since the spectrum found for the replicon by Pytte and Rudnick is now $k^2 - 2yQ^2$ (instead of the massless k^2). We may assume from global stability arguments that the destabilizing terms should not be effective (unlike in Landau Ginzburg systems with long-tailed probability). At most one could expect a first order transition to appear due to the presence of attractive Q powers that dominate for Q large but not too large (in the large Q limit the potential is at most linear). The problem (can we deal with $\mathcal{H}(Q)$ stopped after Q^4) has been taken seriously by Bray and Moore[37,39]. They propose a symmetry breaking that seems to provide a desired answer (restoring the massless mode to lowest order) but which we believe incorrect (the degeneracy weight of their saddle point vanishes with n).

iv) Renormalization group :

The behavior of the cubic Hamiltonian (5.9) has for critical dimension $d_c = 6$ and in the critical region higher order terms (5.13) are irrelevant. Critical exponents have been derived by Harris, Lubensky, Chen and Chen, Lubensky[41]. They give a specific heat

$$C_v = at - bt^{-\alpha} \tag{5.15}$$

$$\alpha = -(1 + 2(6-d) + \dots) \tag{5.16}$$

that shows no cusp and looks linear as one would hope for. How much is to be believed near $d = 3$ is another matter. Besides, the cusp also disappears from the magnetic susceptibility.

* In dynamic terms the replicon χ_R is $\overline{\langle i\hat{\varphi}\varphi\rangle \langle i\hat{\varphi}\varphi\rangle}$, that is also the susceptibility $\delta Q/\delta\Delta_h$ in zero field $\Delta_h=0$ (just like $\langle i\hat{\varphi}\varphi\rangle$ is $\delta\langle\varphi\rangle / \delta h$).

v) Further order parameters :

We have seen in § 4 *iii*) that whether $<\varphi^\alpha\varphi^\beta>$ or $<\varphi^\alpha\varphi^\beta\varphi^\gamma\varphi^\delta>$ shows up first as a spontaneously non-vanishing order parameter is a delicate balance between random attraction and φ^4 repulsion. Similarly here it would be interesting to explore the stability of $<\sigma^\alpha\sigma^\beta>$ against $<\sigma^\alpha\sigma^\beta\sigma^\gamma\sigma^\delta>$. If one assumes that $\mathcal{P}(J_{jk})$ contains also a non vanishing second cumulant, one is in fact pushed into using both fields to discuss the appearance of a T_g . We point out this as a possibly interesting exercise to the reader.

Having dealt with mean field, fluctuations and renormalization group, it would have been appropriate to describe how one relaxes towards equilibrium. However, the long-time behavior evidence from Monte Carlo computations is too controversial[8] to lend itself to a brief summary. We only mention, because we have dealt at length with the SK model, that Kirkpatrick and Sherrington[42] have solved its kinetic equations at and above T_g . The decaying part $\tilde{C}(t)$ of the spin correlation goes like $t^{-1/2}$ (which is an exact result of Ma and Rudnick for the Landau-Ginzburg model). Below T_g Monte Carlo computations show that it is well approximated by the same power law. This slow relaxation arises however from a continuous spectrum of relaxation rates extending down to zero.

6. LOW TEMPERATURE APPROACH AND FRUSTRATION

Given the confusion that exists on the behavior of the presumed ordered phase near T_g , people have tried to understand the ground state and low lying excited states properties. Foremost in this context is the concept of frustration as pointed to by Anderson and developed by Toulouse[43], who in particular emphasized the role played by gauge invariance.

Consider first the ideal spinglass system in dimension $d = 2$, $J^R_{ij} \equiv J\varepsilon_{ij}$

$$\mathcal{H} = - J \sum_{j\ell} \varepsilon_{j\ell}\, \sigma_j\, \sigma_\ell \tag{6.1}$$

where the spins σ (matter field) and the nearest neighbor bonds ε (gauge field) take values ± 1 . The simplest gauge invariant quantity is the frustration on a plaquette $P(ijk\ell)$

$$\Phi_P = \varepsilon_{ij}\, \varepsilon_{jk}\, \varepsilon_{k\ell}\, \varepsilon_{\ell i} \quad . \tag{6.2}$$

If $\Phi_P = +1$ the plaquette is not frustrated (all bonds may be satisfied), if $\Phi_P = -1$ the plaquette cannot have all bonds satisfied, it is frustrated.

In the dual system (which we know to be appropriate for low T) one has spins μ_P on plaquettes (dual sites) and nearest interaction between plaquettes. For a specimen system characterized by a set of frustrated plaquettes $\{f\}$, Fradkin, Huberman and Shenker[44] have shown that

$$\mathop{\mathrm{Tr}}_{\sigma} \exp\left\{J \sum_{j\ell} \varepsilon_{j\ell}\, \sigma_j\, \sigma_\ell\right\} \simeq \mathop{\mathrm{Tr}}_{\mu} \exp\left\{J^* \sum_{pp'} \mu_p\, \mu_{p'}\right\} \prod_{\{f\}} \mu_f \qquad (6.3)$$

Here J^* is the dual coupling. It vanishes as $T \to 0$, being related to $J \equiv \Delta^{1/2}/T$ by

$$e^{-2J^*} = \mathrm{th}\, J \quad . \qquad (6.4)$$

We take the quenched average over a gauge invariant probability law that weights frustration to keep track of it

$$\mathcal{P}\{\varepsilon\} \simeq \prod_{(jk)} \delta(\varepsilon_{jk}^2 - 1) \prod_{P} \exp\{-\frac{\lambda}{2}\, \Phi_P\} \qquad (6.5)$$

We then obtain* [45], for the shift in free energy due to bond randomness

$$\overline{\ln Z(J^R)} - \ln Z(J) = \mathcal{N}^{-1}\left[y^2 \sum_{a,b} \ln \langle\mu_a \mu_b\rangle + y^4 \sum_{abcd} \ln \langle\mu_a \mu_b \mu_c \mu_d\rangle + \dots\right] \qquad (6.6)$$

$$\mathcal{N} = 1 + y^2 \sum_{a,b} + y^4 \sum_{abcd} + \dots + y^N = \frac{1}{2}\left((1+y)^N + (1-y)^N\right) \qquad (6.7)$$

Here averages are computed with the dual Hamiltonian $\mathcal{H}^*$, as in (6.3), and $y = e^{\lambda}$. For $y = 0$ (no frustration) we recover the non-random result, for $y = \infty$ the fully frustrated system and for $y = 1$ the ideal spin glass.

i) Fully frustrated systems :

In attempt to discriminate between frustration and disorder, Villain[46], Alexander and Pincus[47], Derrida, Pomeau, Toulouse, Vannimenus[48] have extensively studied fully frustrated systems. Results are rich. They raise the possibility[47] of a fully frustrated phase separate from the spin glass phase (near the upper critical dimension). Of particular interest is the degeneracy D_o of the ground state.

For Ising systems with $d = 2$, the entropy is extensive, $\ln D_o \sim N$. For example in (6.6) the fully frustrated term gives a free energy $\simeq \ln \langle\mu_a\, \mu_b \cdots \mu_N\rangle$ which has the degeneracy of distributing N/2 dimers on the lattice (for the square lattice it has been given by Kasteleyn[49] as $\exp \frac{G}{\pi} N$ where G is Catalan's constant). The transition temperature is $T = 0$.

For higher but even dimensionality the D_o is finite. For odd dimensionality the entropy is not extensive but behaves as[47] $N^{1/d}$ in f.c.c. systems (but $N^{2/3}$ in s.c. systems[48]), with presumably order in (d-1) dimensions. A Monte Carlo computation by Phani, Lebowitz, Kalos and Tsai[50] shows that the f.c.c. system goes disordered via a first order transition. What happens very close to $T = 0$ is not yet clear.

---

* Directly or via replicas which shows clearly that, properly handled, replicas may be used at or near $T = 0$ with no frustration suppression (see Ref.[18]).

ii) Ideal spin glass :

It is instructive to look at the other end, where only few frustrations are present.

a) Frustration pair : In a specimen system there is a string of flipped bonds between frustrated squares. The string varies from specimen to specimen (Fig.4a,b). For a single frustration the string goes to infinity (Fig.4c). Along a string it is energetically favorable to create a Bloch wall (a closed string encloses a domain of flipped spins).

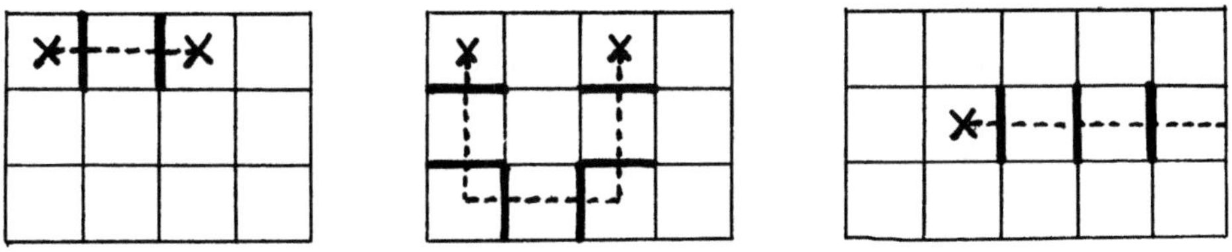

Fig.4 - a , b : Two strings of flipped bonds in two specimen systems with identical frustration configuration.
c : Single frustration square with infinite string.

Whatever the string of flipped bonds that connects the only pair of frustrated squares, the contribution to F/T is the same and proportional to $\ln\langle\mu_o\mu_r\rangle$. In the low T limit ($J^* \to 0$) we have

$$\ln\langle\mu_o\mu_r\rangle \simeq \ln\left\{(J^*)^{|x|+|y|}\, D\,\left[1+O(J^*)\right]\right\} \tag{6.8}$$

$$\simeq -2\,(|x|+|y|)J + \ln D + O(J^*) \tag{6.9}$$

where $r=(x,y)$, $|x|+|y|$ is the length of the shortest path on the dual lattice between 0 and r , D is the degeneracy of the path

$$D = \frac{(|x|+|y|)!}{|x|!\;|y|!} \tag{6.10}$$

longer paths are of order J^* (compared to J). This is Toulouse's[51] rule of minimal strings between frustrated pairs for the ground state.

When going from specimen systems to (quenched) averages over $\mathcal{P}(\varepsilon)$, one has to sum over sites (0,r) and normalize by $\mathcal{N} = 2^{N-1}$. This last factor dwarfs contributions to the entropy or internal energy that come out from (6.9) to let only survive configurations in a region near N/2 frustrated squares[45].

b) Frustration multiplets : With more than one pair of frustrated squares one may have contributions to the entropy that involve correlating any number of them. For example with four sites forming a square with the lattice along its diagonals

$$\ln < \mu_{r_1} \mu_{r_2} \mu_{r_3} \mu_{r_4} > \simeq \ln \left\{ (J^*)^R \, 2D^2 \left[1 + O(J^*) \right] \right\} \tag{6.11}$$

with the total path length

$$R = |x_1 - x_2| + |y_1 - y_2| + |x_3 - x_4| + |y_3 - y_4| = |x_1 - x_4| + |y_1 - y_4| + |x_3 - x_2| + |y_3 - y_2|$$

and $$D = (|x_1 - x_2| + |y_1 - y_2|)! \,/\, |x_1 - x_2|! \, |y_1 - y_2|!$$

The extra factor of two comes from the two possibilities of joining summits in pairs by minimal paths (of total length R).

At the other extreme, when all squares are frustrated, the individual degeneracy of paths between pairs of frustrated squares is $D = 1$, but the degeneracy $\mathcal{D}$ of reorganizing connections (redistributing dimers) may involve anything from two pairs of frustrated squares to all N of them.

c) Zero energy contours and picture :

In its simplest form a zero energy contour appears (Fig.5a) with four frustrated squares forming themselves a square, ($D = 1$, $\mathcal{D} = 2$) . In any specimen the spin at the central site is then allowed to flip freely without changing the ground state energy (two ferro and two antiferro bonds whatever way we connect the squares in pairs) restituting a $\ln \mathcal{D}$ contribution to S.

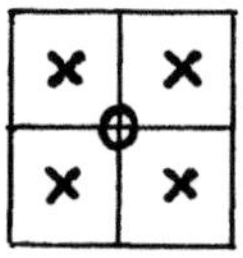

a

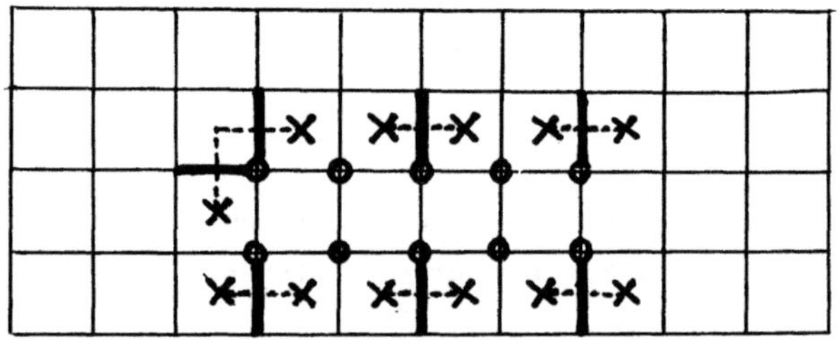

b

Fig.5 - a) Freely flipping spin inside elementary zero energy contour.

b) Freely flipping blob of ten spins inside zero energy contour made of six strings.

More generally if we have a contour (Fig.5b) where half of the perimeter is made of (p) pieces of (minimal) strings connecting (p) pairs of frustrated squares we have a zero energy contour. Imagine all spins aligned outside, spins along the border contour are half frustrated and half aligned whether they (and all other spins inside the contour) are all up and down. The freedom of the blobs of spins, inside zero energy contours, to flip without affecting the ground state energy, is what builds the zero temperature entropy. For evaluation of the entropy it remains easier to count the freedom to connect frustrated pairs by minimal strings than try to determine zero energy contours and spin freedom.

The picture emerging at zero temperature is a sea of blobs with convoluted boundaries, some uncorrelated and flipping freely, others correlated together to flip.

d) Extension to dimension $d = 3$. *Wilson loops* :

The corresponding results for $d = 3$ are then obtained[51,44] if one remarks that

1) in $d = 2$ the spin σ_j is replaced by the spin μ_p on a plaquette or by relabelling, $\mu_{\tilde{j}}$ (dual space site)

in $d = 3$ the spin μ_p on a plaquette is relabelled $\mu_{\tilde{j}\tilde{k}}$ (dual link)

2) in $d = 2$ the n.n. spin σ interaction (link) becomes a n.n. plaquette interaction or by relabelling a n.n. μ interaction (link ; self dual)

in $d = 3$ the $\mu_p \mu_{p'}$ link interaction becomes a $\mu_{\tilde{j}\tilde{k}}\mu_{\tilde{k}\tilde{\ell}}\mu_{\tilde{\ell}\tilde{i}}\mu_{\tilde{i}\tilde{j}}$ plaquette interaction (plaquette ; dual of link)

$$\mathcal{H}^* = - J^* \sum_p \mu\,\mu\,\mu\,\mu \tag{6.12}$$

The frustration expansion (6.6,7) remains identical provided averages over frustrated μ's are taken with the weight $\exp - \mathcal{H}^*$. These averages are only non vanishing if, in dual space, one can pave the product $\prod_{\{f\}} \mu_f$ with plaquettes (6.12). This implies that $\prod_{\{f\}}$ form one or several *closed loops* [43,44,51].

Again let us look at weak concentration of frustrations, i.e. a single Wilson loop[52], in a specimen system. Its contribution to F/T is proportional to

$$\ln < \prod_{\text{loop}} \mu_{\tilde{j}\tilde{k}} >_{J^*} \tag{6.13}$$

which is known[51-54] to undergo a change in behavior at T_c. It asymptotically behaves like the minimal area spanned by the Wilson loop (low T) or like its edge or perimeter (high T). This is of identical nature to the behavior we had in dimension $d = 2$. The $\ln <\mu_{\tilde{j}}\, \mu_{\tilde{k}}>_{J^*}$ term corresponding to (6.13) was behaving like the minimal length of the line spanning $\tilde{j}$, $\tilde{k}$ at low T or like its edge (length independent) at high T.

In the zero temperature limit, one has to cover all the loops present in each term of the frustration expansion, with surfaces of minimum area which is again the Toulouse[43]-Kirkpatrick[55] rule.

7. DEFECT ENERGY

In pure systems topological defects may occur as thermal excitations or as induced by boundary conditions. They have been classified by Toulouse and Kleman[56]. The higher the temperature the easier they are created and the defect energy vanishes as T_c is reached[57].

In frustrated systems we are given beforehand a network of frustrated plaquettes (and their associated minimal strings or surfaces at $T=0$) which act as external sources for defects. When we set spins in such a system, they try to minimize the free energy using their natural defects to adapt to the external conditions. This approach is taken by Toulouse[58] to extend his classification (and propose a macroscopic order parameter for cases with $n \geq 2$ which we do not consider here). Because the frustration network is a source (or a sink) of Bloch walls, we have zero energy contours. Several authors[59-61] have computed, on specimen systems, the defect energy

$$\Delta F = F_A - F_o \qquad (7.1)$$

which is the change in free energy induced by reversing half of the boundary spins in the ground state (free energy F_o) and letting the system relax. Since the ground state is here highly degenerate a most probable ground state is chosen for reference. This defect energy (per unit length, surface) is found to vanish in dimension $d=2$ (but not in $d=3$) as soon as the concentration x of antiferromagnetic bonds exceeds a critical value $x_f \simeq .1$, in particular in the ideal spin glass limit $x=1/2$.

From there Toulouse, Vannimenus and Maillard[62] have suggested that, as the temperature is raised, a transition from spin glass to paramagnetic phase may occur when the zero energy contours become so convoluted that it is impossible to tell from which ground state the system originated. The transition would occur at T_R , the roughening transition[63] temperature, when the width of the boundary of the zero-energy contour reaches the order of the correlation length. In $d=2$, one has $T_R=0$; but in dimension $d=3$, one has $T_R>0$, leaving room to take care of the transition spin glass to paramagnet.

Derrida, Maillard, Vannimenus and Kirkpatrick[64] have given an interesting argument however, to warn not to take a zero defect energy at $T=0$ as necessarily implying that $T_g=0$. Indeed, one has for the defect energy

$$\Delta F = U_A - U_o + T(S_o - S_A) \quad . \qquad (7.2)$$

If $U_A - U_o$ vanishes with T , and if the creation of a defect *lowers* the entropy in a non-vanishing proportion (which they construct examples of), the entropy term in (7.2) may take over. In that case one has a ΔF rising with T, near $T=0$, to vanish eventually at a higher temperature, T_g .

8. ORDER PARAMETERS. CONCLUSION

We now summarize the situation for the order parameter.

In the Landau-Ginzburg case, we have seen the appearance of a gellation or freeze-in transition, with $\langle\varphi^\alpha\varphi^\beta\rangle$ or $\langle\varphi(t)\ \varphi(t')\rangle_{t-t'\to\infty}$ the order parameter supposed to describe the ordered phase. In all cases the system appears blatantly unstable unless the randomness is governed by short tailed probability laws where the instability may become limited to a small region. This is no problem for the pure random potential system where we precisely try to compute instability characteristics (e.g. $\mathrm{Im}\,G = \rho(E)$). In the coupled Landau-Ginzburg Hamiltonian (random temperature), we have to live with that instability and after some short time in the frozen-in phase, we do not know where it would drive us (large cluster paramagnetism ?).

In the more interesting random bond Hamiltonian the corresponding (local) order parameters $\langle\sigma_j^\alpha\ \sigma_j^\beta\rangle$ appears to describe the freeze-in. With the same questions arising : instability but limited perhaps to intermediate values ($\mathcal{H}\{Q^{\alpha\beta}\}$ is positive for large values of $Q^{\alpha\beta}$), uncertainty about the role of order parameters involving more operators.

Looking at the spin glass systems from the low-T side, we have seen above no clear evidence in favor or against a freeze-in transition in $d = 2$ or $d = 3$. Here we have to distinguish between the three models considered, EA (Edwards - Anderson), SK (Sherrington - Kirkpatrick), ISG (Ideal Spin Glass).

The most thoroughly explored (and the less physically relevant), the SK model poses the well defined mathematical problem : how to extend through $T = 0$ a solution known for $T \gtrsim T_g$, and obtain what MC computations reveal, a vanishing entropy at $T = 0$. Dimension is absent in this model. The EA order parameter $\langle\sigma_j^\alpha\ \sigma_j^\beta\rangle$ seems appropriate but one does not know what to do (symmetry breaking in replica space, introduction of further operators) to make sense at low T.

The EA and ISG models have been thoroughly explored near T_g and T=0 and MC computations do not seem to find much difference between the two models. A continuous spectrum of couplings (EA) suppresses exact zero-energy contours by giving rise to a continuous range of activation energies. As a result the EA ground state has a lower degeneracy (the entropy is in N^a, $a \ll 1$, instead of being extensive like for ISG in d=2,3).

Besides, MC computations are confusing since they seem to give consistently an order parameter that decreases on very long time scales, not only for EA or ISG but for SK as well[8], where the reality of such an order parameter seems well established. We have seen the ground state as highly degenerate, a complicated assembly of spin blobs of various sizes, flipping rigidly with no (or almost no) energy cost, even for time scales that become infinite at T=0. In the very words of Kirkpatrick and Sherrington[42], one may view the picture in phase space as a large number of large valleys paved by small local extrema (flipping of small blobs) and separated by hills(flipping of larger blobs). How this picture may coexist with strict time long-range order is not clear to the author. Obviously, it would be very valuable to be able to follow the low temperature behavior of $<\sigma_j^\alpha \sigma_j^\beta>$ or rather $<\sigma_j(s)\ \sigma_j(t)>_{t\to\infty}$ and see what becomes of gellation as one approaches the ground state.

To end up we wish to point out that we have left several very important aspects of the subject. In particular we have ignored xy and Heisenberg systems. We have not even mentioned the giant cluster approach which may very well be the real cue to understanding systems with dilute quenched magnetic impurities.

Finally, we wish to thank A. Blandin, E. Brézin, B. Derrida, C. Itzykson, G. Parisi, J. Rudnick, G. Sarma, G. Toulouse and H. de Vega for discussions from which we benefited. Special thanks go to L. Peliti, in particular, for clearing up level density results and G. Parisi and J. Zittartz for private communication on their own work.

REFERENCES

[1] - M. Fisher, Phys. Rev. 176, 257 (1968).
[2] - P.W. Anderson, "Lectures on Amorphous Systems", Les Houches 1978, Ed. R. Balian, R. Maynard, G. Toulouse (North-Holland, Amsterdam 1979).
[3] - J. Joffrin, "Les Systèmes Désordonnés, Aspect Expérimental", - id -
[4] - S. Kirkpatrick, "Models of Disordered Material", - id -
[5] - T.C. Lubensky, "Thermal and Geometrical Phenomena in Random Systems" - id -
[6] - D.J. Thouless, "Percolation and Localization", - id -
[7] - A. Blandin, J. Physique C6, 1499 (1978).
[8] - K. Binder and D. Stauffer, "Monte Carlo Studies of Systems with Disorder" in Topics in Current Physics, vol.7, Ed. K. Binder (Springer Verlag, Berlin 1979) p.301.
[9] - S.K. Ma, "Modern Theory of Critical Phenomena (Benjamin, Reading, Mass. 1976).
[10] - R. Brout, Phys. Rev. 115, 824 (1959).
[11] - J. Harris, J. Phys. C7, 1671 (1974).
[12] - S. Edwards and P.W. Anderson, J. Phys. F 5, 965 (1975).
[13] - J.A. Mydosh, in Amorphous Magnetism II, Ed. R. Levy, R. Hagasawa (Plenum Press, New York, 1977) p.73.
J.A. Mydosh, J. Magn. Mag. Mat. 7, 237 (1978).
[14] - A. Aharony in Phase Transition and Critical Phenomena, Vol.6, Ed. C. Domb, M.S. Green (Academic Press, London, 1976). p.357.
[15] - D.J. Thouless, Phys. Rep. 13, 93 (1974).
[16] - P.W. Anderson, Phys. Rev. 109, 1492 (1958).
[17] - D. Sherrington and S. Kirkpatrick, Phys. Rev. Lett. 35, 1972 (1975).
[18] - A. Aharony, J. Phys. C11, L457 (1978).
[19] - G. Grinstein, S.K. Ma and G.F. Mazenko, Phys. Rev. B15, 258 (1977).
[20] - S.K. Ma and J. Rudnick, Phys. Rev. Lett. 40, 589 (1978).
[21] - C. De Dominicis and L. Peliti, Phys. Rev. B18, 353 (1978).
[22] - P.C. Martin, E. Siggia and H.Rose, Phys. Rev. A8, 423 (1973).
[23] - C. De Dominicis, Phys. Rev. B18, 4913 (1978).

[24] - B. Halperin, P.C. Hohenberg and S.K. Ma, Phys. Rev. Lett. 29, 1548 (1972) ; Phys. Rev. B10, 139 (1974).
[25] - C. De Dominicis, J. Phys. C1, 247 (1976) ; H.K. Janssen, Z. Phys. 24, 113 (1976).
[25'] - C. De Dominicis, L. Peliti, unpublished.
[26] - C. De Dominicis, J. Rudnick, unpublished.
[27] - F. Tanaka, Prog. Theor. Phys. 60, 380 (1978).
[28] - C. De Dominicis, Nuovo Cimento Lett. 12, 567 (1975) ; C. De Dominicis, E. Brézin, J. Zinn-Justin, Phys. Rev. B12, 4945 (1975).
[29] - J. Cardy, J. Phys. C11, L321 (1978).
[30] - S. Edwards, J. Non-Cryst. Solids 4, 417 (1970).
[31] - I.M. Lifschitz, Soviet Phys. JETP 26, 462 (1968).
[32] - J. Zittartz, S. Langer, Phys. Rev. 148, 741 (1966) and J. Zittartz, private communication.
[32'] - A. Aharony and Y. Imry, J. Phys. C10, L487 (1977).
[33] - E. Abrahams, P.W. Anderson, Licciardello and Ramakrishnan, Phys. Rev. Lett. 42, 693 (1979).
[34] - D.J. Thouless, P.W. Anderson and R.G. Palmer, Phil. Mag. 35, 1792 (1977).
[35] - A. Blandin, M. Gabay and T. Garel, Orsay preprint (1978).
[36] - G. Parisi, communication at 6th MECO Seminar, Trieste 1979.
[37] - A.J. Bray and M.A. Moore, J. Phys. C12, 79 (1979).
[38] - E. Pytte and J. Rudnick, IBM preprint (1978).
[39] - A.J. Bray and M.A. Moore, Phys. Rev. Lett. 41, 1068 (1978).
[40] - A.J. Bray and M.A. Moore, Manchester preprint (1978).
[41] - A.B. Harris, T.C. Lubensky and J.H. Chew, Phys. Rev. Lett. 36, 415 (1976) ; J.H. Chen and T.C. Lubensky, Phys. Rev. B16, 2106 (1977).
[42] - S. Kirkpatrick and D. Sherrington, Phys. Rev. B17, 4384 (1978).
[43] - G. Toulouse, Commun. on Phys. 2, 115 (1977).
[44] - E. Fradkin, B.A. Huberman and S. Shenker, Phys. Rev. B18, 4789 (1978).
[45] - C. De Dominicis and M. Stephen, J. Phys. C11, L969 (1978).
[46] - J. Villain, J. Phys. C10, 1717 (1977).
[47] - S. Alexander and P. Pincus, UCLA preprint (1978).
[48] - B. Derrida, Y. Pomeau, G. Toulouse and J. Vannimenus, ENS-Paris preprint (1979).
[49] - P.W. Kasteleyn, J. Math. Phys. 4, 287 (1963).
[50] - M.K. Phani, J.L. Lebowitz, M.H. Kalos and C.C. Tsai, Phys. Rev. Lett. 42, 577 (1979).
[51] - R. Balian, J.M. Drouffe and C. Itzykson, Phys. Rev. D11, 2637 (1978).
[52] - K.G. Wilson, Phys. Rev. D10, 2445 (1974).
[53] - F. Wegner, J. Math. Phys. 12, 2259 (1971).
[54] - E. Fradkin and L. Susskind, Phys. Rev. D17, 2637 (1978).
[55] - S. Kirkpatrick, Phys. Rev. B16, 4630 (1977).
[56] - G. Toulouse, M. Kleman , J. Physique 37, L149 (1976).
[57] - M. Fisher, J. Phys. Soc. Japan 26, suppl. 87 (1969).
[58] - G. Toulouse, Phys. Rep. 49, 267 (1979).
[59] - J. Vannimenus and G. Toulouse, J. Phys. C10, L537 (1977).
[60] - P. Reed, M.A. Moore and A.J. Bray, J. Phys. C11, L139 (1978).
[61] - S. Kirkpatrick, unpublished and ref. 4 .
[62] - G. Toulouse, J. Vannimenus and J.M. Maillard, J. Physique 38, L459 (1977).
[63] - S.T. Chui and J.D. Weeks, Phys. Rev. B14, 4978 (1976).
[64] - B. Derrida, J.M. Maillard, J. Vannimenus and S. Kirkpatrick, J. Physique 39, L465 (1978).

DYNAMICS OF THE ONE-DIMENSIONAL HEISENBERG SPIN SYSTEM

A. Sjölander

Institute of Theoretical Physics
S-412 96 Göteborg, Sweden

DYNAMICS OF THE ONE-DIMENSIONAL HEISENBERG SPIN SYSTEM

A. Sjölander
Institute of Theoretical Physics
S-412 96 Göteborg, Sweden

I INTRODUCTION

Our interest in low dimensional systems can be traced to two different reasons. Firstly, they show often rather unusual dynamical behaviour, originating from very strong correlation effects. Secondly, our possibility of analysing in detail relevant theoretical models is improved as we go to lower dimensions and this is particularly true in one dimension. A material of particular interest in this respect is $(CD)_3NMnCl_3$ (TMMC) which consists of Mn^{++} ions, lined up along chains, and which shows striking one-dimensional magnetic properties down to 1K[1]. It is found to correspond quite well to an ideal nearest neighbour Heisenberg chain with antiferromagnetic coupling. The Mn^{++} ions have spin 5/2 and this makes that quantum corrections are reasonably small even at low temperatures.

No long range spin order can exist in this system at any finite temperature, but the short range order is found to extend very far. Sharp spin wave resonances appear at low temperatures[2], in spite of lack of long range order, and the line width is found to increase essentially linearly with temperature[3]. This can be understood qualitatively on the basis that we have for small T a strong local ordering of the spins. However, a more detailed understanding of the experimental results requires a proper analysis of the relevant theoretical model, the Heisenberg model in this case.

Various static correlation functions are known exactly[4,5,6] and the value for the inverse correlation length is $\kappa=k_BT/|J|S^2a$, where J is the nearest neighbour exchange constant, S the length of each spin, and a the spin lattice constant. It is also a fact that the Heisenberg model gives sharp spin wave resonances as $T\to 0$[5], but earlier predictions of the linewidth were contradictory[1].

Here, I will report on calculations by G. Reiter and myself on the dynamics of this model. The results on the linewidth and the lineshape are exact to lowest order in temperature. The basic results were reported in Physical Review Letters[7] nearly two years ago, but the full paper on this has been completed just recently and will be published elsewhere. In the intervenient time some new interesting results were obtained and, in particular, extensive computer simulations by Heller and Blume[8] have been completed, both for the ferromagnet and the antiferromagnet. Some comparisons between these results and our asymptotic ones will be presented later on.

I will here only briefly outline our approach, stressing the most important features of the model. We may in certain respects consider T=0 as a critical point, having long range order at T=0. The question of proper hydrodynamics, validity of dynamical scaling etc. are then of particular interest and highly relevant for this conference. The static susceptibility shows the ordinary mean field behaviour near T=0[4], but the dynamics carries some surprises.

II MODEL CALCULATION

We consider the model Hamiltonian

$$H = -\frac{1}{2}\sum_{q'} J_q \vec{S}_{-q}\cdot\vec{S}_q \quad , \tag{II.1}$$

where $\vec{S}_q$ are the Fourier components of the classical spin vectors $\vec{S}_R$ and, for nearest neighbour interaction, $J_q=2J\cos(qa)$. The quantity we will be interested in is the relaxation function

$$\Sigma(q,\omega) = \int_0^\infty dt\, e^{i\omega t} \langle\vec{S}_{-q}(0)\cdot\vec{S}_q(t)\rangle \tag{II.2}$$

and the corresponding spectral function. Using the Zwanzig[9] and Mori[10] projection operator technique, we may write $\Sigma(q,\omega)$ in the form

$$\Sigma(q,\omega) = i\,\langle\vec{S}_{-q}\cdot\vec{S}_q\rangle \Big/ \left[\omega - \frac{\Omega_q^2}{\omega+\Gamma_q(\omega)}\right] \quad . \tag{II.3}$$

Here, $\langle\vec{S}_{-q}\cdot\vec{S}_q\rangle$ as well as Ω_q and $\Gamma_q(t=0)$ are known exactly from earlier work of Fisher[4], Lovesey and Meserve[5], Tomita and Mashiyama[6]. The formal expression for $\Gamma_q(t)$ contains the second time derivative of $S_q(t)$ and we may insert

$$\ddot{\vec{S}}_q(t) = 8J^2\sin(qa/2)N^{-1}\sum_{q'}\{\sin(qa/2)\,h_{q'}(t) + \sin[(q'-q)a/2]$$
$$\times\cos[(q'-q)a]\;e_{q'}(t)\}\,\vec{S}_{q-q'}(t) \quad . \tag{II.4}$$

The new quantities, appearing here, represent two kinds of collective modes and they are defined as

$$e_q(t) = \sum_{q''}\cos(q''a)\,\{\vec{S}_{q''+q/2}(t)\cdot\vec{S}_{-q''+q/2}(t)\}$$
$$h_q(t) = \sum_{q''}\cos^2(q''a)\,\{\vec{S}_{q''+q/2}(t)\cdot\vec{S}_{-q''+q/2}(t)\} \tag{II.5}$$

The first one is simply the Fourier transform of the local energy density and the second one is the corresponding quantity for the next nearest neighbour. We shall by $\delta e_q(t)$ and $\delta h_q(t)$ mean their fluctuation around the corresponding equilibrium values. $\Gamma_q(t)$ becomes for finite q and low temperatures proportional to T, and for the ferromagnet we have asymptotically, as $T \to 0$,

$$\Gamma_q(t) = i\,(8J^2/S^2\kappa a)\,\sin^2(qa/2)\,\langle \delta h_{-q}(0)\,\delta h_q(t)\rangle \;, \tag{II.6}$$

where $\kappa a = k_B T/JS^2$ and $\langle \delta h_{-q}(0)\,\delta h_q(0)\rangle$ is proportional to $(\kappa a)^2$. This yields through (II.3) sharp resonances in $\Sigma(q,\omega)$. It is here sufficient to calculate the time evolution of $\delta h_q(t)$ at absolute zero temperature and this is done by first considering the equation of motion for the two-spin variable $f_q(q',t) = \delta\{\vec{S}_{q'+q/2}(t)\cdot\vec{S}_{-q'+q/2}(t)\}$. For T=0 it satisfies the equation

$$\Big[\frac{\partial^2}{\partial t^2} + \omega_q^2(q_1')\Big]\,\langle f_{-q}(-q_2',0)\,f_q(q_1',t)\rangle - \frac{1}{2}\Big[\delta_{q_1'+q/2,0} + \delta_{-q_1'+q/2,0}\Big]$$
$$\times \sum_{q''} \omega_q^2(q'')\,\langle f_{-q}(-q_2',0)\,f_q(q'',t)\rangle \;=\; 0 \;, \tag{II.7}$$

where

$$\omega_q^2(q_1') = (4JS)^2\,\sin^2(qa/2)\,\sin^2(q_1'a) \;, \tag{II.8}$$

and $\delta_{q,0}$ is the Kronecker symbol. The initial values can be calculated, using results of Tomita and Mashiyama[6]. It is readily checked that

$$\sum_{q'} f_q(q',t) = 0 \;, \tag{II.9}$$

implying that the fluctuations in the spin length are always kept to zero. This turns out to be a very important point in order to recover sharp spin wave resonances at T=0. The solution for $\Gamma_q(\omega)$ contains sums over q_1' and q_2' and these are easily evaluated through residue calculus. We obtain

$$\Gamma_q(t) = i\,(\kappa a/2)\big[4JS\sin(qa/2\big]^2\,\Big\{\,\frac{J_1(\tau)}{\tau} + J_0(\tau)\,\cos^2(qa/2)\Big\} \tag{II.10}$$

where $\tau = 4JS\sin(qa/2)t$ and J_0 and J_1 are Bessel functions. The memory function has for small values of q a very slow decay in time and is furthermore oscillatory. Its Fourier transform, defined as for $\Sigma(q,\omega)$ in (II.2), is

$$\Gamma_q(\omega) = -\,2JS(\kappa a)\sin(qa/2)\big[\zeta - (\zeta^2-1)^{1/2} + (\zeta^2-1)^{-1/2}\cos^2(qa/2)\big] \;, \tag{II.11}$$

where $\zeta = \omega/4JS\sin(qa/2)$. The singular behaviour for $\zeta=1$ reflects simply a singularity in the two-spin density of states, but this has hardly any noticeable effect on the spectral function in the ferromagnetic case. The situation is different in this respect for the antiferromagnet.

With the present results inserted into (II.3) we obtain for low temperatures a spin wave resonance at

$$\omega_q = 4JS\,\sin^2(qa/2)\big[1-(\kappa a/2)\big] \tag{II.12}$$

and the half width at half maximum is

$$\Delta_q = JS(\kappa a)\,\sin(qa)\ , \tag{II.13}$$

except close to $qa=\pi$ where

$$\Delta_q = JS(\kappa a)^{3/2}\ . \tag{II.14}$$

Our results do not give the expected spin diffusion behaviour for $q<<\kappa$ and it is found that (II.11) ceases to hold for $q<\kappa$. This is most easily clarified by referring back to (II.4). We notice there that the coefficient before $h_{q'}(t)$ becomes proportional to q^2, as $q\to 0$, and the fluctuations in the energy density, $\delta e_{q'}(t)$, begin to dominate. However, this contribution enters first in higher order in the temperature and was therefore lost in the asymptotic solution above. An approximate calculation of $\Gamma_q(t)$ beyond lowest order yields for $q<\kappa$ a spin diffusion behaviour for $\Sigma(q,\omega)$ with a singular temperature dependence of the diffusion constant[11],

$$D = \pi JSa^2/2\ \ln(1/\kappa a)\ , \tag{II.15}$$

This contradicts the dynamical scaling hypothesis, which requires that D tends to a finite value as $\kappa\to 0$. We also notice that the important modes are shifted from $h_q(t)$ to $e_q(t)$ as q passes from above κ to below, causing the violation of dynamical scaling.

The antiferromagnetic case is handled in a similar way but the calculations are more lengthy and shall not be discussed here. I only remark that in this case both $\delta e_q(t)$ and $\delta h_q(t)$ are important throughout the whole reciprocal space and there is no correspondence here to what happened above around $q=\kappa$. Again, we find sharp spin wave resonances and the line width is linear in T and independent of the wave vector for $T\to 0$. $\Sigma(q,\omega)$ shows diffusive behaviour for small q and the diffusion constant is

$$D = 2|J|Sa/\kappa\ , \tag{II.16}$$

and dynamical scaling seems to hold. However, the scaling hypothesis is violated to some extent at the zone boundary and this is caused by the singularity of the two-spin density of states.

III COMPARISONS WITH EXPERIMENTAL AND COMPUTER SIMULATION RESULTS

Earlier experiments on TMMC by Hutchings and Windsor[3] and by Shirane and Birgenau[12] have clearly demonstrated a linear temperature dependence of the line width and that the borderline for the existence of spin waves is $q\sim\kappa$. This is in contradiction to some of the earlier model calculations[1]. Extensive computer simulations have been carried out in recent years[8,13,14]. In our earlier publication[7] we presented one figure of $\Sigma(q,t)$, showing how our asymptotic results compare with the computations of Windsor and Locke-Wheaton[13] for a rather high temperature ($\kappa a=0.3$, T=34K for TMMC). We then inserted in (II.3) the exact finite temperature values for $\langle \vec{S}_{-q} \cdot \vec{S}_q \rangle$ and Ω_q, but our asymptotic form for $\Gamma_q(\omega)$ was used. The agreement was within the uncertainty of the computer data over the rather short time region, for which the simulations were done. Since then such calculations over much longer times have been reported by Heller and Blume[8] and some of their most recent results were provided by Professor Heller in private communications. For the ferromagnet they found that the line width varies as

$$\begin{aligned} \Delta_q &= (1.0 \pm .1)\ JS(\kappa a)\ \sin(qa) \quad , \quad qa<\pi/2 \\ &= (1.2 \pm .4)\ JS(\kappa a)^{1.45\pm.2} \quad , \quad qa=\pi \end{aligned} \qquad \text{(III.1)}$$

They did observe a change of the temperature dependence close to $qa=\pi$, and their results seem to agree with ours within the computational uncertainty. The figures below present comparisons between our asymptotic form for $\Sigma(q,t)$ and the computer simulation data at $\kappa a=0.1$ and $qa=\pi/2$ for both the ferromagnet and the antiferromagnet. Again, the exact finite temperature values of $\langle \vec{S}_{-q} \cdot \vec{S}_q \rangle$ and Ω_q are inserted. According to our results, the curve should for $qa=\pi/2$ be identical in the two cases, but it is noticeable

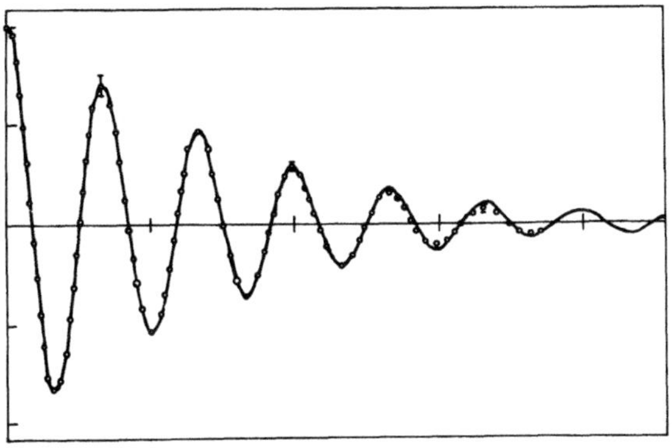

FIG.1. Comparison of present result for $\Sigma(q,t)/\Sigma(q,0)$ (full curve) with computer simulation data (open circles) at $qa=\pi/2$ and $\kappa a=k_BT/JS^2=0.1$ for the ferromagnet. Our asymptotic curve has been folded with an appropriate resolution function (the data were provided by P. Heller and G. Reiter).

from the figures that the "experimental" curve is more strongly damped in the antiferromagnetic case. The agreement between theory and experiments is excellent for the ferromagnet but there are certain discrepancies in the other case. The experimental line width shows a significant wavevector dependence in contradiction to our asymptotic result, but it seems to become less when decreasing the temperature. It leaves, however, open the question concerning the origin of the difference between the ferro- and antiferromagnet.

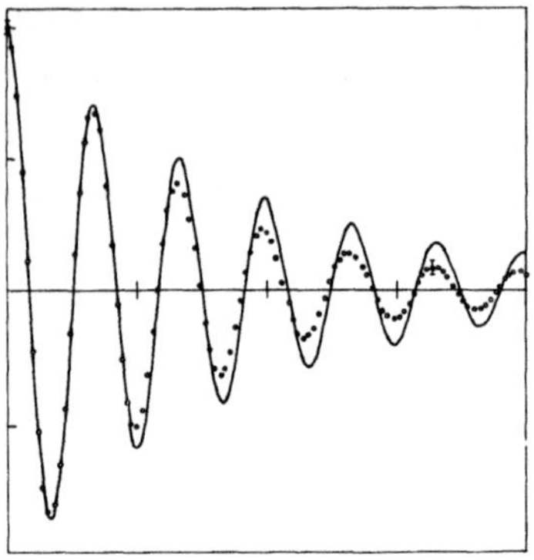

FIG.2. Comparison of present result for $\Sigma(q,t)/\Sigma(q,0)$ (full curve) with computer simulation data (open circles) of Heller and Blume[8] at $qa=\pi/2$ and $\kappa a=k_BT/JS^2=0.1$ for the antiferromagnet. The resolution corrections are negligible in this case (the data were provided by J. Loveluck and G. Reiter).

Another interesting point concerns the long wavelength limit for the ferromagnet, where we found a logarithmic temperature dependence of the spin diffusion constant. Professor Heller has provided preliminary data on this and they are in good agreement with (II.15).

A last remark concerns the good agreement we obtained at high temperatures. This is really not that surprising, for we have to remember that, by inserting the proper values of $<\vec{S}_{-q}\cdot\vec{S}_q>$, Ω_q, and $\Gamma_q(t=0)$, we satisfy automatically the frequency moments of $\Sigma(q,\omega)$ up to the fourth one and $\Sigma(q,t)$ is now quite short range in time.

ACKNOWLEDGEMENTS

I wish to thank Professor P. Heller for providing computer simulation data before publication and Dr. J. Loveluck for useful discussions and for providing material for one of the figures. I am particularly grateful to Dr. G. Reiter, whose collaboration was crucial for obtaining the presented results.

REFERENCES

1. M. Steiner, J. Villain and C.G. Windsor, Adv. in Physics 25, 87 (1976).
2. M.T. Hutchings, G. Shirane, R.J. Birgenau and S.L. Holt, Phys. Rev. B5, 1999 (1972).
3. M.T. Hutchings and C.G. Windsor, J. Phys. C10, 313 (1977).
4. M.E. Fisher, Ann. J. Phys. 32, 343 (1964).
5. S.W. Lovesey and R.A. Meserve, Phys. Rev. Lett. 28, 614 (1972).
6. H. Tomita and H. Mashiyama, Progr. Theor. Phys. 48, 1133 (1972).
7. G. Reiter and A. Sjölander, Phys. Rev. Lett. 39, 1047 (1977).
8. P. Heller and M. Blume, Phys. Rev. Lett. 39, 962 (1977).

9. R. Zwanzig, in Lectures in Theoretical Physics III, eds. W.E. Brittin, B.W. Downs and J. Downs, 1961, Interscience, New York.

10. H. Mori, Progr. Theor. Phys. 33, 423 (1965).

11. Such a logarithmic temperature was first conjectured by Nickels and was also derived through another approach by G. Reiter, Phys. Rev. 1979 (in press).

12. G. Shirane and R.J. Birgenau, Physica 87, 639 (1977).

13. C.G. Windsor and J. Locke-Wheaton, J. Phys. C9, 2749 (1976).

14. M. Blume, G.H. Vineyard and R.E. Watson, Phys. Lett. 50A, 397 (1975).

EXPERIMENTS ON HYDRODYNAMIC INSTABILITIES AND THE TRANSITION TO TURBULENCE

P. Bergé

DPh-G/Dir. CEN Saclay BP no 2 - 91190 Gif-sur-Yvette, France

EXPERIMENTS ON HYDRODYNAMIC INSTABILITIES AND THE TRANSITION TO TURBULENCE

by

P. BERGE
DPh-G/Dir. CEN Saclay BP n°2 - 91190 Gif-sur-Yvette, France

FOREWORD

Many hydrodynamic instabilities fascinated people since a long time and, then, have been subjected to many studies. Even if we restrict our subject to the more known and, may be, the simplest instabilities i.e. Taylor and Rayleigh-Benard instabilities, we cannot give any complete picture of this field. We apologize in advance for the fact that the few features presented here are necessarily very partial and we really cannot account for all the very numerous and beautiful experiments devellopped in this active subject. We wish to refer the reader to some fundamental works and review articles for a broader knowledge of the subjet [1],[2],[3],[4],[5].

INTRODUCTION

In any hydrodynamic instability we are in presence of a competition between destabilizing and stabilizing effects. The detailed balance between these two opposite effects is dependent on an external stress S applied to the system. When the external stress S is higher than a certain threshold S_c, destabilizing effects become dominant and the system becomes unstable.

Let us be more precise and consider the case of the two instabilities considered here, the Taylor instability and the Rayleigh-Benard (R.B.) instability.

The Taylor instability is the problem of the cylindrical Couette fluid flow, let us say when the inner cylinder rotates, the outer one being at rest. Let us suppose furthermore that the spacing d between the two cylinders is very small compared to the radius R and compared to the height L_x of these cylinders (see Fig.1a). The two opposite effects in the fluid under experiment are the *centrifugal force* which tends to produce a radial motion of the fluid and the *viscosity* ν which tends to prevent this motion. A very natural number measuring the balance between these 2 opposite effects is, indeed, the Reynolds number of this experiment

$$R_e = \frac{\Omega R d}{\nu} .$$

Usually and more rigorously we can use for this purpose the *Taylor number* which is, in the limit $d \ll R$ considered above,

$$T_a = 4 \frac{\Omega^2 d^4}{\nu^2} .$$

The Rayleigh-Benard instability is the problem of a fluid confined between two rigid, horizontal, conducting plates, distant of d, when an adverse thermal gradient is applied to this fluid. Here also we suppose $d \ll L_x$, L_x being the greater horizontal extension of the fluid layer .

Two kinds of effects compete: the buoyancy forces tend to create motion of the fluid, when the viscosity and the thermal diffusivity D_T tend to prevent these motions from appearing. The correct number measuring this balance is the *Rayleigh number*

$$R_a = \frac{g \, \alpha \, \Delta T \, d^3}{\nu \, D_T}$$

where g is the gravity acceleration, α the volumic expansion coefficient and ΔT the temperature difference between the two plates, see Fig.1b.

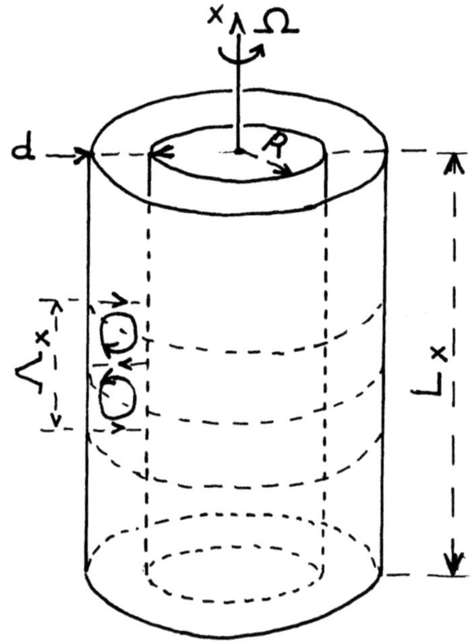

Fig. 1a - Taylor instability

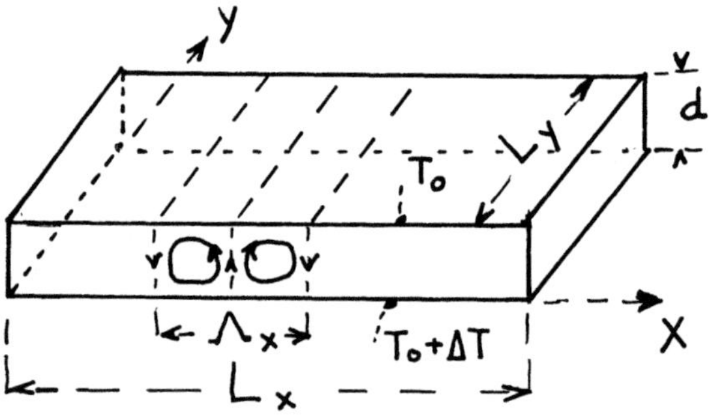

Fig.1b - Rayleigh-Benard instability

Following the terminology used above, the external stress S applied to these two systems is respectively the angular speed Ω and the temperature difference ΔT, more precisely the Taylor number T_a and Rayleigh number R_a. Let us emphasize that all the following results and proposed mechanisms refer always to *absolutely constant* external stress S. (Results obtained under time dependent stress are always much more complicated and, except for some very definite case,[6] rather confusing and not understood.)

For, the expression "by increasing the stress" may be understood as "by a very very slow increase of the stress S and after having allowed the system to reach its new equilibrium state".

This being stated, one can say that "by increasing the stress S" (from zero) the two systems described above become unstable at a perfectly defined value S_c of this external stress. This means that radial (and vertical) velocities set in, in the Taylor instability and vertical (and horizontal) velocities set in, in the Rayleigh-Bénard instability. The existence of these T_{ac} and R_{ac} critical thresholds for these instabilities have been, since a long time, experimentally recognized [7],[8],[9] and theoretically understood following a "linear theory" [10],[11]. A striking, but not fundamental, thing is that the adimensional number measuring these critical thresholds are almost the same

$$T_{ac} \simeq 1724^{*} \quad ; \quad R_{ac} \simeq 1707 \quad .$$

More striking -and more fundamental- is the fact that both systems exhibit -at the critical onset of their instability- a remarkable spatial order i.e. a symmetry breaking of the fluid "properties". In both cases convection sets in under the form of perfectly periodic "rolls". Let us call Λ_x the fundamental wavelength of the convective structure (see Fig.1a and 1b, Λ_x is the periodicity of the structure along x).

Another remarkable similarity between Taylor and Rayleigh-Benard (R.B.) instabilities is the magnitude of this fundamental wavelength Λ_x. In both cases it is very close to twice the distance d

$$\Lambda_x \simeq 2d \quad .$$

More precisely, it is natural to characterize these periodic structures appearing in the fluid by an adimensionalized wave number

$$k_x = \frac{2\pi}{\Lambda_x} d \quad .$$

----

* Value found in Ref.[11] for $\frac{R}{R+d} = 0{,}975$, the corresponding k_{xc} being $k_{xc} = 3{,}127$.

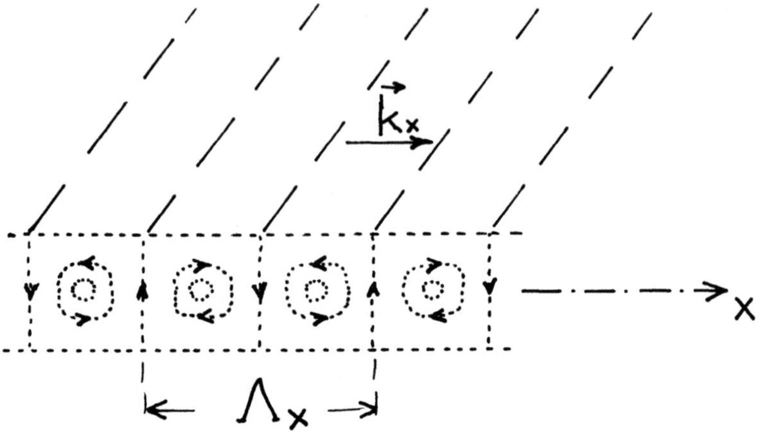

Fig.2 - Schematic representation of convective structures

Careful experiments show, and linear theory confirms, that

$$k_{xc} \text{ Taylor} = 3{,}1_{27} \quad ; \quad k_{xc} \text{ R.B.} = 3{,}1_{17} \quad .$$

Up to now, and as far as the non-linearities of these instabilities are ignored, we are in presence of two very similar phenomena.

IS AN INSTABILITY ANALOGOUS TO A CRITICAL PHENOMENON ?

The presence of a critical onset together with a symmetry breaking phenomenon have since some years incited physicists to make analogies with a critical phenomenon.

Let us remind that the classical Curie point is characterized by the appearance, below a certain critical temperature T_c , of a spontaneous magnetization m. This transition, indeed, corresponds to a lowering of the symmetry of the ferromagnet for $T < T_c$ see Fig.3a.

It is very tempting to make an analogy with the behaviour of the convective velocity see Fig.3b. A natural and powerful approach is, in both cases, that of Landau[12]

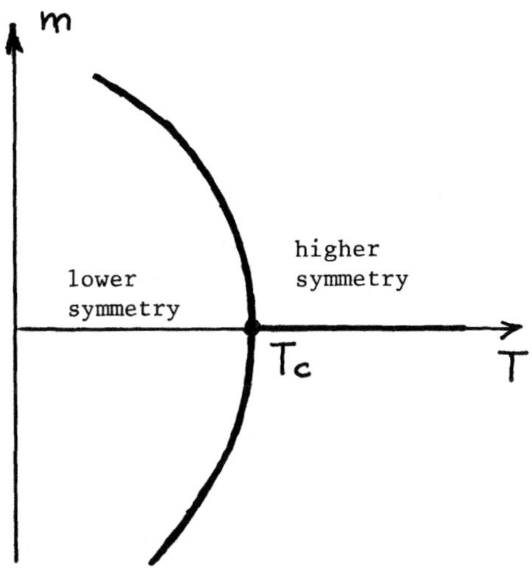

Fig. 3a - Ferromagnet

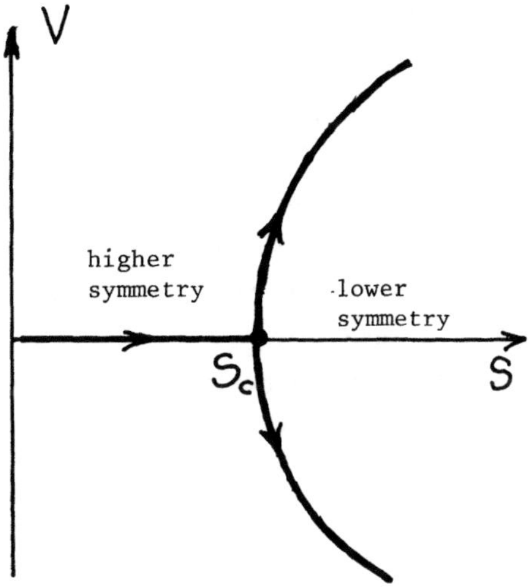

Fig. 3b - Instability

Landau introduced the concept of an order parameter η which measures the degree of "order" of the system. In a ferromagnet $\eta \equiv m$; $\eta \equiv 0$ for $T > T_c$ and $\eta \neq 0$ for $T < T_c$.

Naturally, in the hydrodynamic instabilities considered here, the proper "order parameter" can be the convective velocity V

$$V \equiv 0 \quad \text{for} \quad S < S_c$$
$$V \neq 0 \quad \text{for} \quad S > S_c \ .$$

A much more difficult thing is to define a potential Φ . Let us call this potential, "pseudo-potential" Φ . See Graham in Ref.[5] and in Solvay Conference 1978.

Classically this "pseudo-potential" is expanded in a power serie of the order parameter $\eta \equiv V$; hence

$$\Phi = AV + BV^2 + CV^3 + DV^4 + \dots \qquad (0)$$

For obvious symmetry reasons (see Fig. 3b, the V = f(S) curve represents a so called "normal bifurcation")

$$\Phi = BV^2 + DV^4 \ . \qquad (1)$$

If we remember that $\frac{d\Phi}{dV} \div \frac{dV}{dt}$

we obtain from (1)

$$\frac{dV}{dt} = B'V + D'V^3 \quad , \tag{2}$$

where $B' = \frac{(S-S_c)}{\tau_o}$ in order to account for the fact that $\frac{dV}{dt} < 0$ in the stable state $(S - S_c < 0)$, $\frac{dV}{dt} = 0$ when $S = S_c$ (marginal stability) and $\frac{dV}{dt} > 0$ for $S-S_c > 0$ (see Fig.4) ; τ_o is a characteristic time of the instability and the non-linear term $D'V^3$ tells us that the velocity V has to reach a steady value after the initial period of growing.

Equation (2) predicts a critical slowing down whose characteristic time is $\tau = \tau_o (S-S_c)^{-1}$; this means that

$$\tau \rightarrow \infty \quad \text{when} \quad S \rightarrow S_c \ .$$

Furthermore, the stationary amplitude of V can be determined from (2) by making $\frac{dV}{dt} = 0$; obviously $V = \pm V_o (S-S_c)^{1/2}$.

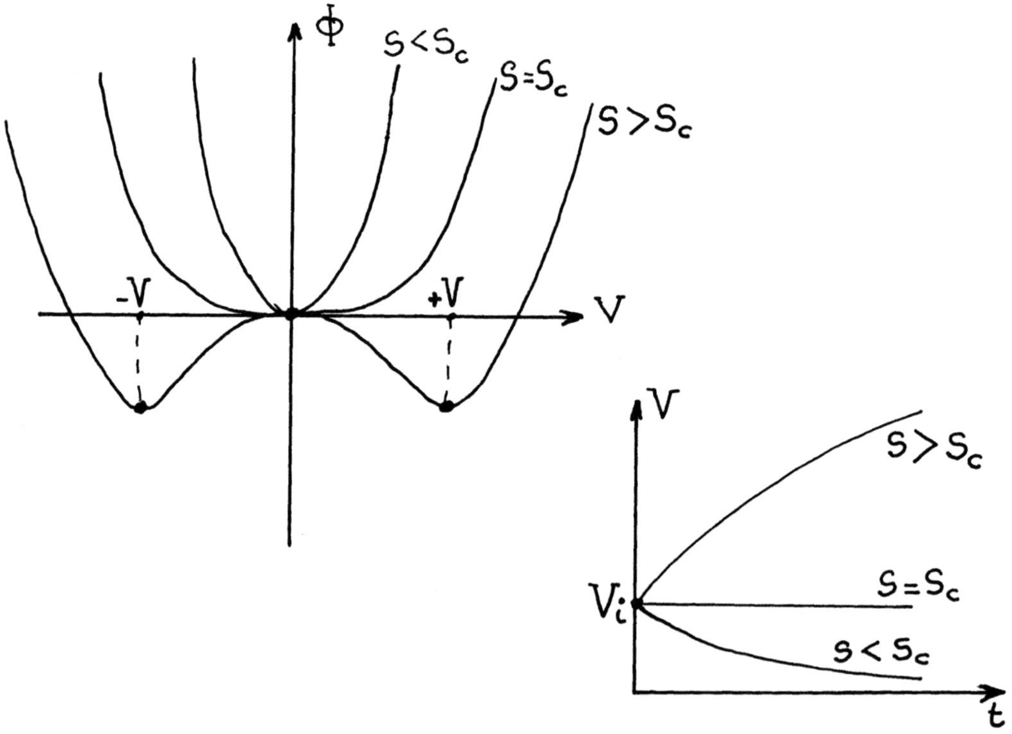

Fig.4 - Landau "picture" of an instability

The critical slowing down as well as the dependence of the "order parameter" V versus $(S-S_c)$ like $(S-S_c)^{\beta}$, with $\beta = \frac{1}{2}$, are two well-known expected properties of a 2nd order phase transition in the frame of a mean-field theory.

The corresponding experimental behaviour in the Taylor [13],[14] as well as in the R.B. [15],[16] instabilities check perfectly well these points : critical slowing down is observed in the dynamical response of the velocity and the steady velocity amplitude varies as the square root of the threshold distance in both cases.

Following Landau's ideas we can also account for spatial effects adding a new term to (2),

$$\frac{dV}{dt} = B'V + D'V^3 + \left(\xi_o^2 \frac{\partial^2 V}{\partial x^2}\right) \frac{1}{\tau_o} .$$

Physically speaking, a local spatial perturbation of the order parameter V will be repercuted in space, along $\vec{k}_x$ direction on distances of order of magnitude

$$\xi = \xi_o (S-S_c)^{-\nu} .$$

$$\xi \to \infty \quad \text{when} \quad S \to S_c ,$$

$\nu = \frac{1}{2}$, and ξ_o is the product of an universal coefficient by d.

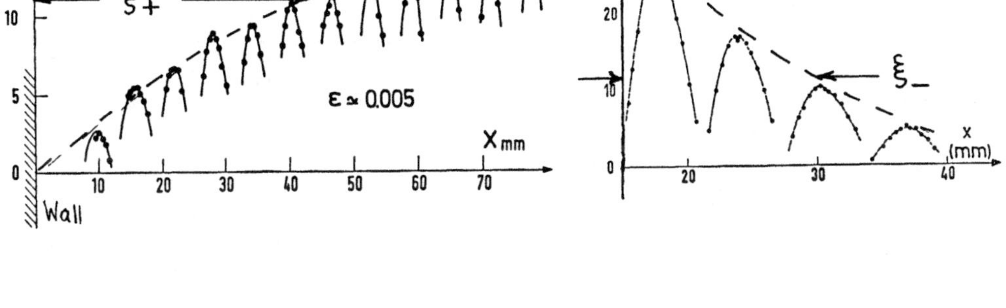

Fig.5 - Spatial "correlation" effects in Rayleigh-Benard instability

$$\left(\varepsilon = \frac{R_a - R_{ac}}{R_{ac}}\right)$$

This point has been precisely checked only in R.B. instability [16] both in supercritical $R_a > R_{ac}$ as well as in subcritical $R_a < R_{ac}$ regime [17] (see Fig.5). As far as we know, in Taylor instability this same phenomenon exists but has not been yet subjected to experimental investigation.

Before leaving the nice and powerfull analogy between the hydrodynamic instabilities and a 2nd order phase transition in the mean-field approach, let us play with the ϕ expression (0). We said above that, in particular, the term C is zero for "symmetry reasons". This situation corresponds, in the R.B. problem,to the so called "Boussinesq" approximation (invariance of all the fluid properties with the temperature, except for the density). This invariance inside the fluid layer may be effectively broken if the viscosity ν or the expansion coefficient α , for example, are temperature dependent. Then (0) must be rewritten as :

$$\phi = BV^2 + CV^3 + DV^4 \qquad C \neq 0 . \tag{3}$$

Physically, the symmetry [V;-V] of the properties of the R.B. convection is broken, inducing a decrease of the symmetry in the convective structure : near the onset the *rolls* are replaced by *hexagons*.

Furthermore from (3) one can deduce

$$\frac{\partial V}{\partial t} = B'V + C'V^2 + D'V^3 , \tag{4}$$

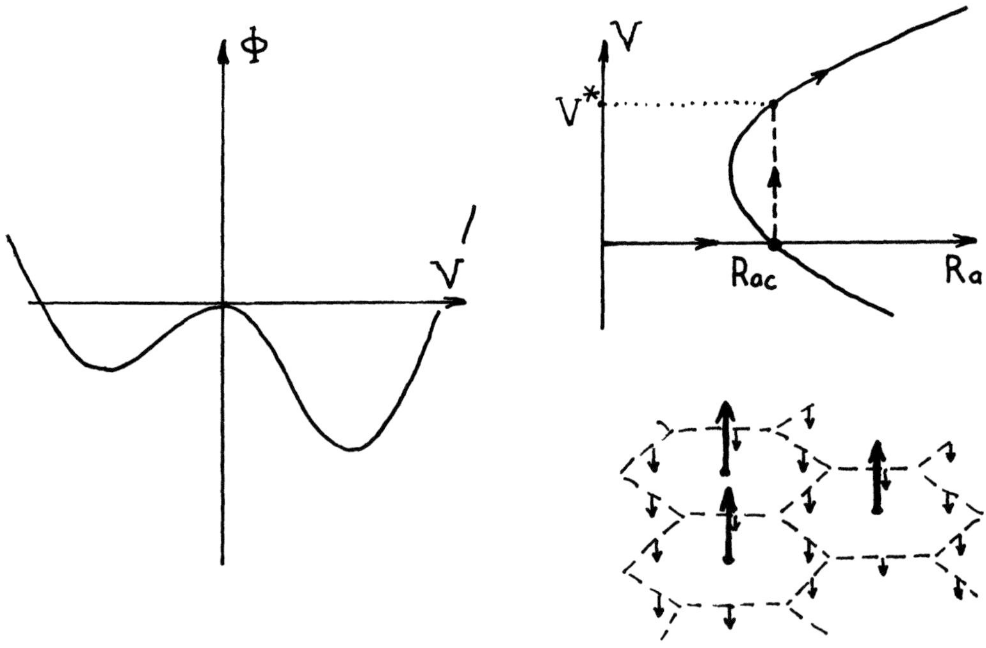

Fig. 6 - Non-Boussinesq convection

the stationary convective solution of (4) is obviously of the form

$$V = V^{*} + V_{1}(R_{a} - R_{ac}) + \dots$$

A new, fundamental, feature appears (see Fig.6): There is a jump V^{*} of the velocity at the onset of the instability. In terms of phase transition, the "transition" corresponding to non-Boussinesq R.B. convection is partially of 1st order. This behaviour was experimentally checked in the case of R.B. convection in water near 4° ; furthermore, and as expected, the convective structure near the onset is hexagonal [18].

SUPERCRITICAL CONVECTIVE PROPERTIES

Up to now we discussed some of the properties of the Taylor and R.B. instabilities near their critical onset.

We found striking similarities between the two kinds of instabilities, let us say in the domain $S_c < S < 1,5\ S_c$. For more supercritical values of S this nice similarity does not hold anymore, the greater the value of S, the more important is the difference between the two kinds of instabilities.

Let us specify that we compare behaviours of the "dissipative" structure in both instabilities for comparable "aspect ratio" $\Gamma_x = \frac{L_x}{d}$, the Rayleigh-Benard convection being installed in a rectangular box whose L_y extension is $d < L_y < L_x$, see Figs.1 and 2. Let us arbitrarily consider a case where Γ_x is of the order of 10.

In Taylor instability, when T_a is very very slowly increased, Λ_x remains absolutely constant, may be up to 100 times the critical onset.

What happens in the Rayleigh-Benard instability ? Only one thing is clear , there is a marked tendency to an *increase* of the wavelength Λ_x by increasing R_a/R_{ac}. But, in a rectangular box of moderate aspect ratio Γ as considered here, the increase is discontinuous (the number of rolls has to remain integer). On the other hand,the Prandtl number $Pr = \nu/D_T$ plays a fundamental role in this process. (Let us assume, for simplicity that $Pr > 10$). Then Λ_x , in R.B. instability, is a complicated, discontinuous, function of Γ_x, Γ_y, $Pr, R_a/R_{ac}$. This wavelength change may be furthermore time-dependent and seems to be very sensitive to uncontrolled boundary conditions as well as to any imperfections in the system. Obviously, complicated hysteretic behaviour accompanies this wavelength change (see Fig.7 for some of the possible behaviours).

This discontinuous, hysteretic, increase of Λ_x versus $\frac{R_a}{R_{ac}}$ has not yet received definitive explanation. However one can find a powerfull approach in [19].

Let us emphasize that this complicated* Λ_x change in R.B. is the main difference

*In the case of the cylindrical cell used in [20] with boundary conditions such that the rolls are axis-symmetric, the central ombilic allows a *continuous* wavelength change.

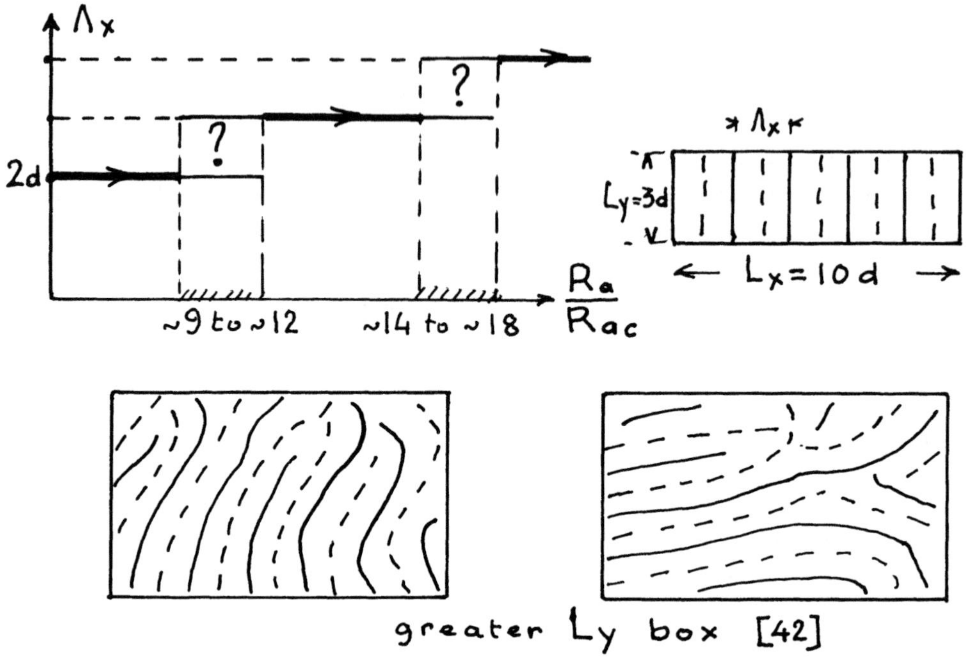

Fig. 7 - Some of the possible behaviours in super-critical R.B. convective structures ($Pr \sim 10^3$)

in the supercritical properties existing between Taylor and R.B. instabilities.

From this difference we can expect that the higher bifurcations toward turbulence will be different and , at least, more simple and **universal in the** Taylor than in the R.B. instability.

VERY SIMPLIFIED PICTURE OF THE TRANSITION TOWARD TURBULENCE IN TAYLOR INSTABILITY

Many experiments have been devoted to this study [21],[22],[23],[24] . The crucial point is that the fundamental structure of wavelength Λ_x remains unchanged in all the studied $\frac{T_a}{T_{ac}}$ domain.

Roughly speaking one can notice three important steps between the onset of the Taylor instability and the turbulent state , by increasing $\frac{T_a}{T_{ac}}$ we can observe what follows:

The first, next step (next "bifurcation") from Taylor steady, mostly two-dimensional, flow corresponds to the appearance of transverse waves superimposed on the horizontal Taylor vortices, see Fig.8. This step is considered as the first appearance of time-dependent effects ; indeed,the corresponding convective velocity measured *at a fixed point* of this new three-dimensional *rotating* structure is time-oscillating with a frequency f_1 .

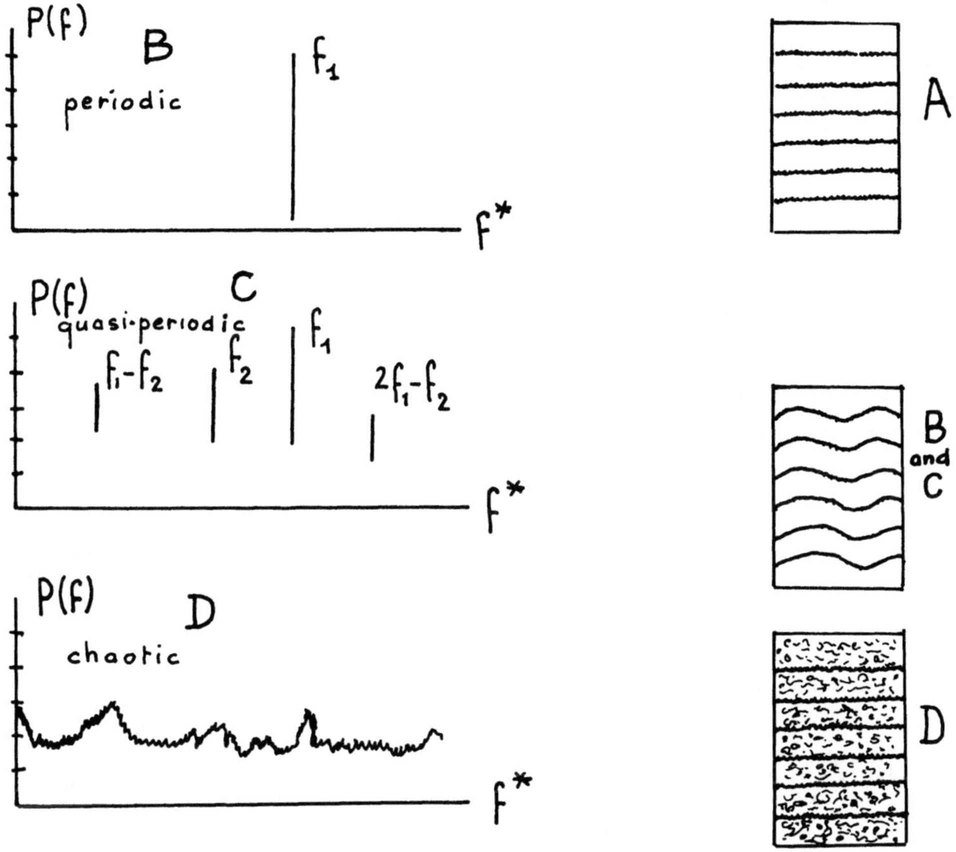

Fig.8 - Transitions toward turbulence in Taylor instability, ref. [34].

The second next step is most generally considered [21] as the appearance of a second new time dependence in the velocity, the frequency of which f_2 is not simply related to f_1. See Fig.8. This flow has been called "*quasiperiodic*" . (The name *biperiodic* would be better) - The last (but not the least !) step is the appearance, in the Fourier frequency spectrum of the velocity, of *noise* or *broad* peaks. (This contrasts with the extremely narrow peaks f_1 and f_2). The corresponding state of the flow is very naturally called *chaotic flow*, see Fig.8.

Even if the above picture is simplified and naive, one can notice that in the particular case of the Taylor instability, the transitions toward turbulence are - in principle - quite simple and natural. This beautiful simplicity is due, in our opinion, to the fact that the *basic convective structure* (Taylor vortices) is almost *conserved* from the onset until the chaos. Let us look - by contrast - to the extremely complicated situation found in the Rayleigh-Benard instability.

TRANSITIONS TOWARD TURBULENCE IN THE RAYLEIGH-BENARD CONVECTION

Before entering in this intricate and highly complicated field let us make a remark. The different few steps toward turbulence,

steady convection → monoperiodic regime →

→ biperiodic regime → chaotic regime

seem to be natural. This sequence is in contradiction with the view of Landau [25] and more coherent with the basic ideas of Ruelle and Takens [26], for example. One can see in Fig. 9 illustration of the above-described cascade toward turbulence. But we have to ask ourselves if this "natural" behaviour is *unique* ; we will see that this is absolutely *not* the case !

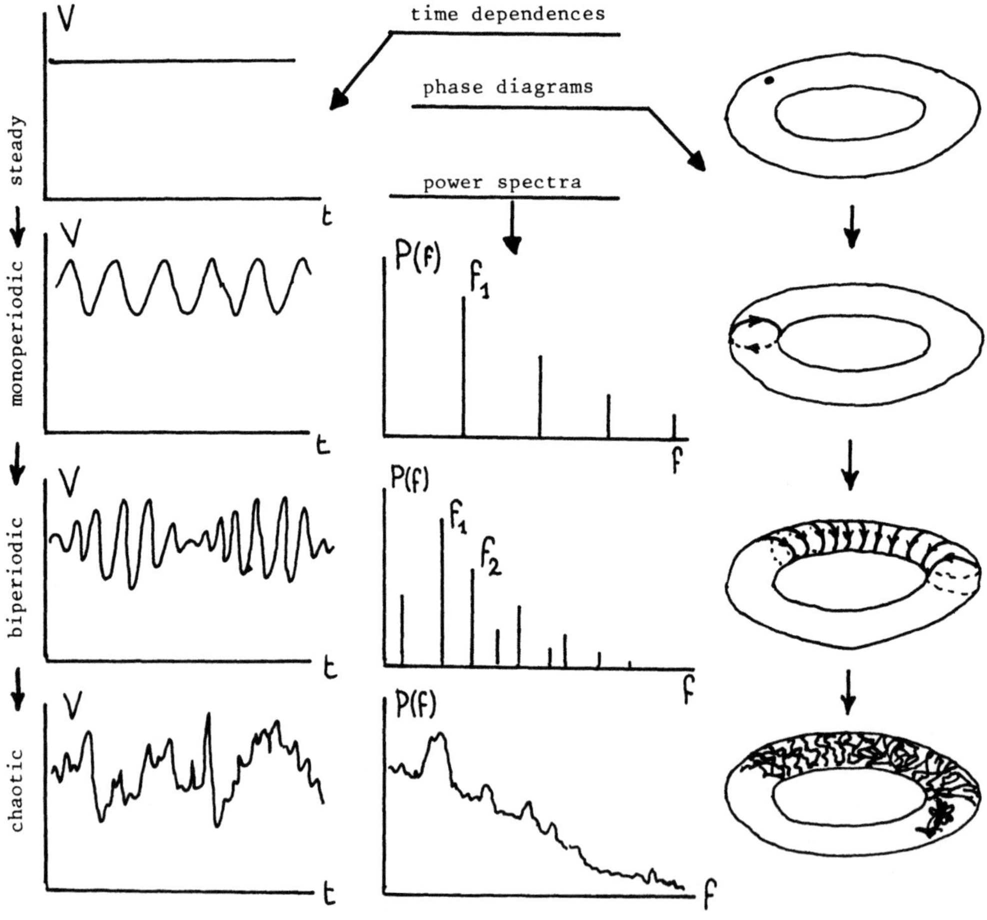

Fig. 9 - Some illustrations of a cascade toward turbulence ... one of the possible cascades !

Let us return to the specific case of R.B. convection in a rectangular box ($\Gamma_x = 10$ $\Gamma_y = 3$) containing near R_{ac} a perfect structure of 10 two-dimensional rolls. (similar to the Taylor vortices). By increasing R_a the convective amplitude grows and furthermore, Λ_x has a *tendency to increase* [23] (with a *perfect* structure we have a *discontinuous* increase, see Fig.7).

The first new step after R_{ac} is obviously the transition - at a certain threshold R_{aII} - from the two-dimensional structure to a new three-dimensional one (a new set of rolls develops at R_{aII} perpendicularly to the normal rolls existing just above R_{ac}). In our opinion, this transition corresponds to that occurring in the Taylor problem when wavy vortices appear. But the instructive difference is that in the R.B. case one cannot predict the value of R_{aII}. The R_a number is *not the only relevant* parameter ; we have to know also the actual Λ_x value. Anyway the crucial point is : the R_{aII} threshold is unambiguously related to Λ_x (for a given box Γ_x, Γ_y and a given Prandtl number $Pr > 10$). See Fig.10 for some experimental behaviours[27].

This means that the first step (R_{aII}) toward turbulence (after R_{ac}) can be understood only if the fundamental structure is well-defined and has a *known* Λ_x. Remember from above that the actual Λ_x may be *a posteriori* measured (if there is an optical access to the structure) but *not predicted* (without external trick).

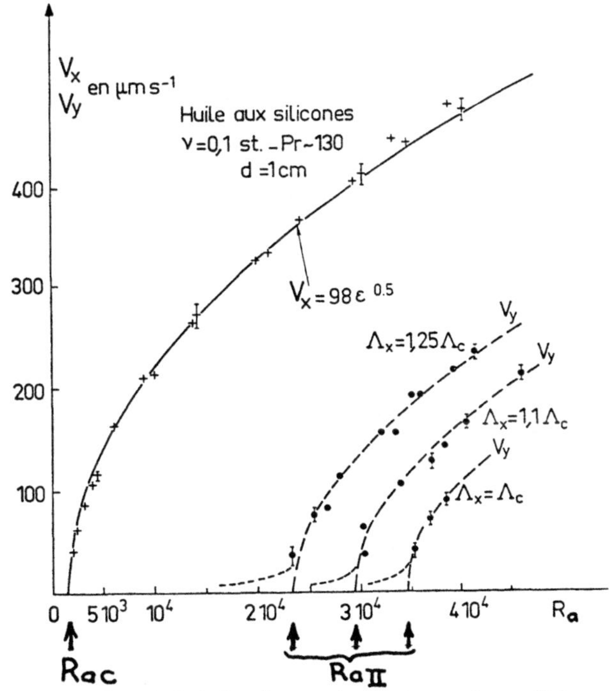

Fig.10 - Experimental behaviour of the R_{aII} *transition*

What happens at a further increase of R_a/R_{ac} ? The answer is simple but, in some sense, quite unexpected : the next step (which corresponds to the first appearance of a time-dependent effect) is *directly* the transition to the *chaotic regime*.

In the rectangular cell with $\Gamma_x = 12$, $\Gamma_y = 4$ and $Pr \simeq 130$ this chaos appears at R_{aT} such that $R_{aT}/R_{ac} \sim 20$ *. This chaotic time dependence of the velocity is identified through the appearance of a broad, very low-frequency spectrum P(f) of the time-dependent velocity, see on Fig.11 . This chaos is clearly (but retrospectively !) due to time-dependent Λ_x change *correlated* to structure motion and rearrangement as schematically pictured in Figures 7 and 12.

Let us try to give tentative explanations and models for these structure motions giving rise to this kind of chaotic regime.

A crucial point is the following : the Λ_x increase (as required in the R.B. convection) implies some *structure "compressibility"* to be achieved almost continuously and without great energy expense. Let us call this Λ_x change "soft change".

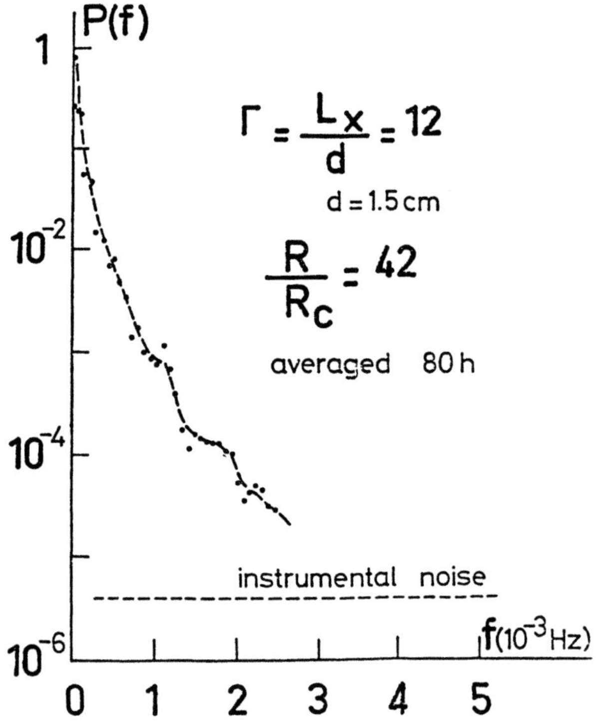

Fig.11 - Noise power spectrum of the velocity in a "large box" (R.B. convection), ref. [28].

----

* We believe that this R_{aT} threshold is not as well defined as R_{ac} or R_{aII} .

For example, the compressibility of axially symmetric rolls in a cylindrical container under conditions of reference [20] is almost infinite due to the presence of the central ombilic, see Fig.12a. On the contrary, in the case of a rectangular box of small Γ_x, Γ_y containing a *perfect* structure of two-dimensional rolls, this structure is almost incompressible *in its whole* - the Λ_x increase is then discontinuous implying a roll disappearance with a high energy barrier ("hard change"). A good compressibility in a rectangular box may exist, but only due to the *presence of defects*. See as a first example Fig.12b ; dislocation-like defects allowing dilatation(i.e. global Λ_x increase) just by the motion of the dislocation edge A or B. The energy barrier in that case is obviously much smaller (if any) than that corresponding to the "hard" discontinuous case. Then defects allow "soft Λ_x change" ;

one can easily understand why Γ is of prime importance in the occurrence of this last mechanism (for a given Pr number, the lower the Γ, the higher the R_a number for which dislocation defects can appear). Other mechanisms allowing a soft Λ_x change exist (see Fig.12c). The important point is the following :

All these mechanisms for a "soft increase" of Λ_x in *rectangular geometry* imply the existence of some kind of *degeneracy in the structure* to account for a finite compressibility.

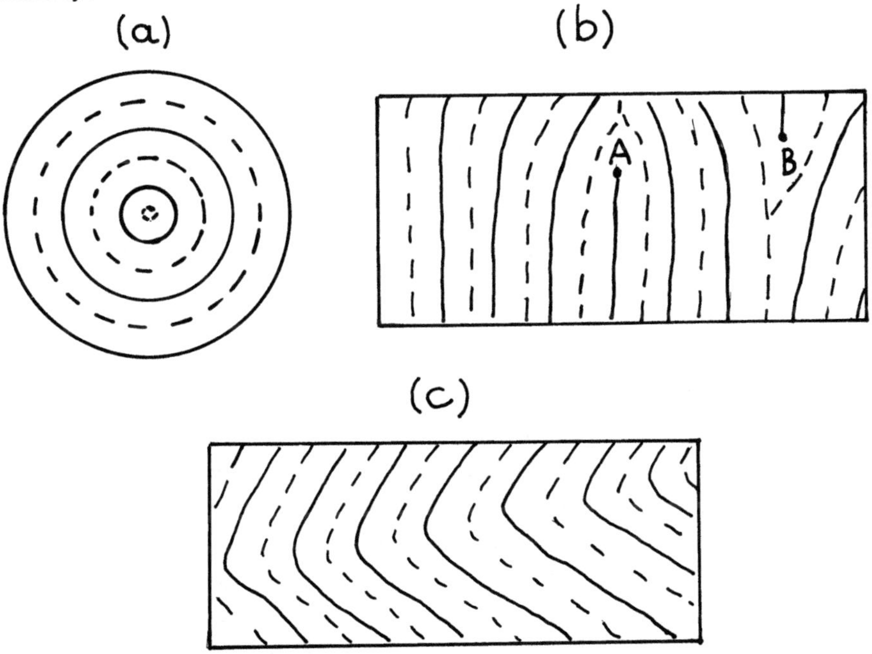

Fig.12 - Defects in supercritical R.B. convective structures, refs. [29] ,[20] .

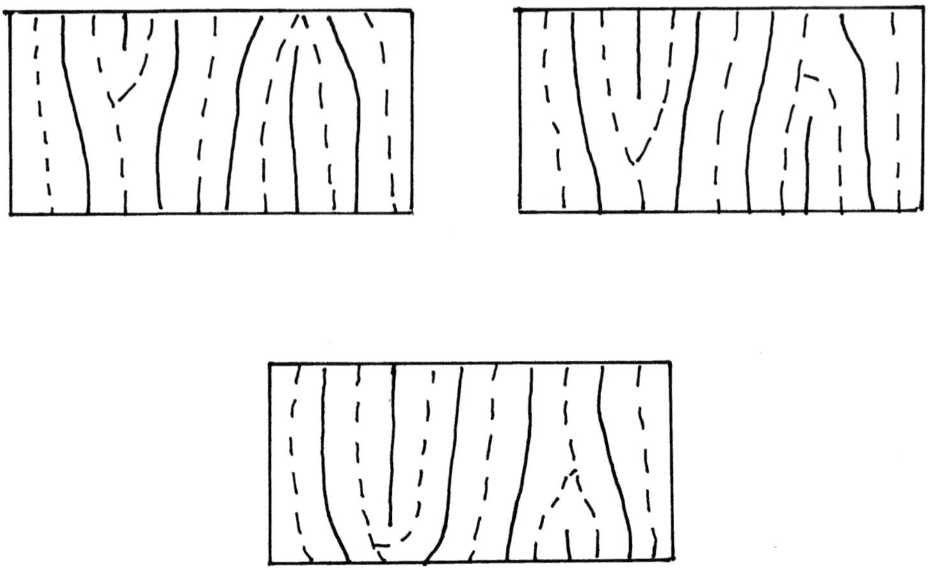

Fig.13 - Degeneracies in R.B. structures...one example

The presence of such a degeneracy *unlock* obviously the structure (see Fig.13), allowing slow erratic motion of this structure. As a (first) consequence, the velocity (measured at a fixed point) will fluctuate strongly and in a chaotic manner. But this chaotic regime, specific of the R.B. convection, is *basically different* from that observed in the Taylor vortices. Let us tentatively call this kind of turbulence "phase" turbulence or "structure" turbulence. But let us emphasize that this kind of turbulence is also *true turbulence*. Obviously unlocked structure will produce chaos in any local property ; but *furthermore*, the local structure change (i.e. local Λ_x fluctuations) does produce convective *amplitude* change as well as anharmonicity mode exchange [30]. As a (second) consequence even the macroscopically averaged convective properties are chaotically time-dependent. See heat flux measurements [31],[32]. One can easily understand that the appearance of this "structure" turbulence is very sensitive to the aspect ratio (the greater the box, the lower $R_a/R_{ac} = R_{aT}$ value for which it can appear) and also to the Prandtl number (the lower the Pr number, i.e. the viscosity, the lower the $R_a/R_{ac} = R_{aT}$ threshold)[19]. A naive but natural way to prevent "structure" turbulence in a *rectangular container* is to lower the Γ value in order to stabilize the structure through boundary effects, and then inhibit the Λ_x increase. Let us work, now, in a box which is so small that *only one* Λ_x (2 "x" rolls) can be present ($\Gamma_x = 2$). Also in order to have *only* 2 "y" rolls, let us choose $\Gamma_y \simeq 1.5$. See Fig.14a, the so called "small box" [27].

Even in such a small box we cannot claim that the structure is imposed (blocked) but we can reasonably expect that - at least - "phase" motion will be strongly reduced. We expect to have *artificially* introduced a phase and Λ_x stability which is very similar to that *naturally* present in the Taylor instability [23].

Effectively, and as far as the structure is composed of two symmetrical three-dimensional rolls, we get the striking and expected feature : until more than *one* (or maybe a few) *hundred times* R_{ac} *the convection remains absolutely steady !* The "structure" or "phase" turbulence has then been effectively blocked.

The next step (first time-dependence appearance) consists in the occurrence of perfectly coherent monoperiodic oscillations of the velocity. One can see on Fig.14b a power frequency spectrum of the velocity. Note the presence of harmonics of f_1 and the fact that the highest peak is more than 10^6 times the (instrumental) noise. Experimental identification of the mechanism of these highly coherent oscillations is described elsewhere [33].

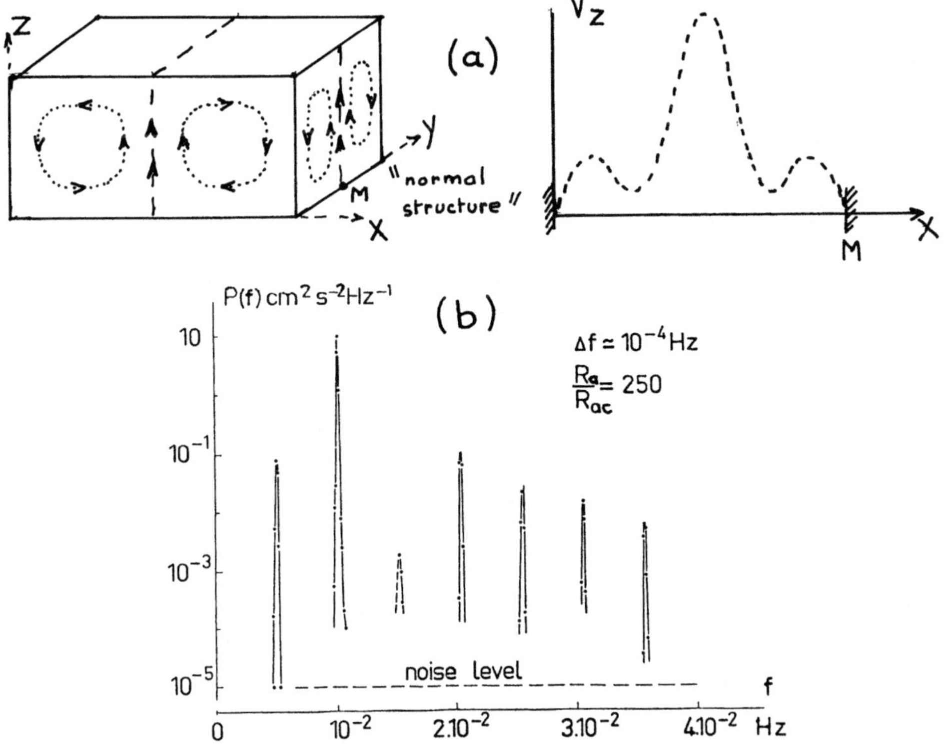

Fig.14 - Power spectrum of the oscillating velocity in R.B. convection

Another further step *may be* the appearance of a second frequency f_2 which is *generally** not simply related to the previously described f_1 ("biperiodic" regime or "quasi periodic" regime). For more details see [21][32][34].

Naturally, the last step in the R.B. "phase locked" convection is the appearance of turbulence (chaotic regime). This turbulence manifests itself by a broad low-frequency spectrum centered at zero frequency together with the broadened oscillation peaks previously described [27]. Summarizing, we found in the R.B. problem in a rectangular small box and with Prandtl number fluid Pr > 10 the usual sequence toward turbulence which was found in Taylor instability. But we have to emphasize that this "simple" behaviour is *one of the possible* behaviours in a small box : *the key of the problem is the actual convective structure.* A "normal" structure (Fig.14a) gives this behaviour. Many other kinds of structure may exist with different parity and different amount of higher harmonics. Indeed, the number of degrees of freedom for structures is strongly reduced in a "small box" ; but keep in mind that many kinds of structure are however possible with corresponding *completely different* time-dependent behaviours.

In order to point out the *extreme complexity* of the time-dependent convection *even* in a small box, let us mention two striking, unexpected and *apparently* confusing behaviours.

In a *given small* box filled with a *given fluid* and at a *given* (supercritical !) R_a number, one can obtain :

- steady convection
- chaotic convection
- mono-or biperiodic convection ,

depending on the *actual* convective structure present in the fluid.

Even more confusing situations are the following experimental results :

Two *identical twin cells* (small boxes) filled with the same fluid and inserted between the *same massive* copper plates (then submitted to the *same* R_a number with the *same thermal* (*history*) are generally exactly under the *same* convective regime. Sometimes however, we have found the following situation :

One may be oscillating, the other being either turbulent *or* stationary !

These confusing behaviours are just mentionned to emphasize - for the last time - the fact that any behaviour cannot be either predicted or understood from the only knowledge of R_a, Γ_x, Γ_y, Pr ...etc... etc... if one does not specify the actual structure which is sensitive to so small uncontrolled perturbations (even in a highly careful experiment) that they remain far away from the attention of the standard R.B. experimentalist !

---

* Indeed, the R_a dependences of f_1 and f_2 being independent, f_1/f_2 ratio cannot always be incommensurate ! On the contrary, f_1 and f_2 can be beautifully coupled [35].

Anyway the R.B. experimentalist can have empirical "recettes de cuisine" in order to get "simple" structure in a small box, and in that case a simple mechanism of transition toward turbulence can be understood.

Acknowledgments

It is a pleasure to thank all my colleagues of the friendly-named "bande à Bergé" by the organizer of this conference [36] for various help. Indeed, my major gratitude is addressed to M. Dubois to whom most of the original results and understanding on R.B. convection are due . I am also deeply indebted to Y. Pomeau and J. Wesfreid for numerous help and constant illuminating discussions. Lastly, I thank E. Cotteverte and F. Lefèvre for their great patience and care in the typing of this manuscript.

REFERENCES

[1] - S. Chandrasekhar, "Hydrodynamic and Hydromagnetic Stability" (Clarendon, Oxford 1961).
[2] - C. Normand, Y. Pomeau and M.G. Velarde, Rev. Mod. Phys. 49, 581 (1977).
[3] - E.L. Koschmieder, Adv. Chem. Phys. 26, 177 (1974).
[4] - G.Z. Gershuni, E.M. Zhukhovitskii, "Convective Stability of Incompressible Fluids", Keterpress Entr., Jerusalem 1976.
[5] - "Fluctuations, Instabilities and Phase Transitions", edited by T. Riste (Plenum, New York 1975).
[6] - J.P. Gollub and S.V. Benson, Phys. Rev. Lett. 41, 948 (1978).
[7] - R.J. Schmidt and S.W. Milverton, Proc. Roy. Soc.(London)A 152, 586 (1935).
[8] - G.I. Taylor, Phil. Trans. Roy. Soc.(London) A223, 289 (1923).
[9] - P.L. Silveston, Forsch. Ing. Wes. 24, 29-32 and 59-69 (1958).
[10] - A. Pellew and R.V. Southwell, Proc Roy. Soc.(London)A 176, 312 (1940).
[11] - P.H. Roberts, Proc. Roy. Soc.(London)A 283, 550 (1965).
[12] - L.D. Landau and E.M. Lifshitz, "Statistical Physics", 2nd Edition (Addison-Wesley, Reading, Mass. 1969).
[13] - R.J. Donnelly and K.W. Schwarz, Proc. Roy. Soc.(London)A 283, 531 (1965).
[14] - J.P. Gollub and M.H. Freilich, Phys. Fluids 19, 618 (1976).
[15] - R.P. Behringer and G. Ahlers, Phys. Lett. 62A, 329 (1977) ;
Y. Sawada, Phys. Lett. 65A, 5 (1978) ;
C. Allain, H.Z. Cummins and P. Lallemand, J. Physique Lettres 39, L473 (1978).
[16] - J. Wesfreid, Y. Pomeau, M. Dubois, C. Normand and P. Bergé, J. Physique 39, 725 (1978).
[17] - J. Wesfreid, P. Bergé and M. Dubois, Phys. Rev. A, March 1979 ;
J.C. Legros, J. Wesfreid and J.K. Platten, Int. J. Heat Mass Transfer , to be published.
[18] - M. Dubois, P. Bergé and J. Wesfreid, J. Physique 39, 1253 (1978).
[19] - Y. Pomeau and P. Manneville, submitted to Phys. Rev. Lett.
[20] - E.L. Koschmieder and S.G. Pallas, Int. J. Heat Mass Transfer 17, 991 (1974).
[21] - J.P. Gollub and H.L. Swinney, Phys. Rev. Lett. 35, 927 (1975).
[22] - A. Bouabdallah, G. Cognet (I.N.P.L. - Nancy, France), to be published.
[23] - E.L. Koschmieder, to be published in Proc. Solvay Conference (1978).
[24] - R.W. Walden and R.J. Donnelly, Phys. Rev. Lett. 42, 301 (1979).
[25] - L.D. Landau and E.M. Lifshitz, "Fluid Mechanics" (Pergamon, London 1959).
[26] - D. Ruelle and F. Takens, Commun. Math. Phys. 20, 167 (1971).
[27] - L. de Sèze and Y. Pomeau, J. Physique Colloque C5, 39 and C5, 95 (1978) ;
M. Dubois and P. Bergé, Synergetics, Vol.III - "Far from equilibrium", Bordeaux, Sept. 1978, published by Springer and Verlag (1979) ;
[28] - P. Bergé and M. Dubois, to be published.
[29] - F.H. Busse and J.A. Whitehead, J. Fluid Mech. 47, 305 (1971) ;
J.A. Whitehead, J. Fluid Mech. 75, 715 (1976) ;
F.H. Busse, Rep. Prog. Phys. 41, 1929 (1978).
[30] - M. Dubois, C. Normand and P. Bergé, Int. J. Heat Mass Transfer 21, 999 (1978) ;
F.H. Busse, J. Math. Phys. 46, 140 (1967).
[31] - G. Ahlers and R.P. Behringer, Phys. Rev. Lett. 40, 712 (1978).
[32] - A. Libchaber and J. Maurer, J. Physique Lettres 39, L369 (1978).
[33] - P. Bergé and M. Dubois, Optics Comm., to be published.
[34] - P.R. Fenstermacher, H.L. Swinney, S.V. Benson and J.P. Gollub, in "Bifurcation Theory in Scientific Disciplines, edited by O.G. Gurel and O.E. Rössler (N.Y. Academy of Sciences, 1978).
[35] - A. Libchaber and J. Maurer, to be published.
[36] - C.P. Enz, private communication.
[37] - R. Graham, Phys. Rev. A10, 1762 (1974).
[38] - H. Haken, Rev. Mod. Phys. 47, 67 (1975).
[39] - K. Stork and U. Müller, J. Fluid Mech. 71, 231 (1975).
[40] - P. Atten and J.C. Lacroix, J. de Méc. (1979), to be published.
[41] - K. Bühler, K.R. Kirchartz and H. Oertel Jr., Acta Mechanica 31, 155 (1979).
[42] - R. Krishnamurti, J. Fluid Mech. 42, 309 (1970) and 60, 285 (1973).

THEORY OF HYDRODYNAMIC INSTABILITIES

M.G. Velarde

Departamento de Fisica de Fluidos, Universidad Autonoma de Madrid
Cantoblanco (Madrid) Spain

THEORY OF HYDRODYNAMIC INSTABILITIES

Manuel G. Velarde
Departamento de Fisica de Fluidos
Universidad Autonoma de Madrid
Cantoblanco (Madrid) SPAIN

Introduction

Three levels of description can be used to deal with (thermo) hydrodynamic instabilities and non-equilibrium phase transitions at large. We have :

i) The microscopic theory which starts from a Master Equation of the system's evolution,

ii) The semimicroscopic , Langevin-type of approach, and

iii) The purely phenomenological description which starts with the macroscopic equations (Navier-Stokes ,etc) or suitably truncated versions of them (Landau approach, Lorenz model, etc). With the introduction of a Landau-Ginzburg"potential" (or using the nowadays current ideas of elementary catastrophe theory) a rather simplified and compact description is achieved. As Graham /1,2/ has recently reviewed quite in depth the field I shall be concerned here only with an illustration of the Landau method /3/to describe for you some old and new predictions in the stability analysis of a binary fluid layer heated from below or above. The cross-transport Soret effect, operating in the fluid layer (the effect of an impurity or an "external field") leads to a rather varied phenomenology. Besides the impurity extra field, a true external field can be considered, should the impurity be magnetically or electrically responsive or should the fluid be a nematic liquid crystal. Recently, it has been shown /4/ that the Soret-driven convection and the Rayleigh-Bénard convection of a homeotropic layer heated from below share a number of common features, at least to a qualitative level of description. The Soret parameter is replaced in the nematic by the anisotropic property of the material constants of the fluid /5/.

The Landau theory seems enough to accurately describe thermo-hydrodynamic instabilities within present-day experimental error bars. Non-classical corrections simply improve in extremely small regions or lead to extremely small corrections, at least for the cases studied until now /1-3/. Promising work remains to be done for the two-component Rayleigh-Bénard-Soret problem or for the liquid crystal case. However, as correlation lengths are of macroscopic size , side boundary effects may eventually prevent from seeing non-classical corrections.

The basic features of the Landau theory /1-3/ are :

i) The existence of an order parameter. In thermo-hydrodynamic instabilities this can be either the velocity field, or the temperature field or the mass-fraction of one of the components or the Nusselt convective heat transport. Considering the (convective) velocity field we have the amplitude and the spatial phase in a roll-type of convective structure /See Bergé's report to this Conference/. According to the evolution of the order parameter,transitions can be continuous (soft- or second-order type) or hard-excitation type, first-order transitions with hysteretic phenomena and metastability(subcritical instability,as called in the Fluid Mechanics jargon).

ii) The appearance of symmetry restoring fluctuations,

iii) There is "critical slowing down" at continuous transitions where we have "exchange of stabilities", and

iv) The appearance of "long range ordering" as determined by a macroscopic correlation length.

Elementary exposition of the Landau approach

i) Continuous transitions

For a Newtonian and Boussinesquian single-component layer (standard oils, water far from four degrees,etc which are the analogue of the simplest homogeneous magnet) we have a continuous transition at the onset of Rayleigh-Bénard convection /3/ with "exchange of stabilities" /6/. A velocity distrubance , $\tilde{V}$, from the motionless regime (the paramagnet) is expected to evolve in time according to the simple law $d\tilde{V}/dt \sim \tilde{V}$, or more properly, using convenient scales and dimensionless quantities

$$\tau_o \, d\tilde{V}/d\tau = \tilde{V}/\tau_r = \epsilon\tilde{V}, \text{ with } \epsilon = (R-R_c)/R_c \qquad (1)$$

where τ_r denotes some relaxation time scale, and R_c is the critical Rayleigh number for the onset of steady convection. Thus $\tau_r = \tau_o \epsilon^{-1}$, with τ_o being a convenient time unit (scale). Critical slowing down is described in Figure 1, where for the sake of completeness we have plotted together the results found for both the single component , and the two-component layer at different values of the Soret separation.

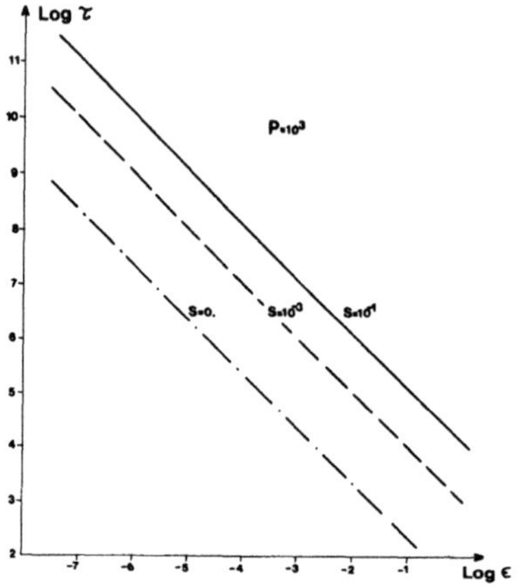

Fig. 1. Characteristic time for a layer thickness d = 0.6 cm, with d^2/κ = 320 s where κ denotes the thermal diffusivity of the liquid. S accounts for the Soret separation parameter. Values : S = 0 , τ_o = 21 s ; S = 10^{-3}, $\tau_o = 10^3$ s; S = 10^{-1}, τ_o = 15 x 10^3 s. P = 10^3 denotes the Prandtl number.

Further in the developped non-linear roll-type regime Eq.(1) needs to be corrected. We have

$$\tau_o \, d\tilde{V}/d\tau = \epsilon \tilde{V} + \alpha \tilde{V}^2 + \beta \tilde{V}^3 \tag{2}$$

However, the parameter α vanishes due to the symmetry of the problem. For a Newtonian and Boussinesquian fluid heating the layer from below is equivalent to cooling it from above !

We introduce a Landau potential through the relation

$$d\tilde{V}/d\tau = \delta\Phi/\delta\tilde{V} \tag{3}$$

The steady states of the system (whether convective or motionless) correspond to the extrema of the potential. Up to an additive constant we have

$$\Phi = \frac{1}{2}\tau_o^{-1}\,\epsilon\,\tilde{V}^2 + \frac{1}{4}\tau_o^{-1}\,\beta\,\tilde{V}^4 \qquad (4)$$

Like in the description of the simplest magnet we have no odd term in the Landau potential. We have the standard continuous, second-order type of transition as illustrated in Fig. 2.

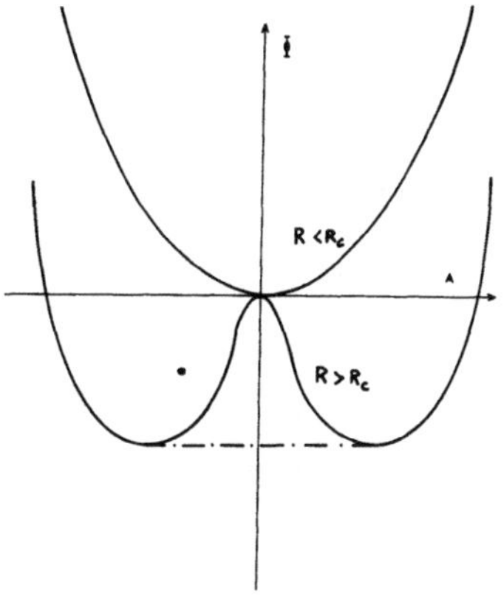

Fig. 2. The Landau potential for the Rayleigh-Bénard transition with "exchange of stabilities". Linear stability analysis is enough to delineate the onset of steady convection.

Figure 3 gives a typical evolution of the order parameter.

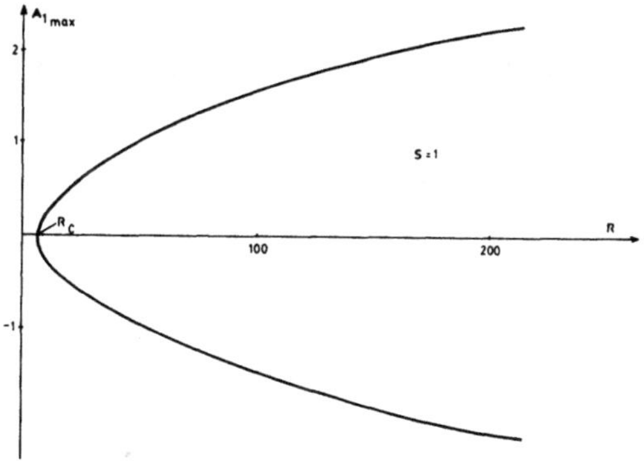

Fig. 3. Order parameter (maximum velocity amplitude) versus Rayleigh number. S = 1 denotes a value of the Soret separation. For vanishing S the behavior is qualitatively identical.

Like in magnets , spatial inhomogeneities can be included in our Landau picture by merely defining a Landau-Ginzburg potential. For the simplest case of a roll structure where $\tilde{V} = \tilde{V}_{max} \sin(kx + \varphi)$, it suffices to introduce the following functional

$$\Phi = \int dx \left[a\tilde{V}^2 + b\tilde{V}^4 + c \left\| \nabla V \right\|^2 \right] \qquad (5)$$

Theory has been developped /7/ and experiments have been conducted/7,8/ which again show that the Landau approach explains fairly accurately the data. As a matter of fact the onset of convection fits very precisely with the point at which the influence of the vertical boundaries of the layer shows up at the center in a circular container or when such "influence" or "correlation" length reaches half the value of the horizontal length in a square or rectangular container.

ii) First-order transitions: Non-Boussinesquian effects

Non-Boussinesquian properties in Rayleigh-Bénard convection, like variations of the material parameters with temperature, the properties of water in the vicinity of four degrees with anomalous variation of its thermal expansion coefficient,etc ,play a similar role to symmetry-breaking properties in a magnet. In the simplest case the Landau potential which incorporates a non-Boussinesquian property is

$$\Phi = a\tilde{V}^2 + b\tilde{V}^3 + c\tilde{V}^4 \quad (c \text{ positive}) \qquad (6)$$

where due to the odd term there is the possibility of hard-mode excitation (first-order transitions). A number of authors have carried out nice experimental work disclosing the hystertic phenomena and the metastability at the transition. For a recent paper see /9/. I shall come back to this when discussing the Landau approach to the two-component Rayleigh-Bénard problem.

Landau approach to the Rayleigh-Bénard-Soret instability: First-, second-order transitions and tricritical point

A large body of literature exists on the role of the Soret effect in the stability of a horizontal two-component fluid layer heated from below or above /10-12/. Here I shall consider only the case of a Newtonian and Boussinesquian layer.

According to the simplest approach to hydrodynamic instability /3/, for given boundary conditions, say dynamically free, conducting and permeable to matter transport, the evolution of a roll-type of steady convection (in the vicinity of the transition point) can be expressed in terms of a model description of the non-linear fields. The minimal set of modes needed for a relevant description involves one for the velocity field, two for the temperature distribution , and two for the mass-fraction field of one of the two components /13/. This indeed is a highly truncated representation of the convection. Such a truncation, however, provides a rather varied phenomenology which compares well, to a qualitative level, with the known experimental facts/12,14/. The five modes are the following/13,15/

$$\pi^2 k \psi = A_1 \sin \pi kx \sin \pi z \qquad (7.a)$$

$$\pi^2 k\, T = A_3 \cos \pi kx \sin \pi z + A_2 \sin 2\pi z \qquad (7.b)$$

$$\pi^2 k\, N_1 = A_5 \cos \pi kx \sin \pi z + A_4 \sin 2\pi z \qquad (7.c)$$

where ψ, T and N_1 denote the stream function (velocity potential), the temperature distribution and the mass-fraction of the heavier component. k is the roll wave number at the onset of convection and x,z denote horizontal and vertical coordinates respectively. The amplitudes A_i (i = 1,...,5) are unknowns whose spatiotemporal evolution is governed by the Navier-Stokes equations, the Fourier and Fick equations, and the continuity equation. We have

$$d\, A_1/dt = R\, k(A_3 + SA_5) + \pi^3 A_1 (1 + k^2)^2 \qquad (8.a)$$

$$d\, A_2/dt = \tfrac{1}{2} A_1 A_3 - 4\pi^2 A_2 \qquad (8.b)$$

$$d\,A_3/dt = \pi k\,A_1 + \pi^2(1 + k^2)A_3 + A_1A_2 \tag{8.c}$$

$$d\,A_4/dt = \tfrac{1}{2}\,A_1A_5 + 4\,\pi^2\,r_D\,(A_2 - A_4) \tag{8.d}$$

$$d\,A_5/dt = \pi^2(1 + k^2)\,r_D(A_3 - A_5) - A_1A_4 - \pi kA_1 \tag{8.e}$$

where, as before, R is the Rayleigh number and S measures the influence of the Soret cross-transport in the layer. r_D is the ratio of mass to heat diffusivity. Clearly, the system (8) is the natural extension of the Lorenz model /3/ to the two-component problem.

The steady solutions of (8) are the motionless regime (all A_i vanishing) and the solutions corresponding to the quartic equation

$$A_1^4 + A_1^2\left[27(1 + r_D^2)\,\pi^4/4 - R\right]16/9 + 64\,\pi^4\,r_D^2\left[27\,\pi^4/4 - R(1 + S + S/r_D)\right]/3 = 0 \tag{9}$$

According to the convention (3) we have the following Landau potential

$$\Phi = A_1^6/6 + A_1^4\left[27(1 + r_D^2)\,\pi^4/4 - R\right]4/9 + 32\,A_1^2\,\pi^4 r_D^2\left[27\,\pi^4/4 - R(1 + S + S/r_D)\right]/3 \tag{10}$$

which is a sixth-order polynomial, very much like in the description of metamagnets or Helium mixtures /16/ and the laser/17/.

Numerical calculations using (9) and (10) yield the following predictions (see Fig. 4). There is a line of first-order transitions in the region where according to linear stability theory/12/ overstability is to be expected. As this first-order line remains below the locus of overstable modes (bifurcation without exchange of stabilities) the fluid layer should depart from the motionless state either via a hard excitation of finite amplitude or through a transient oscillation whose amplitude exponentially grows until the finite amplitude solution is attained. This transition should exhibit metastability and the typical hysteretic phenomena of first-order transitions. These predictions agree well with experimental findings reported in /11/, and do not necessarily disagree with the observations given in /18/. It is rather surprising, however, that some linear oscillations can be steadily observed in a Rayleigh-Bénard experiment, for such a linear mode of instability is generally a precursor of a non-linear (finite amplitude) precursor, and thus, should belong to a metastable branch or a branch of solutions with quite a narrow stability domain. On the other hand, the kind of (inverted) bifurcation /3/ just described for the two-component problem has been observed in recent experiments with homeotropic liquids heated from below /5/.

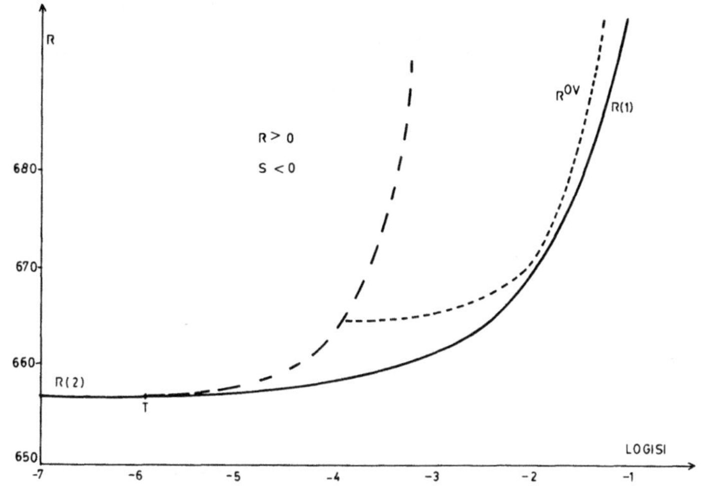

Fig. 4. Phase diagram-according to non-linear stability analysis-for the five-mode problem (8). The quadrant corresponds to heating the layer from below and the denser component migrating to the warmer boundary. For the remaining quadrants, our non-linear analysis reproduces the predictions already given in /10-12/. T denotes the tricritical point. To the left of T, the heavy line corresponds to second-order, continuous transitions, R(2), as also predicted by linear analysis. To its right, the broken line also belongs to the linear prediction of second-order transitions. The dotted line R^{OV}, corresponds to (linear) overstability whereas the heavy line, R(1), is the line of first-order subcritical transitions.

Another important prediction is that the first-order line meets the second-order one at a tricritical point/16/ where three coexisting "phases" become identical. Figures 3,5 and 6 describe the behavior of the "order parameter", A_1, at various values of the Soret separation, S. As indicated already, Fig. 3 corresponds to a typical second-order transition, in similar manner as when S vanishes. Fig. 5 is an (inverted) bifurcation/3/, or first-order transition for a value of S below the Soret region where linear theory predicts overstability. Fig. 6 also corresponds to an inverted bifurcation at $S = 10^{-3}$, and thus to a hard-mode excitation where according to linear theory, instability proceeds through a complex eigenvalue (Overstability). Notice that the Rayleigh number for overstability, R^{ov}, is smaller than the linear critical Rayleigh number for the onset of steady convection, though R^{ov} remains below the onset of non-linear steady convection.

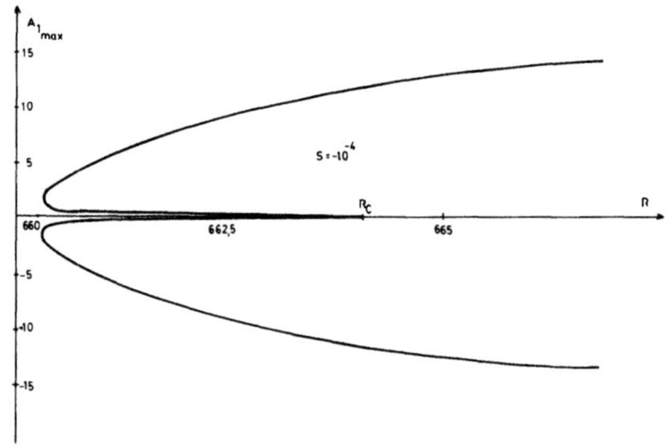

Fig. 5. Behavior of the order parameter, maximum value of A_1 plotted, in a first-order transition to steady convection.

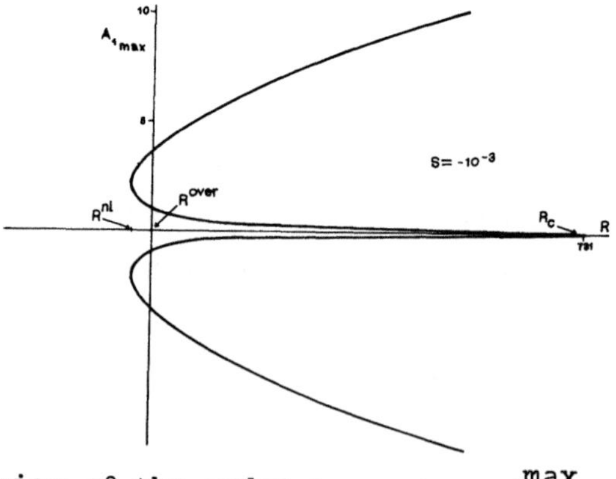

Fig. 6. Behavior of the order parameter, A_1^{max}, in a first-order transition where, according to linear theory, overstability is predicted.

In conclusion, careful experiments are to be encouraged in either the two-component Rayleigh-Bénard problem or in the nematic liquid case where a rather rich phenomenology is expected. As Landau theory seems valid, indeed, for the description of the main features of non-equilibrium phase transitions, theory and experiments on these two problems might lead to a better understanding of "tricritical" phenomena and all that.

The results reported here heavily rely upon the computer work of my collaborator J. Carlos Antoranz, to whom I express my sincere thanks.

References

1. R. Graham, Report at the XVII Conseil Solvay de Physique(to appear).
2. R. Graham, Proceedings Geilo Winter School, edited by T. Riste, FLUCTUATIONS, INSTABILITIES AND PHASE TRANSITIONS, Plenum, N.Y., 1975.
3. C. Normand, Y. Pomeau and M.G. Velarde, Revs.Mod.Phys., 49(1977)581.
4. H.N.W. Lekkerkerker, J.Physique-Lett.(Paris), 38(1978)L-277.
5. E. Guyon, P. Pieranski and J. Salan, J. Fluid Mech., to appear.
6. S. Chandrasekhar, HYDRODYNAMIC AND HYDROMAGNETIC STABILITY, Clarendon, Oxford, 1961.
7. E. Wesfreid, Y. Pomeau, M. Dubois, C. Normand and P. Bergé, J.Physique(Paris), 39(1978)725.
8. E. Wesfreid, Report at this CDC-79 Conference (unpublished).
9. M. Dubois and P. Bergé, Proceedings of an International Meeting, edited by A. Pacault and C. Vidal, SYNERGETICS(FAR FROM EQUILIBRIUM), Springer-Verlag, N.Y., 1979.
10. M.G. Velarde and R.S. Schechter, Phys. Fluids, 15(1972)1707.
11. D.T.J. Hurle and E. Jakeman, J. Fluid Mech., 47(1971)667. See also Advs. Chem. Phys., 32(1975)277.
12. R.S. Schechter, M.G. Velarde and J.K. Platten, Advs.Chem. Phys., 26 (1974)265.
13. G. Veronis, J. Marine Res., 23(1965)1.
14. M. Giglio and A. Vendramini, Phys.Rev. Lett., 39(1977)1014.
15. J.C. Antoranz and M.G. Velarde, Phys. Lett., A, submitted.
16. See, for instance, J.M. Kincaid and E.G.D. Cohen, Phys.Reports(Phys. Lett., C)22(1975)57.
17. J.F. Scott, Optics Comm., 15(1975)343.
18. J.K. Platten and G. Chavepeyer, J. Fluid Mech., 60(1973)305.

FULLY DEVELOPED TURBULENCE AND RENORMALIZATION GROUP

P.L. Sulem, J.D. Fournier and A. Pouquet
C.N.R.S., Observatoire de Nice, France

FULLY DEVELOPED TURBULENCE AND RENORMALIZATION GROUP

P.L. Sulem, J.D. Fournier and A. Pouquet
C.N.R.S., Observatoire de Nice, France

1. INTRODUCTION.

The word turbulence is used to describe diverse phenomena. In this paper, we shall restrict ourselves mostly to incompressible flows described by the Navier-Stokes (NS) equation

$$\begin{cases} \frac{\partial u}{\partial t} + u \cdot \nabla u = - \nabla p + \nu \nabla^2 u \\ \nabla \cdot u = 0 \qquad \text{(+ boundary and initial conditions)} \cdot \end{cases} \tag{1}$$

In this equation, u denotes the velocity, p the pressure and ν the kinematic viscosity. The relative importance of the non-linear terms to the viscous one is characterized by the Reynolds number

$$R = \frac{\ell_o \; v_o}{\nu} \quad , \tag{2}$$

where ℓ_o is a scale characteristic of the geometry of the flow and v_o a typical velocity difference across a distance of order ℓ_o. For small values of the Reynolds number, the flow is laminar. As the Reynolds number is increased, there may be a complicated sequence of instabilities leading eventually to a turbulent flow in which the velocity correlation decays to zero in the limit of long time separation (Lorenz, 1963; Martin, 1976; Gollub and Swinney, 1975; Normand, Pomeau and Velarde, 1977). Asymptotically, in the limit of infinite Reynolds number, the regime of fully developed turbulence is achieved. The flow has a highly chaotic behaviour with universal statistical properties in sufficiently small scales. These small scales are nearly homogeneous, isotropic and independent of the large scale motions, the instability of which is the source of the turbulent energy. The last property permits one to achieve the mathematical description of the flow by replacing the source of turbulent energy in the large scales by a statistically defined external force.

From a theoretical point of view, fully developed turbulence is essentially an unsolved problem. The main understanding we have is based on phenomenological analysis suggested by experiment. Most of the analytic theories of turbulence developed during the last decades can be viewed as semi-heuristic closures which reformulate the phenomenology of turbulence in a more quantitative way. They are models rather than approximations in the sense that they do not appear to be a given order expansion in terms of a small parameter. An important exception is provided by the Renormalization Group (RG). The success of this technique in critical phenomena, which also display universal self-similar properties, has raised some hope for turbulence. Up to now the RG has succeeded only in studying the large scale motions of a randomly stirred fluid.

We present in this paper a brief survey of these different approaches to turbulence. Special attention is paid to the Renormalization Group. For more details about the other approaches, the reader is referred to the review papers of Kraichnan (1975a), Orszag (1977), Nelkin (1978), Rose and Sulem (1978) and to the book of Leslie (1973). A comprehensive survey of the physics of turbulence is given in Monin and Yaglom (1971, 1975).

2. THE INERTIAL RANGE : SELF SIMILARITY AND UNIVERSALITY.

The experimental result on three-dimensional turbulence which has been the focus of theoretical interest is the existence of a range of eddy sizes $\{\ell;\ \ell_o \gg \ell \gg \ell_d\}$, called inertial range where energy production and dissipation are negligible and where energy is only "cascading" at a constant rate from large to small scales. In this range, the energy spectrum

$$E(k,\ t) \sim k^2 \int e^{i\vec{k}.\vec{r}} < u(\vec{x},\ t)\ .\ u(\vec{x}+\vec{r},\ t) > d\vec{r} \tag{3}$$

exhibits a self similar behaviour

$$E(k) \propto k^{-m} \tag{4}$$

with an exponent m close to the value 5/3 predicted by Kolmogorov (1941) on the basis of a phenomenological analysis. The inertial range is limited in the small scales by a dissipation range where dissipation copes with non-linear transfer. This latter range is pushed to smaller and smaller scales as the viscosity tends to zero.

Three parameters are associated with the equations of motion: the kinematic viscosity ν, the characteristic stirring length ℓ_o at which turbulent energy is put into the fluid and the rate of energy input per unit mass $\bar{\varepsilon} \sim v_o^3/\ell_o$. Kolmogorov 1941 theory (in short K41) assumes that in the cascade all detailed statistical information about the source of energy in the large scales is lost except the injection rate $\bar{\varepsilon}$, because it equals the transfer rate (and also the dissipation rate). Since the dissipation is negligible in the inertial range, the viscosity does not enter the energy spectrum, and dimensional analysis yields

$$E(k) \sim \bar{\varepsilon}^{\,2/3} k^{-5/3} \quad . \tag{5}$$

This spectrum can also be derived by a more dynamical argument using a simple phenomenological model (Frisch, Sulem and Nelkin, 1978).

The K41 theory predicts also the form of higher order correlation functions in the inertial range, for example the normalized structure function

$$S_p(\ell) = \frac{<|u(x) - u(x+\ell)|^p>}{<|u(x) - u(x+\ell)|^2>^{p/2}} \quad . \tag{6}$$

Since the energy transfer rate $\bar{\varepsilon}$ is the only external parameter which plays a role in the K41 theory, a simple dimensional analysis indicates that $S_p(\ell)$ should be independent of p and ℓ. This prediction is actually contradicted by the experimental results which show that $S_p(\ell)$ increases with $1/\ell$ and p (Van Atta and Park, 1972; see also Monin and Yaglom, 1975, chap. 25 for review). This departure is believed to be related to the fact that K41 does not take into account the statistical fluctuations of the energy transfer. The experimental evidence of the intermittency of the flow, i.e. the concentration of the small-scale structures on smaller and smaller fractions of the available space as the eddy size decreases (Batchelor and Townsend, 1949; Kuo and Corrsin, 1971, 1972) is in turn believed to be a consequence of such fluctuations. Amplification of these fluctuations along the cascade has been numerically observed on a model based on a wave packet description of the NS equation (Siggia, 1977).

Kolmogorov (1962), Obukhov (1962) and after them Novikov and Stewart (1964), Yaglom (1966), Mandelbrot (1974, 1976), Frisch et al. (1978) have proposed modifications of the K41 theory to take intermittency into account. The energy cascade is described as a breakdown process of eddies into smaller eddies which fill a lesser portion of the space with a spatial distribution prescribed by the model. The statistics of the velocity field at scale $1/k$ then depends upon the number of steps required to generate eddies of size $1/k$ starting from eddies of size ℓ_o, and the energy spectrum becomes

$$E(k) \sim \bar{\varepsilon}^{\,2/3} k^{-5/3} (k\,\ell_o)^{-B} \quad . \tag{7}$$

The exponent B is not determined by dimensional arguments. It can be expressed (differently according to the models) in terms of the Hausdorff dimension of the "fractal" set on which dissipation concentrates in the limit of zero viscosity (Mandelbrot, 1976; Frisch et al., 1978).

The universal self-similar behaviour of turbulence in the limit of infinite Reynolds number and large wavenumbers has been compared with the universality of critical phenomena when the temperature approaches the critical temperature in the limit of small wavenumbers (Nelkin, 1974, 1975; de Gennes, 1975). In this comparison K41 has been considered as the analog of the Landau mean field theory of phase transitions in the sense that they both neglect statistical fluctuations of the energy flux and of the order parameter, respectively (Siggia, 1977; Nelkin, 1978). In critical phenomena for spatial dimensions d larger than a crossover dimension d_c the mean field theory becomes exact. It has thus been asked whether there exists a dimension above or below which intermittency disappears (Nelkin, 1975; de Gennes, 1975). Up to now there is no such evidence, even in the limit of infinite dimension (Fournier, Frisch and Rose, 1978). Notice that in critical phenomena, when the mean field theory is exact, departures of the Hamiltonian from a quadratic functional are irrelevant, and consequently the evolution equations are linear. Within such an analogy the K41 energy cascade would be describable in a linear framework, which seems rather doubtful.

The notion of universality itself does not have the same meaning for fully developed turbulence as for critical phenomena . In turbulence, universality means that the sufficiently small scales are independent of the detailed initial conditions. For a Hamiltonian system this can plausibly correspond to the fact that the state of thermal equilibrium is characterized only by the isolating integrals. On the other hand, in critical phenomena universality means that certain properties of the system in the vicinity of the critical point, such as the critical exponents, are independent of the detailed form of the Hamiltonian. One is then led to compare universality in critical phenomena with the sensitivity of turbulence to changes in the NS equation. Following Kraichnan (1975a), let us consider a modification of the coupling coefficients in the non-linear terms of the NS equation which preserves the integral of motion. This affects in general the locality of the interactions in Fourier space and thus the step size in the cascade. If the build-up of intermittency depends indeed on the effective number of steps, then the build-up and the value of the exponent B correcting the K41 spectrum will be affected by this alteration.

3. ANALYTIC THEORIES OF TURBULENCE.

We present in this chapter the statistical approaches to turbulence made in a spirit somewhat akin to mode-coupling formalism (Kawasaki, 1968, 1969), and which were developed around the Direct Interaction Approximation (DIA) of Kraichnan (1958, 1959). There exist several derivations of the DIA. One of them makes use of the formalism developed in quantum electrodynamics, which leads to the Dyson-Schwinger equations. As adapted to classical systems (Martin, Siggia and Rose, 1973;

Pythian, 1977) it leads, for homogeneous turbulence, to three coupled equations for the velocity correlation, the response function and the three-point vertex function. The DIA can be obtained by replacing in those equations the renormalized vertex by the bare vertex. It is tempting to associate the renormalization of the response function to the idea, developed in hydrodynamics at the end of the 19th century, that the nonlinear interactions can be partly described by a turbulent viscosity. Kraichnan (1958) noticed that the master equations of the DIA are exact consequences of a "stochastic model" obtained by coupling many replicas of NS equations to each other with random coupling coefficients. This insures the realizability of the DIA, for example the positivity of the energy spectrum. For the study of homogeneous turbulence, the DIA has however the defect of being non-invariant under random Galilean transformations (Kraichnan, 1964). This leads to a spurious description of the advection of the small scales by the large scales. The DIA predicts in particular an inertial energy spectrum

$$E(k) \sim (\bar{\varepsilon}\, v_o)^{1/2}\, k^{-3/2}\,, \qquad (8)$$

where v_o is the characteristic velocity of the most energetic eddies. To cure this deficiency, Kraichnan (1965, 1977) proposed to use a Lagrangian description. The theory is then invariant under random Galilean transformations but the resulting equations are quite involved. An alternative non-Eulerian framework leading to K41 scaling has recently been proposed by Horner and Lipowsky (1979). For both these theories the realizability is not insured.

In view of these difficulties, a class of phenomenological modifications of the DIA has been proposed (Orszag and Kruskal, 1968; Kraichnan, 1971a, 1971b).
This leads to realizable Galilean invariant equations, consistent with K41 theory and thus neglecting a priori intermittency. These equations can also be viewed as semi-heuristic closures providing a quantitative reformulation of the K41 theory (see Orszag, 1977; and Rose and Sulem, 1978, for review). The simplest of these models, the Eddy Damped Quasi Normal Markovian (EDQNM) approximation leads to an integro-differential equation for the energy spectrum which, contrary to the primitive NS equation, can be numerically integrated at very high Reynolds numbers. The reason is that the spectrum, being an averaged quantity, has a gentle variation with wavenumber and can be adequately computed using a logarithmic discretization. In this way Reynolds numbers up to 10^6 have been achieved. Such closures give the possibility to study evolution problems. Moreover, their integration in various dimensions has confirmed or disclosed the existence of other self-similar regimes than the celebrated direct cascade of Kolmogorov.

In three-dimensions, the numerical calculations indicate that for arbitrary initial conditions, the energy spectrum evolves indeed toward the K41 spectrum which is a stationary solution in the limit of zero viscosity (André and Lesieur, 1977). Comparisons with direct numerical simulations of the NS equation at moderate Reynolds number and with experiments indicate that the closures give an adequate qualitative description of the evolution of the energy spectrum at those Reynolds numbers (Herring and Kraichnan, 1972).

In two dimensions, the non-linear terms of the NS equation have the property of conserving the vorticity ω = curl u of each fluid element as it follows the velocity field . A special consequence is the conservation of the enstrophy

$$\Omega = \langle \mathrm{curl}\ u)^2 \rangle = \int_0^\infty k^2\, E(k)\, dk \ , \tag{9}$$

which excludes the possibility of significant energy transfer towards the small scales. For large Reynolds numbers the enstrophy cascades to the small scales with a -3 energy spectrum (with a possible logarithmic correction). If energy is injected into the fluid, it cascades to the large scales with a -5/3 energy spectrum (Kraichnan, 1967; Leith, 1971; Pouquet, Lesieur, André and Basdevant, 1975). A description of the inverse energy cascade by a hierarchical clustering of discrete vortices has been proposed by Aref (1978). Intermittency corrections are expected to decrease the 5/3 exponent in this cascade (Kraichnan, 1975b, Frisch et al., 1978).

The significant difference between turbulence in two and three dimensions, in particular in connection with the direction of the energy cascade (ultraviolet in three dimensions and infrared in two dimensions), leads one to consider turbulence in non-integer dimension $2 < d < 3$ (Frisch, Lesieur and Sulem, 1976; Fournier and Frisch, 1978). The enstrophy does not go over continuously into another conserved quantity; nevertheless, the energy cascade is in the infrared direction for $2 \lesssim d < d_c$ ($d_c \gtrsim 2.05$) and in the ultraviolet direction for $d > d_c$. For $2 < d < d_c$, in addition to the infrared -5/3 energy cascade, there appears a fluxless ultraviolet $-m(d)$ range with m varying with dimension from 3 to 5/3. Similar results were obtained by Bell and Nelkin (1977, 1978) using a simple phenomenological model.

4. RENORMALIZATION GROUP FOR NAVIER-STOKES EQUATION.

The self-similarity of the inertial range has suggested a renormalization group approach to turbulence (Nelkin, 1974). Actually, no RG calculation starting from the Navier-Stokes equation has succeeded to date in treating the inertial range. The main difficulty lies in the fact that a cascade does not seem to be obtainable by a perturbation of a linear Langevin equation. In contrast, the infrared properties of a randomly stirred fluid have been studied with the RG, both in the formulation of

Wilson (1971) and of field theory. We discuss mainly in this chapter the first approach which has been implemented in its dynamical version (Ma and Mazenko, 1975) by Forster, Nelson and Stephen (1976, 1977; to be referred as FNS). The field theoretical approach is discussed in chapter 5.

Let the NS equation be written in Fourier space with wavenumbers extending from 0 to an ultra-violet cut-off Λ. The Fourier components of the velocity and forcing are separated in two groups: the first one for which $\Lambda e^{-\ell} < k \leq \Lambda$ is denoted by a subscript $>$, and the second one ($k \leq \Lambda e^{-\ell}$) by $<$. The parameter ℓ is positive and the limit $\ell \to \infty$ is eventually taken. Without any approximation, reference to velocity Fourier components $u^{>}$ can be suppressed by the following procedure: using the NS equation, $u^{>}$ can in principle be written in terms of $u^{<}$ and of the forces $f^{>}$; the expression for $u^{>}$ is then substituted in the equation of motion for $u^{<}$. In the resulting equation for $u^{<}$, variables are then rescaled to make it look as much as possible like the original equation. This defines the RG transformation.

It turns out that in the problem studied by FNS, there is a crossover dimension d_c (depending to some extent on the properties of the forcing) above which the iteration of the RG transformation makes the non-linear terms essentially negligible. For d slightly less than d_c one can solve perturbatively in powers of $\varepsilon = d_c - d$. In addition, the new couplings generated by the RG transformation turn out to be irrelevant variables as $\ell \to \infty$, and they can often be neglected at least for the leading approximation. Instead of defining the crossover in terms of space dimensionality, it is also possible to keep the space dimension fixed and to consider a crossover for the forcing spectrum (Fournier, 1977). This permits one to study problems which cannot be analytically continued in non-integer dimensions, for example helical turbulence.

The suppression of the small scales in the NS equation cannot be carried out exactly. But, since iteration of the RG transformation makes the strength of the non-linear terms tend either to zero or to $\varepsilon^{1/2}$, a perturbative calculation in powers of the non-linear terms is feasible, at least for sufficiently large ℓ. The generality of the results obtained in this way is subordinated to a universality hypothesis.

In the case of the forced NS equation, to the leading approximation, the elimination of the small scales produces a renormalization of the viscosity and possibly of the forcing but not of the vertex. A phenomenological interpretation can be given to the renormalization of these coefficients: the small scale motions generally produce an energy sink for the large scales described by an "eddy viscosity", and also an energy source, described by an internal forcing or "eddy noise" (Rose, 1977).

It appears that indefinite iteration of the RG transformation where only second order diagrams are retained, generates the very diagrams the summation of which gives the DIA. It follows that near the crossover, and in the limit $k \to 0$, the DIA is indeed an $O(\varepsilon^2)$ - approximation. This is not necessarily true any more in magnetohydrodynamics (see chapter 6).

The infra-red behaviour of a forced turbulence depends to some extent on the properties of the external forcing. The internal forcing, generated by the elimination of the small scales, has a spectrum proportional to k^{d+1} and there are two broad classes of universality according to whether it is negligible or not when compared to the external forcing. When the internal forcing is negligible or comparable to the external one, the fluid is at thermal equilibrium and the energy spectrum obtained corresponds to an equipartition of energy among the Fourier modes (models C and A of FNS) . The velocity probability distribution is gaussian and there is a fluctuation - dissipation theorem. This property permits other approaches to the problem, such as a Fokker-Planck description (Enz, 1978). On the other hand, when the external forcing is dominant (model B of FNS, model R of Fournier 1977), the fluid is in contrast far from equilibrium. Let us give for this case some details on the procedure. We assume that the forcing is a Gaussian white noise with a spectrum $F(k) \sim Dk^{-r}$ with $-(d + 1) < r$. After elimination of the small scales, the variables are rescaled according to :

$$\tilde{k} = e^{\ell} k \; ; \; \tilde{\omega} = e^{z\ell} \omega ;$$

$$\tilde{u}(\tilde{k}, \tilde{\omega}) = e^{-\tau\ell} u^{<}(k, \omega) \; ; \; \tilde{f}(\tilde{k}, \tilde{\omega}) = e^{(z-\tau)\ell} f^{<}(k, \omega). \tag{10}$$

It is convenient to choose the scaling factors z and τ in such a way that the rescaled velocity satisfies a NS equation with viscosity and forcing intensity being unchanged. For the j-component of the velocity, this equation reads :

$$(-i\tilde{\omega} + \nu\tilde{k}^2)\, \tilde{u}_j(\tilde{k}, \tilde{\omega}) = D^{1/2}\, \tilde{f}_j(\tilde{k}, \tilde{\omega}) + i\lambda(\ell)\tilde{k}_m \left(\delta_{jn} - \frac{\tilde{k}_j \tilde{k}_n}{\tilde{k}^2}\right)$$

$$\times \int d\Omega \, d\vec{q} \; \tilde{u}_m(\vec{p}, \Omega) \; \tilde{u}_n(\vec{q}, \omega - \Omega), \tag{11}$$

$$\tilde{k} = \vec{p} + \vec{q}$$

$$|\vec{p}|, |\vec{q}| < \Lambda$$

where the coupling coefficient $\lambda(\ell)$ satisfies, to the leading approximation $O(\varepsilon^2)$,

$$\frac{d\,\lambda(\ell)}{d\ell} = \lambda(\ell)\left[r + 3 - C(d)\,\frac{D}{\nu^3}\,\lambda^2(\ell) \right] \quad ; \tag{12}$$

C(d) is a positive constant depending on space dimension only. Clearly, $r = -3$ is a crossover for the behaviour of the coupling $\lambda(\ell)$ as $\ell \to \infty$. Indeed, for $\varepsilon = r + 3 < 0$ $\lambda(\ell)$ tends exponentially to zero as $\ell \to \infty$ and the nonlinear terms are negligible. The scaling factors are then found to be

$$z = 2 \quad ; \tau = d + \frac{\varepsilon}{2} \quad . \tag{13}$$

For $\varepsilon > 0$, the coupling coefficient $\lambda(\ell)$ has a finite limit of order $\varepsilon^{1/2}$. Near the crossover, the velocity field for wavenumbers $\tilde{k}$ of order unity can then be computed perturbatively. The scaling factors become

$$z = 2 - \frac{\varepsilon}{3} \quad ; \tau = d + 1 \; . \tag{14}$$

For $\varepsilon = 0$, the coupling $\lambda(\ell)$ tends to zero as $1/\ell$ and logarithmic corrections are to be taken into account.

For negative or small positive ε, one can discard the nonlinear terms in equation (11). Returning to the unscaled variables, one then obtains the effective equation, valid in the limit $k \to 0$;

$$(-\,i\omega + \nu(k)\,k^2)\,u(\vec{k},\omega) = D^{1/2}\,f(\vec{k},\omega) \tag{15}$$

where the renormalized viscosity is given by $\nu(k) \sim k^{z-2}$. On the trivial side of the crossover ($\varepsilon < 0$), the damping $\nu(k)$ is, as expected, independent of wavenumber, whereas for $\varepsilon > 0$ it behaves as $k^{-\varepsilon/3}$. From the Langevin equation (15), one can easily deduce the statistical properties of the velocity field and in particular the energy spectrum

$$E(k) \sim \frac{k^{1-\varepsilon}}{\nu(k)} = \begin{cases} k^{1-\varepsilon} & \text{if } \varepsilon < 0 \quad (16a) \\ k^{1-2\varepsilon/3} & \text{if } \varepsilon > 0 \; . \quad (16b) \end{cases}$$

The renormalization group thus indicates that for small positive ε, the effect of small on large scales may be represented by a renormalized viscosity of the form $\nu(k) \sim (E(k)/k)^{1/2}$. This coincides with the usual expression of the "eddy viscosity" obtained on phenomenological grounds (without the restriction of small ε) when the transfer is local (Heisenberg, 1948). In this context, the spectrum (16b) results from an equilibrium between external forcing and transfer.

5. FIELD THEORETICAL APPROACH.

The phenomenological interpretation of the spectra obtained with the renormalization group suggests that their validity is not restricted to the neighbourhood of the crossover. The same results have also been obtained using the EDQNM (Fournier and Frisch, 1978) or a variational method (Lücke, 1978). Anticipated by Forster et al., the generality of such results has been established systematically by de Dominicis and Martin (1978) who have used the RG as formulated in field theory. This method, whose utilization in static critical phenomena is now standard (Brezin, Le Guillou and Zinn-Justin, 1977), has recently been implemented for dynamical critical phenomena (de Dominicis and Peliti, 1978). Let us briefly recall the procedure in the case of the Navier-Stokes equation.

One considers a generating functional for the various correlation and response functions of the velocity field (Martin, Siggia and Rose, 1973; Phythian, 1977). One evaluates the superficial divergences which, in the absence of ultraviolet cut-off Λ, appear in a perturbative calculation of the correlation functions. Such a power counting exhibits the same crossovers which obtain in the Wilson formalism (Abarbanel, 1978; de Dominicis and Martin, 1978). One then introduces, in front of the coupling constants (and possibly of the fields), renormalization factors which can be made to depend on a parameter μ (which appears to be a wavenumber). These factors are chosen in order to suppress the logarithmic divergences arising at the crossover when the cut-off Λ becomes infinite. The physical quantities (e.g. the correlation functions) calculated from the renormalized generating functional must be independent of the parameter μ, at least in the range of wavenumbers $\mu \ll \Lambda$. In other words, the total derivative with respect to μ of the velocity correlation function must vanish identically: this identity is called a Callan-Symanzik equation. By solving this linear partial differential equation by means of the characteristics method one obtains the scaling laws of the correlation functions. The equations of the characteristics are the analog of the recursion equations of the Wilson formalism. Those equations are usually obtained perturbatively. There are three renormalization factors to determine, those of viscosity, forcing and vertex. Actually, the forcing is either not renormalized (model R) or renormalized in the same way as the viscosity, because of the fluctuation-dissipation theorem (models A and C). A Ward identity linked to Galilean invariance insures that the vertex is not renormalized. The existence of a stable non-trivial fixed point for the equation of the paramater playing the role of a Reynolds number then determines the renormalization of the viscosity without recourse to perturbative calculations. Thus the results are not restricted to small ε , provided that no neglected operators become relevant.

When the external forcing is not renormalized (model R), an analysis based on the normal dimensions of the operators restricts a priori $\varepsilon = r + 3$ (where r is the forcing exponent) to be smaller than 4, a value which corresponds to a $-5/3$ energy spectrum. The relationship between this limit spectrum and the K41 spectrum for the energy cascades - where both forcing and dissipation are negligible - is still unclear. It must nevertheless be stressed, as done by Martin and de Dominicis (1978), that the anomalous dimension corrections to the operators, which vanish when $\varepsilon = 0$, could modify the domain of validity of the calculation. In this context one may recall that the semi-heuristic EDQNM approximation indicates that in dimensions close to 2 ($d < d_c \approx 2.05$) the validity condition for the FNS result (which demands that the energy transfer be negative) is more stringent than $\varepsilon < 4$ (see Fournier and Frisch, 1978, fig. 1b).

6. INFLUENCE OF NEW COUPLING.

In this chapter we look at the modifications of the results obtained in chapter 4 when a new coupling enters the NS equation. An example is provided by (three-dimensional) helical turbulence, i.e. a turbulence which is invariant under translations and rotations but not under plane reflections. In this case the helicity $H = \langle u . \text{curl } u \rangle$, which is an invariant of the inviscid NS equation, does not vanish. This leads us to consider in parallel with the energy injection and the energy spectrum, the helicity injection and the helicity spectrum. It appears that this symmetry breakdown is never more than a marginal perturbation, in the sense that it does not modify the crossover and the energy spectra found in the purely isotropic case (Pouquet, Fournier and Sulem, 1978).

We have also considered the infrared properties of turbulence in a conducting fluid governed by the magnetohydrodynamic (MHD) equations when the velocity u is coupled to the magnetic field b through the Lorentz force $b . \nabla b$:

$$\left\{ \begin{aligned} & \frac{\partial u}{\partial t} + u . \nabla u = - \nabla p + \nu \nabla^2 u + b . \nabla b + f^V \\ & \frac{\partial b}{\partial t} + u . \nabla b = b . \nabla u + \eta \nabla^2 b + f^M \\ & \nabla . u = 0 \quad ; \quad \nabla . b = 0 , \end{aligned} \right. \tag{17}$$

where η denotes the magnetic diffusivity. We shall restrict ourselves to the non-helical case since in MHD, helicity has a destabilizing influence on the large scale magnetic fields leading to turbulent dynamo (see Moffatt, 1978, and references therein). Such an effect does not appear at this point to be tractable by the RG procedure. In the context of non-helical MHD turbulence one may investigate the sensitivity of the velocity field to the Lorentz force. To deal with this problem in the RG framework it seems necessary to introduce, in addition to the kinetic forcing f^V, a (somewhat academic) magnetic forcing f^M. In three dimensions, if the magnetic forcing is strong enough in comparison with the kinetic forcing, then even for $\varepsilon > 0$ the advection is negligible, and the dynamics of the velocity field is governed by an equilibrium between Lorentz force and viscous damping. This situation corresponds to a new stable fixed point in parameter space (referred to as the magnetic fixed point). To be specific, let us assume a kinetic forcing spectrum $F^V(k) \propto k^{3-\varepsilon}$ and a magnetic forcing spectrum $F^M(k) \propto k^{3-\varepsilon^M}$ with $\varepsilon^M = a\varepsilon$. We concentrate on $\varepsilon > 0$. For $a < 1/4$, only the non-magnetic fixed point is stable and the Lorentz force is not relevant. For $a > 1.16$, this fixed point is unstable and only the magnetic fixed point is stable. Between these two values there is a universality breakdown in the sense that the two fixed points are simultaneously stable. When the Lorentz force is not relevant, the magnetic field behaves as a passive vector and, similarly to the passive scalar studied by FNS, the effective large-scale Prandtl number is universal. The magnetic energy spectrum is given by

$$E^M(k) \sim k^{1 - (a - 1/3)\varepsilon} \quad . \tag{18}$$

A peculiar result is that for $1 < a < 1.16$, and for initial values of the parameters in the attractive domain of the non-magnetic fixed point, the magnetic energy dominates the kinetic energy in the large scales and, nevertheless, the Lorentz force is negligible in comparison with the inertial force. The stability of the non-magnetic fixed point in this range is due to the renormalization of the Lorentz force vertex by the RG procedure. No invariance principle prevents in effect such a renormalization. Consequently, the DIA which consists in neglecting the vertex corrections is not a priori exact in the MHD problem in the limit $k \to 0$.

7. CONCLUDING REMARKS.

Most of the properties of fully developed turbulence, in particular the cascade phenomenon, are not to date apprehended within a systematic theory. The closure techniques have provided a semi-heuristic description of turbulence compatible with Kolmogorov 1941 phenomenology and as such do not describe the intermittency. The experimental investigation of intermittency is mostly concerned with its effect on high-order velocity correlation functions. The geometrical aspect of intermittency is still fuzzy. One would like to be confronted with a map of the region with strong velocity gradients. This demands the development of visualization techniques and of precise measurements of velocity gradients.

Another "experimental" approach involves resorting to computers. The direct numerical simulation of isotropic turbulence generally based on spectral methods (Orszag and Patterson, 1972). It is only in two dimensions that the present calculations have achieved Reynolds numbers large enough (10^4) to display an inertial range, here the enstrophy cascade (Orszag, 1976). The interactions between the Fourier modes in the enstrophy cascade are non-local, and this probably makes the intermittency of small-scale two-dimensional turbulence very different from the three-dimensional case (Kraichnan, 1975b). In two-dimensional MHD turbulence, in contrast, the enstrophy conservation is broken by the Lorentz force, and there is an energy cascade towards the small scales (Pouquet, 1978). A calculation of Orszag and Tang (1979) displays intermittency, and further investigations are under way. As for three-dimensional NS turbulence, the Reynolds numbers presently achieved (of order of a hundred) are too moderate to demonstrate intermittency unambiguously (Siggia and Patterson, 1978). Higher Reynolds number simulations are needed. This demands the development of new codes, possibly using moving mesh schemes, in order to refine the calculation in the regions with strong velocity gradients.

REFERENCES.

ABARBANEL, H.D.I. (1978) Character of homogeneous turbulence. Preprint. FERMILAB - PUB. 78/28 - THY.

ANDRE, J.C. and LESIEUR, M. (1977) J. Fluid Mech. 81, 187.

AREF, H. (1978) Dynamics of point vortices and the inverse cascade. Preprint. Dept. Physics. Cornell University.

BATCHELOR, G.K. and TOWNSEND, A.A. (1949) Proc. Roy. Soc. A 199, 238.

BELL, T.L. and NELKIN, M. (1977) Phys. Fluids 20, 345.

BELL, T.L. and NELKIN, M. (1978) J. Fluid Mech. 88, 369.

BREZIN, E., LE GUILLOU, J.C. and ZINN-JUSTIN, J. (1977) in "Phase Transition and Critical Phenomena". C. Domb and M.S. Green eds. Academic Press.

DE DOMINICIS, C. and MARTIN, P.C. (1979) Phys. Rev. A 19, 419.

DE DOMINICIS, C. and PELITI, L. (1978) Phys. Rev. B 18, 353.

DE GENNES, P.G. (1975) in "Fluctuation, Instability and Phase Transition". (Proc. NATO Advance Study Institute, Jeilo, Norway). T. Riste ed. Noordhoff, Leiden, series B, p. 1.

ENZ, C.P. (1978) Physica 94 A, 20.

FORSTER, D., NELSON, D.R. and STEPHEN, M.J. (1976) Phys. Rev. Lett. 36, 867.

FORSTER, D., NELSON, D.R. and STEPHEN, M.J. (1977) Phys. Rev. A 16, 732.

FOURNIER, J.D. (1977) Quelques méthodes systématiques d'approximation en turbulence homogène. Thèse de 3ème Cycle. Université de Nice.

FOURNIER, J.D. and FRISCH, U. (1978) Phys. Rev. A 17, 747.

FOURNIER, J.D., FRISCH, U. and ROSE, H.A. (1978) J. Phys. A 11, 187.

FRISCH, U., LESIEUR, M. and SULEM, P.L. (1976) Phys. Rev. Lett. 37, 895.

FRISCH, U., SULEM, P.L. and NELKIN, M. (1978) J. Fluid Mech. 87, 719.

GOLLUB, J. and SWINNEY, H. (1975) Phys. Rev. Lett. 35, 927.

HEISENBERG, W. (1948) Z. Phys. 124, 628.

HERRING, J.R. and KRAICHNAN, R.H. (1972) in "Statistical Models and Turbulence" p. 148. M. Rosenblatt and C. Van Atta eds. Lecture Notes in Physics 12, Springer Verlag.

HORNER, H. and LIPOWSKY, R. (1979) On the theory of turbulence : a non-Eulerian renormalized expansion. Preprint. Inst. für Theoretisch Physik. Universität Heidelberg.

KAWASAKI, K. (1968) Progr. Theor. Physics 39, 1133.
KAWASAKI, K. (1969) J. Phys. Soc. Japan 26 Suppl. 160.
KOLMOGOROV, A.N. (1941) C.R.Ac.Sc. U.R.S.S. 30, 301.
KOLMOGOROV, A.N. (1962) J. Fluid Mech. 13, 82.
KRAICHNAN, R.H. (1958) Proc. 2nd Symp. on Naval Hydrodynamics, p. 29. R. Cooper ed. ACR-38. Office of Naval Research. Dept. Navy. Washington.
KRAICHNAN, R.H. (1959) J. Fluid Mech. 5, 497.
KRAICHNAN, R.H. (1964) Phys. Fluids 7, 1723.
KRAICHNAN, R.H. (1965) Phys. Fluids 8, 575; erratum 9, 1884.
KRAICHNAN, R.H. (1967) Phys. Fluids 10, 1417.
KRAICHNAN, R.H. (1971a) J. Fluid. Mech. 47, 513.
KRAICHNAN, R.H. (1971b) J. Fluid. Mech. 47, 525.
KRAICHNAN, R.H. (1975a) Avances in Math. 16, 305.
KRAICHNAN, R.H. (1975b) J. Fluid Mech. 67, 155.
KRAICHNAN, R.H. (1977) J. Fluid Mech. 83, 349.
KUO, A.Y. and CORRSIN, S. (1971) J. Fluid Mech. 50, 285.
KUO, A.Y. and CORRSIN, S. (1972) J. Fluid Mech. 56, 447.
LEITH, C.E. (1971) J. Atm. Sc. 28, 145.
LESLIE, D.C. (1973) Development in the Theory of Turbulence. Clarendon Press (Oxford).
LORENZ, E.N. (1963) J. Atm. Sc. 20, 130.
LÜCKE, M. (1978) Phys. Rev. A 18, 282.
MA, S. and MAZENKO, G.F. (1975) Phys. Rev. B 11, 4077.
MANDELBROT, B.B. (1974) J. Fluid Mech. 62, 331.
MANDELBROT, B.B. (1976) in "Turbulence and Navier-Stokes Equation". p. 121 R. Temam ed. Lecture Notes in Mathematics 565, Springer Verlag.
MARTIN, P.C. (1976) J. Phys. (Paris) Suppl. 37, C1-57.
MARTIN, P.C. and DE DOMINICIS, C. (1978) The Long Distance Behaviour of Randomly Stirred Fluids. Preprint. Havard University.
MARTIN, P.C., SIGGIA, E.D. and ROSE, H.A. (1973) Phys. Rev. A 8, 423.
MOFFATT, H.K. (1978) 'Magnetic Field Generation in Electrically Conducting Fluids. Cambridge Monographes of Mechanics and Applied Mathematics. Cambridge University Press.
MONIN, A. and YAGLOM, A.M. (1971) vol. 1; (1975) vol. 2. Statistical Fluid Mechanics : Mechanics of Turbulence. English edition J.L. Lumley ed. MIT Press (Cambridge).
NELKIN, M. (1974) Phys. Rev. A 9, 388.
NELKIN, M. (1975) Phys. Rev. A 11, 1737.
NELKIN, M. (1978) in Proc. 13th IUPAP Conf. on Stat. Phys. (Haifa, August 1977) p. 235. D. Cabib, C.G. Kuper and I. Riess eds. Annals Israel Phys. Soc. vol. 2.
NORMAND, C., POMEAU, Y., VELARDE, M.G. (1977) Rev. Mod. Phys. 49, 581.
NOVIKOV, E.A. and STEWART, R.W. (1964) Izv. Akad. Nauka SSSR, Ser. Geofiz. 3, 408.
OBUKHOV, A.M. (1962) J. Fluid Mech. 12, 77.
ORSZAG, S.A. (1976) Proc. 5th Int. Conf. on Numerical Methods in Fluid Dynamics, p. 32. I.A. Van de Vooren and P.J. Zandbergen eds. Lecture Notes in Physics 59, Springer-Verlag.
ORSZAG, S.A. (1977) in Fluid Dynamics, p. 235. Les Houches Summer School of Theoretical Physics (1973). R. Balian and J.L. Peube eds. Gordon and Breach.
ORSZAG, S.A. and KRUSKAL, M.O. (1968) Phys. Fluids 11, 43.
ORSZAG, S.A. and PATTERSON, G.S. (1972) Phys. Rev. Lett. 28, 76; also in "Statistical Models and Turbulence" p. 127. M. Rosenblatt and C. Van Atta eds. Lecture Notes in Physics 12, Springer Verlag.
ORSZAG, S.A. and TANG, C.M. (1979) J. Fluid Mech. 90, 129.
PHYTHIAN, R. (1977) J. Phys. A 10, 777.
POUQUET, A., FOURNIER, J.D. and SULEM, P.L. (1978) J. Phys. Lett. (Paris) 39, L 199.
POUQUET, A., LESIEUR, M., ANDRE, J.C. and BASDEVANT, C. (1975) J. Fluid Mech. 72, 305.
POUQUET, A. (1978) J. Fluid. Mech. 88, 1.
ROSE, H.A. (1977) J. Fluid Mech. 81, 719.
ROSE, H.A. and SULEM, P.L. (1978) J. Phys. (Paris) 39, 441.
SIGGIA, E.D. (1977) Phys. Rev. A 15, 1730.
SIGGIA, E.D. and PATTERON, G.S. (1978) J. Fluid Mech. 84, 567.

VAN ATTA, C.W. and PARK, J. (1972) in "Statistical Models and Turbulence" p. 402. M. Rosenblatt and C. Van Atta eds. Lecture Notes in Physics 12. Springer Verlag.
WILSON, K.G. (1971) Phys. Rev. B 4, 3174; B 4, 3184.
YAGLOM, A.M. (1966) Sov. Phys. Dokl. 11, 26.

CRITICAL DYNAMICS FAR FROM EQUILIBRIUM

K. Kawasaki and A. Onuki

Department of Physics, Faculty of Science
Kyushu University, Fukuoka, 812 Japan

CRITICAL DYNAMICS FAR FROM EQUILIBRIUM

Kyozi Kawasaki and Akira Onuki

Department of Physics, Faculty of Science
Kyushu University, Fukuoka, 812 Japan

Systems near critical points are known to be very sensitive to even a small amount of disturbances. Thus if the critical point is approached with external disturbances of fixed sizes that bring the system out of equilibrium, we encounter a variety of new problems of critical dynamics which cannot be handled by the usual linear response type treatments. Some examples of such a new kind of critical dynamics will be described which include critical fluids under steady incompressible flow, nonlinear relaxation of the order parameter, kinetics of first order phase transitions near critical points, and the superconducting transition under electric field.

1. Introduction

During the past fifteen years we have witnessed a phenomenal growth of our understanding of both static and dynamic aspects of critical phenomena which culminated in the introduction of renormalization group ideas [1). We now understand the nature of singularities in thermodynamic properties at the critical point and the spatial and temporal behavior of fluctuations occurring in the equilibrium state. As far as the dynamics is concerned, however, much of the past studies have been limited to the problems where deviations from equilibrium are so small as to be treated with the linear response scheme [2). On the other hand, the characterization of a critical point as a point of marginal stability suggests that the linear regime where the linear response theory is adequate is very limited, and is indeed more so as the critical point is approached more closely. This is illustrated by considering the average order parameter m(t) when it started out with $m(0) = m_0$ for simple dynamical systems such as the TDGL model without conservation laws in the absence of a conjugate field and above T_c. There is then a simple scaling law [2)

$$m(t) = (\Delta T)^{\beta} f[(\Delta T)^{\nu z} t, (\Delta T)^{-\beta} m_0] \quad (1.1)$$

where $\Delta T = T - T_c$ and z is the dynamic critical exponent. The linear regime is then limited to $m_0 \ll (\Delta T)^{\beta}$ where (1.1) can be linearized in m_0. Therefore, even when m_0 is fixed to some very small value, as ΔT tends to zero we will leave the linear regime at some stage.

Thus there is a whole new field of critical dynamics where a small but finite deviation must be taken into accout*) and which only recently has started to be explored. Here the types of problems one encounters will be much more diversified than the usual linear critical dynamics which is concerned with the linear regime. Thus it is clearly presumptuous at this stage to try to enumerate all the possible types of problems. On the other hand, by looking at the limited number of cases where detailed studies are now becoming available, one can nevertheless recognize the kinds of phenomena expected in this new field of critical dynamics.

Here we wish to illustrate the situation by discussing several typical examples which we do not of course claim to exhaust all the types.

2. Steady State Problems

Here we will be concerned with systems in nonequilibrium steady states brought about by constant dissipative perturbations.

2-1. Critical Fluid under Steady Incompressible Flow

Consider a critical fluid subject to uniform shear flow with the velocity field $\vec{u} = Dy\vec{e}_x$, $\vec{e}_x$ being a unit vector. Here one notices the existence of two characteristic times, namely, the inverse of the rate of shear D^{-1} which is the distortion time by the flow and the decay time of critical fluctuations in the absence of shear $\tau = [16\bar{\eta}\xi^3/k_B T]$, where $\bar{\eta}$ is the shear viscosity near equilibrium and $\xi = |\Delta T/T_c|^{-\nu}\xi_0$ is the equilibrium correlation length. As one approaches the criticality with a fixed rate of shear one encounters a regime where $D\tau$ exceeds unity and large scale critical fluctuations are expected to be greatly affected by the flow within their life times (Fig. 1). Thus the problem is characterized by the relevent dimensionless parameter $D\tau = (k_c\xi)^3$ which grows near the criticality where $k_c = (16\bar{\eta}/k_B T)^{1/3} D^{1/3}$. In this new regime, $D\tau > 1$, typical critical fluctuations with size greater than k_c^{-1} acquire strongly anisotropic needle-like shapes extending in the direction of flow, while fluctuations with size smaller than k_c^{-1} are little affected by the flow because they dissipate thermally before the distortion mechanism by the flow becomes effective. The above-mentioned modification in the fluctuation spectrum drastically changes the character of critical singularity [3,4,5].

*) Here we exclude cases which can be treated by finite order power series expansion in deviations from equilibrium.

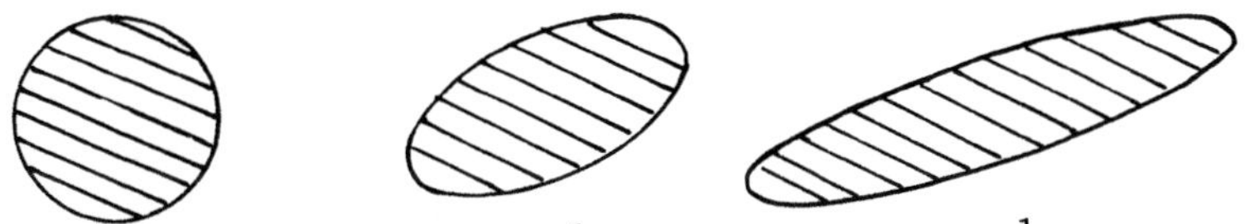

Fig. 1 Time-development of a local disturbance with size greater than k_c^{-1}

In the steady state the space-time correlation function of order parameter fluctuations $s(\vec{r},t)$ satisfies the relation [3]

$$\langle s(\vec{r},t)s(\vec{r}',t')\rangle = \langle s(\vec{r}-\vec{r}'-Dy'\vec{e}_x(t-t'),t-t')s(\vec{0},0)\rangle \tag{2.1}$$

The equal time correlation function has the translational invariance in space. The invariance property is lost for $t \neq t'$, however. This results in a Doppler shift broadening in dynamic light scattering as we shall see below [6].

Our starting model is the following fluctuating mode-coupled hydrodynamic equation

$$\frac{\partial}{\partial t}s = -\vec{u}(\vec{r})\cdot\vec{\nabla}s - \rho_0\vec{\nabla}\cdot(s\vec{v}) + \lambda_0\nabla^2\frac{\delta H_0}{\delta s} + \theta \tag{2.2}$$

$$\frac{\partial}{\partial t}\vec{v} = -\rho_0\left(s\vec{\nabla}\frac{\delta H_0}{\delta s}\right)_\perp + \bar{\eta}_0\nabla^2\vec{v} + \vec{\zeta} . \tag{2.3}$$

The first term of the right hand side of (2.2) represents the convection of the order parameter by the average flow. In the absence of this term, (2.2) and (2.3) are the stochastic fluid model familiar in critical dynamics [2]. Here $s(\vec{r},t)$ and $\vec{v}(\vec{r},t)$ are the local order parameter and the (transverse) velocity field, respectively, and θ and $\vec{\zeta}$ are the corresponding random forces of microscopic origin. λ_0 and $\bar{\eta}_0$ are the bare kinetic coefficients. ρ_0 is the mode coupling strength which is usually set equal to 1. H_0 is the usual Wilson Hamiltonian plus the kinetic energy of flow of the form

$$H_0 = \int d\vec{r}\left[\frac{1}{2}(\tau+\tau_{0c})s^2 + \frac{1}{2}(\vec{\nabla}s)^2 + \frac{u_0}{4!}\Lambda^\varepsilon s^4 - hs + \frac{1}{2}\frac{\bar{\rho}}{k_BT}\vec{v}^2\right] \tag{2.4}$$

where τ is proportional to $T-T_c$, Λ is the upper cut-off wave number, h is the chemical potential conjugate to the order parameter and $\bar{\rho}$ is the mass density of fluid. We assume the familiar fluctuation-dissipation relations between θ, $\vec{\zeta}$ and λ_0, $\bar{\eta}_0$, which may be expected to be hardly affected by the shear flow since the fluctuations involved are of microscopic nature.

First we consider a linearized version of (2.2)-(2.4) where ρ_0, u_0 and τ_{0c} are set equal to zero. This provides a good approximation above four dimensions.

Peculiar effects of shear are already apparent in this case. The variables s and $\vec{v}$ are now uncoupled and the shear flow only affects the former. (The steady state variance of the velocity field is almost unchanged for realistic values of D[3]). The variance $C_{\vec{k}} = \langle |s_{\vec{k}}|^2 \rangle$ taken on the critical isochore, $\tau \geq 0$ and $h = 0$, which will be denoted as $C_{\vec{k}}^{(0)}$ in this case, obeys the equation

$$[k^2(k^2+\tau) - \frac{1}{2}k_c^4 k_x \frac{\partial}{\partial k_y}] C_{\vec{k}}^{(0)} = k^2 \tag{2.5}$$

where $k_c \equiv (D/\lambda_0)^{1/4}$. This equation can be solved under the condition that $C_{\vec{k}}^{(0)}$ is positive definite for any $\vec{k}$. The effect of shear is important only in the case of strong shear $k_c > \tau^{1/2}$ and in the small wave number region $k < k_c$. Therefore we consider this case in the following. (For $k_c < \tau^{1/2}$ or for $k > k_c$ the effect of shear is small and the variance is almost unchanged from the Ornstein-Zernike form). The variance $C_{\vec{k}}^{(0)}$ becomes anisotropic and has different functional forms in several domains of small k ($< k_c$). Its overall behavior can be well represented by the following approximate expression:

$$C_{\vec{k}}^{(0)} \simeq [\tau + c\, k_c^{8/5} |k_x|^{2/5} + k^2]^{-1} \tag{2.6}$$

where $c = (5/2)^{2/5}\,\Gamma(3/5)^{-1} \simeq 0.97$. Fourier-transforming to the coordinate space we obtain for $g(\vec{r}) \equiv \langle \delta s(\vec{r})\, \delta s(\vec{r}=0) \rangle$ with $\delta s \equiv s - \langle s \rangle$ the following behavior at $\tau=0$ in strong shear regime (here g is again denoted by $g^{(0)}$):

$$g^{(0)}(\vec{r}) = k_c^{d-2}\, g^*(k_c \vec{r}) \tag{2.7}$$

with

$$g^*(\vec{R}) \propto \begin{cases} |\vec{R}_\perp|^{-(d+2)} & \text{for } |\vec{R}_\perp| \gg |X|^{1/5} \text{ and } 1 \\ |X|^{-(d+2)/5} & \text{for } |\vec{R}_\perp| \ll |X|^{1/5},\ |X| \gg 1 \end{cases} \tag{2.8}$$

where d(>2) is the dimensionality of space, and X and $\vec{R}_\perp$ are the components of $\vec{R}$ parallel and perpendicular to the direction of flow. This peculiar behavior is due to the fact that fluctuations can be transported for a long distance without much decay by the shear flow along the x-direction in the region $|\vec{R}_\perp| \gg |X|^{1/5}$.

Let us now turn to the complete nonlinear model (2.2)-(2.4), where we are faced with tough problems of critical phenomena. The easiest way to deal with them is to try an ε-expansion where $\varepsilon = 4-d$ and we know from the usual treatment of critical phenomena[1, 2] that $\rho_0 \sim \varepsilon^{1/2}$ and u_0, $\tau_{0c} \sim \varepsilon$. The shear flow does not enhance critical fluctuations and hence the nonlinear couplings among fluctuations can be neglected above four dimensions as in the usual cases near equilibrium. The rate of shear D is a new relevent parameter and can be regarded as a kind of a new thermodynamic variable. Various physical quantities such as the variance of critical fluctuations and the equation of state depends on D strongly for $k_c\xi > 1$. Not enough attention has been paid to nonlinear effects of dissipative disturbances near the critical point, although much work has been done on nonlinear effects of static perturbations added to the Wilson Hamiltonian. The introduction of nonlinearities brings in coupling between order parameter and velocity field fluctuations that lead to renormalizations of λ_0 and $\overline{\eta}_0$. Technically, the procedure is similar to the usual renormalization group (RG) treatment of critical dynamics[1,2] except for one important difference. In the usual RG treatment the recursion relations for the static parameters such as u and τ do not involve dynamic parameters like λ and $\overline{\eta}$ where subscripts 0 are omitted to denote renormalized quantities. Namely, static and dynamic aspects of the problem are separable. This is not so in our problem far from equilibrium since the steady nonequilibrium state is determined by both static and dynamic parameters. This produces some complications but no unsurmountable difficulty. The main results in the case of strong shear to first order in ε are summarized below:

(1) The critical temperature $T_c(D)$ is shifted downward below its equilibrium value $T_c(0)$ as

$$T_c(D) = [1 - 0.0832\,\varepsilon\,\overline{\tau}_s(D)]\,T_c(0) \qquad (2.9)$$

where $\overline{\tau}_s(D)$ is the cross-over reduced temperature at which $k_c\xi = 1$ holds and is given by $(k_c/\Lambda)^{2-\varepsilon/3}$ to order ε.

(2) The equation of state for $k_c\xi > 1$ is of the mean field functional form but strongly depends on k_c :

$$h/m = (T - T_c(D))/T_c(0) \cdot \Lambda^{2-\varepsilon/3} k_c^{\varepsilon/3} + \frac{1}{2} u^* m^2 k_c^{\varepsilon} \qquad (2.10)$$

where h is the chemical potential and m the average order parameter. This relation indicates that the critical exponents β and γ are given by the mean field values 1/2 and 1, respectively.

(3) The critical divergence of the kinetic coefficients λ and $\overline{\eta}$ are suppressed for $k_c\xi > 1$ as

$$\lambda = \lambda^*(k_c/\Lambda)^{-18\varepsilon/19} \qquad (2.11)$$

$$\eta = \eta^*(k_c/\Lambda)^{-\varepsilon/19} \qquad (2.12)$$

where λ^* and η^* are the critical amplitudes for diverging transport coefficients near equilibrium. Here we have omitted corrections to λ^* and η^* of order ε.

(4) The variance for $k < k_c$ is given by

$$C_{\vec{k}} = k_c^{-2}\, C^*(\vec{k}/k_c,\ 1/k_c\xi_s) \qquad (2.13)$$

where

$$\xi_s = [T - T_c(D)]^{-1/2} T_c(0)^{1/2} (k_c/\Lambda)^{-\varepsilon/6} \Lambda^{-1} \quad (2.14)$$

is the new correlation length of critical fluctuations in the direction perpendicular to the flow. The scaling function $C^*(\vec{\ell},\delta)$ is well represented by the following:

$$C^*(\vec{\ell}, 1/k_c\xi_s) = [\xi_s^{-2} k_c^{-2} + c|\ell_x|^{2/5} + \ell^2]^{-1} \quad \text{for } \ell < 1 \quad (2.15)$$

The Fourier transform of (2.15) to the coordinate space reflects the spatial extension of critical fluctuations; they are elongated along the direction of flow with the width ξ_s in the directions perpendicular to the flow and with the length $k_c^4 \xi_s^5$ in the direction of flow.

(5) The structure function of light scattering $I(\vec{k},\omega)$ which is the Fourier transform of the space-time correlation function $\langle\delta s(\vec{r},t)\, \delta s(\vec{r}', t')\rangle$ in the scattering volume is written as

$$I(\vec{k},\omega) = C_{\vec{k}}\, S_{\vec{k}}(\omega) \quad (2.16)$$

The line space function S depends strongly on the direction of $\vec{k}$ and on the size of the scattering volume in contrast to the usual cases without flow. S is written using (2.1) as

$$S_{\vec{k}}(\omega) = \frac{1}{L_y} \int_{-L_y/2}^{L_y/2} dy\, G_{\vec{k}}(\omega - Dk_x y) \quad (2.17)$$

where L_y is the length of the scattering volume along the y-axis and $G_{\vec{k}}(\omega)$ is the line shape function in the limit of small scattering volume. The peak width of $G_{\vec{k}}(\omega)$, denoted by $\Delta_{\vec{k}}$, is strongly anisotropic and $G_{\vec{k}}(\omega)$ is not Lorentzian in wide regions of $\vec{k}$ corresponding to the distortion of fluctuations by the shear flow. $S_{\vec{k}}(\omega)$ coincides with $G_{\vec{k}}(\omega)$ only for small k_x such that $D|k_x|L_y < \Delta_{\vec{k}}$, as can be seen from (2.17). For $D|k_x|L_y > \Delta_{\vec{k}}$ the line shape is broadened as

$$S_{\vec{k}}(\omega) \simeq \begin{cases} 2\pi/2D\, L_y\, |k_x| & \text{for } |\omega| < DL_y|k_x| \\ 0 & \text{otherwise} \end{cases} \quad (2.18)$$

This can be explained as a Doppler shift broadening due to velocity gradients which will be readily observable experimentally.

These results demonstrate that the strong anisotropy of critical fluctuations as indicated by (2.15) suppresses the effects of critical fluctuations on various quantities like the transport coefficients. Namely, for $k_c\xi > 1$ and $k < k_c$ the variance is dependent only on k_x in most regions of $\vec{k}$:

$$C_{\vec{k}} \simeq k_c^{-8/5} |k_x|^{-2/5} \qquad \text{for} \quad k_c^4 |k_x| > k^5 \quad \text{and} \quad \xi_s^{-5} . \tag{2.19}$$

The Fourier transform of a function of $\vec{k}$ almost independent of k_y and k_z is sharply peaked on the x-axis. Therefore, (2.19) shows that the critical fluctuations with sizes greater than k_c^{-1} are elongated in the direction of flow. One can also show that this weakening of the critical singularity reduces the critical dimensionality from 4 to $12/5 = 2.4$. This can be seen by considering the fluctuation contribution to λ. There one encounters integrals like

$$\lambda \sim \int d^{(d)} \vec{q} \, \frac{1}{q^2} C_{\vec{q}} \tag{2.20}$$

By putting $C_{\vec{q}} \simeq k_c^{-8/5} |q_x|^{-2/5}$ the contribution from $q < k_c$ becomes

$$\lambda \sim \int_0^{k_c} dq \, q^{d-3} |q_x|^{-2/5} \sim \int_0^{k_c} dq \, q^{d-17/5} , \tag{2.21}$$

which diverges logarithmically at $d = d_c = 12/5$.

It is interesting to compare the suppression effects found here with an analogous situation found in anisotropic Ising-like ferromagnets with dipolar interactions[7]. There the Fourier transform of the spin correlation function $\langle s^x(\vec{r}) \, s^x(\vec{0}) \rangle$ takes the form

$$\langle s^x_{\vec{k}} \, s^x_{-\vec{k}} \rangle = [\xi^{-2} + k^2 + g(k_x/k)^2]^{-1} \tag{2.22}$$

where g measures the strength of dipolar interaction. Here wave vectors of the critical fluctuations giving rise to the dominant renormalization contributions are restricted to be within narrow regions almost perpendicular to the x-axis (namely, $|k_x/k|$ must be small). In this way the number of important fluctuations is decreased and the critical dimensionality is reduced from 4 to 3. In both the problems critical fluctuations are elongated along a particular direction to attain the lengths $g^{-1/2} \xi^2$ and $k_c^4 \xi_s^5$ for the cases of dipolar interaction and strong shear flow, respectively.

The levelling-off of initially diverging shear viscosity to a finite value in the presence of a finite shear flow was noted sometime ago by Oxtoby[8], which, in fact, is an example of non-Newtonian flow in critical fluids. There is a long history of the study of this problem starting with Botch and Fixman[9] in 1965 who predicted rather large non-Newtonian effects reflecting the current view advocating strong critical divergence of shear viscosity. Oxtoby made an extensive use of the method developed in Ref. 10 and concluded that

$$\eta(D)/\eta(D{=}0) \simeq 1 - \frac{8}{45\pi^2} \ln\left(\frac{\eta \xi^3 D}{0.45 \, k_B T}\right) \qquad \text{(strong shear regime)} \tag{2.23}$$

where ξ in the right hand side will cancel with that in $\eta(D=0)$ as $\xi \to \infty$. The overall behavior of $\Delta \equiv 1-\eta(0)/\eta(D=0)$ is shown in Fig. 2 as a function of $\lambda = \eta\xi^3 D/k_B T$

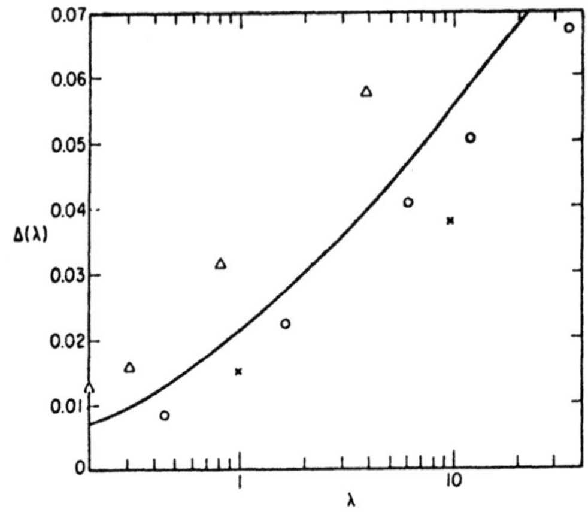

Fig. 2 The shear dependent part of viscosity as a function of the dimensionless rate of shear. The solid line is the theoretical prediction. After Ref. 8.

together with some experimental values of capillary viscometers. Recent theory of critical divergence of shear viscosity near equilibrium critical point which predicts power law singularity[2] suggests that (2.23) may well be replaced by

$$\eta(D)\,/\,\eta(D{=}0) \simeq \left(\frac{\eta\xi^3 D}{0.45\,k_B T}\right)^{-8/45\pi^2} \tag{2.24}$$

although this is not explicitly demonstrated yet. However, the non-Newtonian aspect of the shear flow is rather weak and is difficult to make precise measurements.

Our model can be extended to more general flow fields as follows. Assuming that the flow is incompressible and its velocity gradients are constant over the system one can rewrite the velocity field $\vec{u}(\vec{r})$ in (2.2) as

$$\vec{u}(\vec{r}) = \overleftrightarrow{D}\cdot\vec{r} \qquad \text{with} \qquad \sum_{i=1}^{d} D_{ii} = 0\,, \tag{2.25}$$

where $\overleftrightarrow{D} = \{D_{ij}\}$ is a tensor with the dimension of frequency. If $\overleftrightarrow{D}$ is constant over the system the equal time correlation function has the translational invariance in space and the steady state variance on the critical isochore can be obtained in the linear approximation in the form

$$C_{\vec{k}}^{(0)} = \int_0^\infty dt \exp\left[-\int_0^t d\zeta \lambda_0 \{\tau\vec{k}(\zeta)^2 + \vec{k}(\zeta)^4\}\right] \lambda_0 \vec{k}(t)^2\,, \tag{2.26}$$

where the time-dependent wave vector $\vec{k}(t)$ is defined by

$$\vec{k}(t) = \exp[\frac{1}{2} t \overleftrightarrow{D}^T] \cdot \vec{k}, \qquad (2.27)$$

$\overleftrightarrow{D}^T$ being the transposed matrix of $\overleftrightarrow{D}$. Various possible flows can be classified into two groups as follows :

(1) One is the case where at least one eigenvalue of $\overleftrightarrow{D}$ is zero or pure imaginary. This condition in three dimensions reduces to det $\overleftrightarrow{D} = 0$. For small wave vectors which are the corresponding eigenvectors of $\overleftrightarrow{D}^T$ or their linear combinations $C_{\vec{k}}^{(0)}$ grows as $\tau \to 0$ and instability occurs for $\tau < 0$. Below four dimensions the critical temperature is lowered below its equilibrium value and forms a critical surface parametrized by the independent elements of $\overleftrightarrow{D}$.

(2) The second is the case where all real parts of the eigenvalues of $\overleftrightarrow{D}$ are non-vanishing. Here $\vec{k}(t)$, (2.27), always grows or decays exponentially in time and $C_{\vec{k}}^{(0)}$ remains finite even if τ is pushed to negative values. This means that fluctuations are suppressed in such a way that no linear instability occurs. This might imply that the phase transition becomes first order. Analogous situations can be found in superconducting systems near the transition point under electric field, as will be discussed below[13].

The nature of critical singularity appears to depend on the types of flow and a simple shear flow constitutes only a special case. There is even a possibility of non-universal behavior. Full analysis of this interesting problem is currently under way. It is interesting to note that sensitive dependence of polymer conformations in dilute solutions on the types of flow was also discussed by de Gennes[11]. Similar circumstances seem to exist in two-dimensional fluids where the velocity field correlation function decays as $r^{-2/3}$ in a simple shear flow, for example[12].

2-2. Superconducting Transition under Electric Field

Another problem of considerable interest dealing with a nonequilibrium steady state near the critical point is the superconducting transition under a finite electric field[13,14]. Here one is interested in the "equation of state" relating the temperature, the electric field E and the current density J. Near the transition the equation of state may take the following scaling form[15]

$$E = J^x \chi(\Delta T / J^\lambda) \qquad (2.28)$$

with the critical exponents x and λ which are the analogs of δ and $1/\beta$, respectively. The scaling function $\chi(z)$ reduces to a finite number as $z \to 0$ and behaves as z^μ as $z \to \infty$. The requirement that $E \propto J$ as $\Delta T / J^\lambda \to \infty$ imposes the condition

$$x - \mu\lambda = 1 \qquad (2.29)$$

A mean field type theory based on the TDGL equation in which $\vec{\nabla}$ is replaced by $\vec{\nabla} - (2ie\hbar/c)\vec{A}$, $\vec{A}$ being the vector potential due to the electric field, yields the mean field exponents $x=3$, $\mu=1$ and $\lambda=2$ for films whereas the latest experiment for NbN films[15] indicates that $\mu=3.6\sim4.2$, $x=2.8\sim3.2$, $\lambda=0.4\sim0.6$. The value $x=3$ is in agreement with other experiments but the values of $\mu\sim1$ or $\mu\sim2$ are also reported[16]. The problem discussed here has one point in common to the critical fluid under incompressible flow: dissipative disturbances substantially suppress critical fluctuations.

3. Transient Phenomena

Consider a system in thermal equilibrium near a critical point. Suppose we suddenly change some of the thermodynamic variables like temperature and pressure to other values still in the neighborhood of the critical point. Although amounts of change can be very minute, this can give rise to rather drastic effects if it takes place sufficiently near the criticality. One can also imagine, for instance, periodic variations of thermodynamic variables although we shall here limit ourselves to sudden changes.

We now divide transient phenomena into those involving only a single phase, and those where two coexisting phases in the ordered state are involved.

3-1. Nonlinear Order Parameter Relaxation

Here we take up nonlinear relaxation of the order parameter which was briefly described in Sec. 1. Much attention has been paid to the nonlinear relaxation of the spatially uniform average order parameter m(t) or the total energy after sudden removal of magnetic field or temperature change for systems described by the time-dependent Ginzburg-Landau (TDGL) equation without the consevation law or the single-spin-flip kinetic Ising models which are believed to exhibit the same dynamic critical behavior[17]. The relaxation time of the overall nonlinear relaxation process may be defined by an expression such as[17]

$$\tau_m = \int_0^\infty dt\, m(t) / m(0) \tag{3.1}$$

As T_c is approached, the critical slowing-down takes place, so that

$$\tau_m \propto (\Delta T)^{-\Delta_m} \tag{3.2}$$

If, as ΔT approaches zero, the initial magnetic field is also reduced to maintain[18] $|m(0)| \lesssim \text{const.}(\Delta T)^\beta$ the scaling law (1.1) tells us that (also note that $m(t)\propto m(0)$ for very small m(0))

$$\Delta_m = \Delta_m^{(\ell)} = \nu z \tag{3.3}$$

On the other hand, if m(0) is kept fixed as $\Delta T\to 0$, we have a different situation. The initial state where $|m(0)| >> \text{const.}(\Delta T)^\beta$ can be considered to be relatively far from the critical point. Thus the initial relaxation process takes place within a relatively short time until m(t) decreases towards smaller values near the critical point, after that the system will forget its initial memory, the value of m(0)[19]. If this initial "noncritical" relaxation process does not contribute substantially to $\int_0^\infty dt\, m(t)$, it behaves as $(\Delta T)^{\beta-\nu z}$. Hence we have[19, 20]

$$\Delta_m = \Delta_m^{(n\ell)} = \nu z - \beta \ . \tag{3.4}$$

Thus

$$\Delta_m^{(n\ell)} = \Delta_m^{(\ell)} - \beta \tag{3.5}$$

$\Delta_m^{(\ell)}$ and $\Delta_m^{(n\ell)}$ have been referred to as linear and nonlinear relaxation time exponents, respectively, and (3.5) has been first explicitly written down by Rácz[19] although Suzuki[17] had earlier noticed possible distinction between the two regimes. One can extend the argument given here to the relaxation of variables A other than the order parameter and write down the exponent relations between linear and nonlinear relaxation time exponents $\Delta_A^{(\ell)}$ and $\Delta_A^{(n\ell)}$, respectively, as follows[20];

$$\Delta_A^{(n\ell)} = \Delta_A^{(\ell)} - \beta_A \tag{3.6}$$

where β_A is the exponent defined by

$$A(T) - A(T_c) = \text{const.}(\Delta T)^{\beta_A} \ . \tag{3.7}$$

In particular, for the internal energy E we have $\beta_E = 1-\alpha$.

Attempts have been made to verify the relations (3.5) and (3.6) which were obtained above on the heuristic grounds, and to explore how generally valid are these relations from various directions: (i) examine simple systems such as the one-dimensional kinetic Ising model[21] and the dynamical spherical model[22], (ii) use approximate models such as the mean field[19] one and the dynamical droplet model[18] (iii) the series expansion method[21, 23] (iv) Monte-Carlo simulation for a kinetic Ising model[18, 24] (v) experimental verification[24']. These results seem to indicate that (3.5) and (3.6) with $A = E$ are valid for purely dissipative systems, and not necessarily for non-dissipative cases. For instance, the best estimates by series method for the two-dimensional Ising model are $\Delta_m^{(\ell)} = 2.125 \pm 0.01$ and $\Delta_m^{(n\ell)} = 2.00 \pm 0.04$[21].

Further, one would like to know the actual scaling function f of (1.1) not just the exponents, the Δ's. For this purpose one needs to know the equation of motion satisfied by m(t). Such an equation was obtained for the time-dependent Ginzburg Landau (TDGL) model [eqs. (2.2) and (2.4) without D, ρ_0 and ∇^2 in front of $\delta H_0/\delta s$] near the four dimension up to $O(\varepsilon)$ which is nonlinear in m(t) and contains memory effects[25]. This takes the following form:

$$\frac{\partial\hat{m}}{\partial\hat{t}} = -F(\hat{m}) + \frac{\varepsilon}{12}\hat{m}(\hat{t})\int_0^\infty d\hat{s}\,\hat{s}^{-2}\{\exp[-2\hat{\gamma}(\hat{t})\hat{s}] - \exp[-2\int_0^{\hat{s}} d\hat{s}'\hat{\gamma}(\hat{t}-\hat{s}')]\} \tag{3.8}$$

where

$$\hat{\gamma}(\hat{t}) = 1 + \frac{8\pi^2}{3}\varepsilon\hat{m}(\hat{t})^2 \tag{3.9}$$

$$F(\hat{m}) = \hat{m}\{1 + \frac{8\pi^2}{9}\varepsilon\hat{m}^2 + \frac{\varepsilon}{6}(1 + \frac{8\pi^2}{3}\varepsilon\hat{m}^2)[\ln(1 + \frac{8\pi^2}{3}\varepsilon\hat{m}^2) - 1]\} \tag{3.10}$$

and $\hat{m}$ and $\hat{t}$ are the appropriately scaled order parameter and the time proportional to $m/(\Delta T)^\beta$ and $(\Delta T)^{z\nu}t$, respectively.

It is important to note that the values of $\Delta^{(n\ell)}$ do not necessarily reflect the time scale of dynamical processes occurring in the nonlinear regime. For instance, the mean field solution in the nonlinear regime still contains the time in the combination $(\Delta T)^{z\nu}t$. Rather, $\Delta^{(n\ell)}$ reflects the particular definition of the nonlinear relaxation time like (3.1) especially the presence of m(0) in the denominator. Indeed if one defines the nonlinear relaxation time by [26)]

$$\tau_m' = [\int_0^\infty dt\, t\, m(t)/m(0)]^{1/2} \tag{3.11}$$

(3.5) will be changed to [27)]

$$\Delta_m^{(n\ell)'} = \Delta_m^{(\ell)} - \frac{1}{2}\beta . \tag{3.12}$$

3-2. Kinetics of First Order Phase Transition

Here we would like to take up the kinetics of first order phase transitions that take place near critical points. Far from the critical point kinetics of first order phase transitions involves fluctuations of microscopic sizes. That is, the nucleation process can be started by critical nuclei with radii of, say, a few Å [28)]. Also a spinodal decomposition begins with fluctuations with wave lengths of, say, 50 Å [29)]. There one often has to worry about microscopic details which can depend on individual systems under consideration that enter, for example, the calculation of the statistical weight of a critical nucleus. On the other hand, near the critical point the smallest characteristic length is ξ, the correlation length of critical fluctuations, and the smallest characteristic time is the time τ_ξ that governs the order parameter dynamics, both of which are enormouly large on microscopic scales. One thus expects nice features that characterize studies of critical phenomena like scaling, universality, etc. In particular, the continuum models such as the TDGL and other continuum stochastic model systems described earlier retain their validity.

Here we have a unique opportunity of attacking the difficult and complex problems of the kinetics of first order phase transitions in their purest forms without being bothered by inessential details. In the following we briefly review this promising area with plenty of room for further exploration.

3.2.1 Nucleation

Transformation of phase by nucleation is a very old subject traced back to Maxwel, Gibbs and Thomson. Quantitative treatments are mostly based on the classical theory of Becker and Döring in 1935 as elaborated by Frenkel and others. This theory predicts the following homogeneous nucleation rate per unit volume I_n for a supersaturated vapor [28-30]:

$$I_n = J_n \exp(-W/k_B T) \tag{3.13}$$

where J_n is the "trial frequency" and is a very large number, and W is the free energy of a critical nucleus given by

$$W = \frac{16\pi\sigma^3}{3(\delta P)^2} \tag{3.14}$$

Here δP is the pressure difference of new and old phases, and σ is the surface tension. This classical theory and its modifications have been successfully applied to various nucleation phenomena[30].

On the other hand, since around 1962 it has been realized that things do not always go well with the classical theory [31]. Indeed we now know that as one approaches the critical temperature, the classical theory predicts nucleation rates which are rather too large. That is, the supersaturated vapor is much more stable than predicted by the classical theory. The detailed experimental study of supercooling near the critical point has been performed recently by Huang <u>et al</u>.[32] and the result is shown in Fig. 3. The classical theory (3.14) yields with $\sigma \propto (\Delta T)^{2(2\beta+\gamma)/3}$ and $\delta P \propto \delta T(\Delta T)^{\beta(\delta+1)-1}$, δT being the amount of supercooling and with the scaling relation $\gamma = \beta(\delta-1)$, the following near the critical point:

$$W/k_B T = -w \cdot (\Delta T/\delta T)^2 \tag{3.15}$$

The finite coefficient w is estimated to be 0.7 for CO_2. In Fig. 3 is indicated by the dashed line the prediction of the classical theory for the scaled supercooling $\Delta T/\delta T$ for nucleation that corresponds to $W/k_B T = 70$. The observed supercooling reach close to 3 times the prediction for $\Delta T/T \simeq 10^{-3}$. The increased supercooling then amounts to an increase of the nucleation rate by roughly 10^{28} times! Considerations of inhomogeneous nucleation and the translation-rotation degrees of freedom of the critical muclei [30] tend to make the situation worse.

There have been alternative theoretical approaches to the nucleation. Cahn and Hilliard used the Landau-Ginzburg type free energy functional to obtain the droplet free energy which turned out to be smaller than that of the classical theory [29].

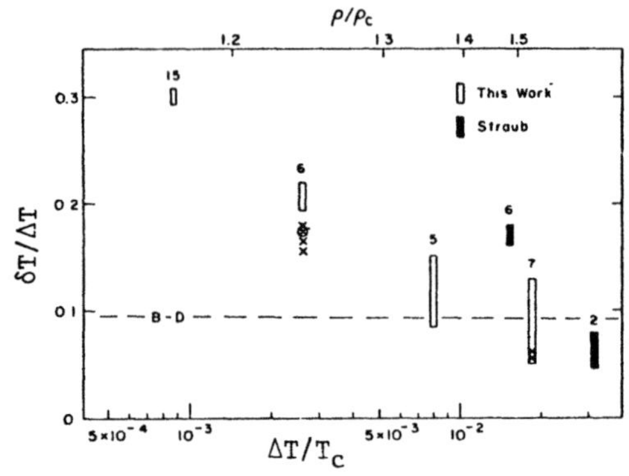

Fig. 3 Supercooling in CO_2. The dashed line is the prediction of the classical theory due to Becker and Döring. After Ref. 32.

This theory is improved by Sarkies and Frankel[33] by adopting the free energy function of Fisk and Widom[34] which is more appropriate near the critical point, obtaining the droplet free energy which becomes distinctly larger than that of Cahn and Hilliard near the critical point but still appears to be short of explaining the large discrepancies with the experiments. Attempts have been made to use Fisher's droplet model, which, however, predicts a much larger nucleation rate than the classical theory[35]. Binder proposed to take into account the process of growth of supercritical nuclei after nucleation in order to explain the experimental results [28, 35]. Mou and Lovett[36] constructed a theory where not only the droplet radius but also its time derivative enter as independent dynamical variables. At this moment the problem of nucleation near the critical point is full of controversy without any convincing explanation of the large gap between theory and experiments.

In connection with the nucleation problem, it is worth pointing out a possible usefulness of modern ideas of critical phenomena[1] in understanding the nature of metastable states. For instance the equation of state for the case with a scalar order parameter obtained by the usual ε-expansion[37] can be smoothly continued into the metastable region of the thermodynamic parameter up to the van der Waals spinodal. We conjecture this to be true for any finite order ε-expansion. This, then, will provide a natural way of obtaining the equation of state in metastable states, which complements the equation of state of a supersaturated vapor obtained by excluding clusters over a certain finite size[30]. Whatever there are the peculiarities associated with metastable states such as finite life times should then come from infinite order terms in the expansion. Indeed the recent developments[38] of large order perturbation theory are closely related to the study of metastable states in statistical mechanics.

3.2.2 Spinodal decomposition

Previously, kinetics of phase separation that takes place when a system is suddenly brought into the region of unstable thermodynamic parameters have been studied in such systems as alloys and glasses with small diffusion constants[29, 39]. In ordinary fluids, phase separation in unstable states takes place almost instantly because of high mobility of fluid molecules. On the other hand, in the immediate vicinity of a critical point, there occurs an enormous reduction of the diffusion constant which becomes proportional to $(\Delta T)^{\nu}$. Thus kinetics of phase separation becomes experimentally accessible[40, 41, 42].

In some sense the spinodal decomposition (SD) is easier to treat theoretically as compared to the nucleation because the former deals with instability against infinitesimal disturbances whereas a finite amplitude disturbance (critical nucleus) is needed to trigger a nucleation. Thus the earliest theory of SD is formulated as a problem of linear stability[29, 39]. Thus the initial thermal fluctuations are amplified exponentially with time. That is, with $I_k(t) \equiv \langle |s_k(t)|^2 \rangle$ we have

$$I_k(t) = \exp [R_k t] I_k(0) \tag{3.16}$$

with

$$R_k \equiv 2 \lambda_0 k^2 (|\tau| - k^2) \tag{3.17}$$

Here we have started from the stochastic model (2.2) and (2.4) without D, ρ_0, (the TDGL model with conserved order parameter) to represent the order parameter dynamics of a spinodally decomposing system below T_c ($\tau < 0$) and have linearized in fluctuation amplitudes. This linear theory of SD is useful only to provide a very crude idea of what SD is. The fact is that to author's knowledge there appears to be no single example (actual or computer experiments) which establishes the existence of a regime described by the linear theory[43]. It is obvious that the linear theory gets progressively worse as the time goes on since amplitudes of fluctuations are predicted to grow very fast. Thus, there now exist many theoretical attempts to take into account the nonlinearities of the problem[44)-46]. In case of solids such a nonlinearity is provided by the u_0 term in (2.4). The problem of similar nature occurs also for the TDGL model without conservation laws which applies to the kinetics of order-disorder transitions in alloys, structual phase transitions, strongly anisotropic magnets, etc.[47] Although this sort of nonlinear continuum problems is familiar in statistical physics and quantum field theory, there is a special difficulty here. The nonlinearity is essential here in bringing a system out of a uniform single phase to states with two coexisting phases (the so-called double peak structure), the very processes that we want to describe. Thus the standard techniques such as the random phase or decoupling approximation schemes which presuppose the validity of a Gaussian form for the probability distribution functional of the order parameter do not make sense.

As an example of the approaches which go beyond such decoupling-type schemes we mention the work of Langer, Baron and Miller (LBM)[45] for the TDGL model with a conserved scalar order parameter which appear to reproduce the time development of fluctuation spectra obtained by Monte Carlo simulations of the spin-exchange kinetic Ising model[48] reasonably well despite inadequacies of the method that have been noted recently[46, 49].

The crux of the method resides in the simple ansatz expressing the two-point probability distribution function of the local order parameter in terms of the one-point distribution function.

On the other hand the results do not agree with the observed behavior of the fluctuation spectra of critical fluid mixtures where good experimental data free from secondary complications are now becoming available. Here however, hydrodynamic effects have to be properly taken into account. Several years ago[50] the stochastic fluid model which includes such effects was proposed. The model is expressed by a Fokker-Planck type equation for the order parameter distribution functional $P(\{s\},t)$ of the form,

$$\frac{\partial}{\partial t} P(\{s\}, t) = \mathcal{L}\{s\} P(\{s\}, t) \tag{3.18}$$

where

$$\mathcal{L}\{s\} = \mathcal{L}_0 + \mathcal{L}_{hd} \tag{3.19}$$

$$\mathcal{L}_0 = -\lambda_0 \int d\vec{r} \frac{\delta}{\delta s(\vec{r})} \nabla^2 \left[\frac{\delta}{\delta s(\vec{r})} + \frac{\delta H_0}{\delta s(\vec{r})} \right] \tag{3.20}$$

$$\mathcal{L}_{hd} = 2 \int\int d\vec{r}_1 d\vec{r}_2 \frac{\delta}{\delta s(\vec{r}_1)} \vec{\nabla}_1 s(\vec{r}_1) \cdot T(\vec{r}_1 - \vec{r}_2) \cdot \vec{\nabla}_2 s(\vec{r}_2) \left[\frac{\delta}{\delta s(\vec{r}\ )} + \frac{\delta H_0}{\delta s(\vec{r}\ )} \right] \tag{3.21}$$

$$T(\vec{r}) = \frac{1}{8\pi\eta} \left(\frac{1}{r} + \frac{1}{r^3} \vec{r}\vec{r} \right) . \tag{3.22}$$

$\mathcal{L}_{hd}$ expresses the long-range hydrodynamic interaction between distant order parameter fluctuations mediated by velocity field fluctuations, which is familiar in polymer solutions. In the absence of $\mathcal{L}_{hd}$, (3.18) reduces to the TDGL model treated by LBM. Near equilibrium the hydrodynamic term produces the now familiar renormalization of the transport coefficient λ_0[2]. This horribly-looking model equation can be handled if we note that the nonlinearity contained in (3.20) is short-ranged (point-like) whereas that of (3.21) is long-ranged. Hence we regard the LBM result for the solid model without $\mathcal{L}_{hd}$ as a reference and study the effects of $\mathcal{L}_{hd}$ by applying the random-phase-approximation which should be valid for long range interactions. The results of such a treatment[51] are shown in Figs. 4 and 5 together with experimental results[41]. Although some of the excellent agreements obtained may very

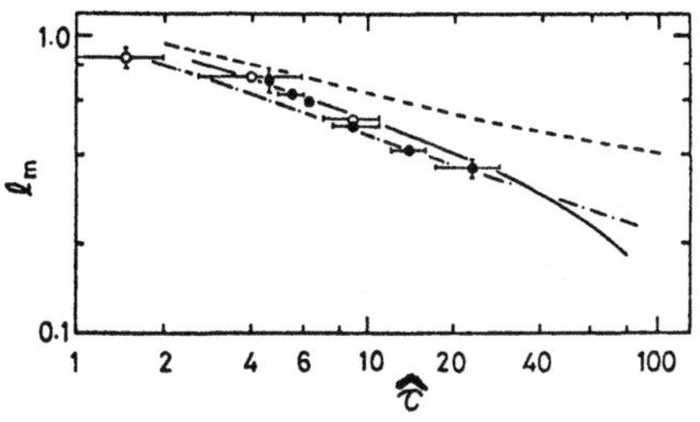

Fig. 4 Time dependence of the peak position k_m where $I_k(t)$ reaches a maximum. Here $\ell_m = |\tau|^{-1/2}k_m$ and $\hat{\tau} = \lambda\tau^2 t$. The solid line is the theoretical prediction of Kawasaki and Ohta[51]. The dashed curve is that of LBM. The dash-dotted curve represents the $\hat{\tau}^{-1/3}$-law of K. Binder and D. Stauffer [Phys. Rev. Lett. 33 1066 (1974)]. After Ref. 51.

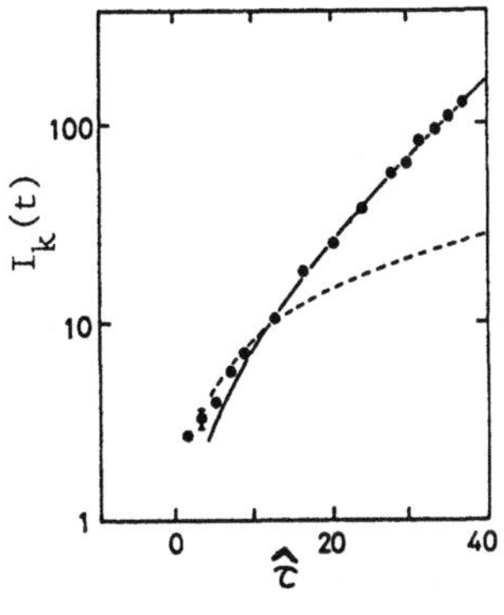

Fig. 5 Time dependence of the peak value of $I_k(t)$. The ordinate is in an arbitrary unit. The lines are the predictions of the same theories as explained in Fig. 4. After Ref. 51.

well be fortuitous, we believe that this comparison establishes the importance of taking hydrodynamic effects into consideration in spinodal decomposition of critical fluids. At late stages much of the local velocity fluctuations can be incorporated into renormalization of transport coefficient λ_0, as in the nucleation problem[52]. However, the situation at the late stage[42] appears to be too complicated to be properly understood at this moment where gravitational effects and various hydrodynamic instabilities may take important parts.

4. Concluding Remarks

In the preceding sections we have presented several concrete examples that demonstrate a wide variety of types of problems pertaining to critical dynamics far from equilibrium. These examples, however, are very limited and we have merely scratched the rim of this potentially rich field.

We have not touched upon various important hydrodynamic problems of critical fluids[53] and superfluids near the critical point[54]. There is also an interesting question of what happens if the liquid He^4 is suddenly quenched below criticality[55]. The same question can be asked for any system having broken continuous symmetry below criticality.

Finally there is a growing interest in critical phenomena that are expected to occur near points of instability far from equilibrium.[56] We have not mentioned them here since we are concerned exclusively with critical phenomena in its more traditional sense. The whole purpose of this brief review is to call attention to the existence of a poorly explored area of critical dynamics even near thermodynamic critical points.

References

1) S. K. Ma, Modern Theory of Critical Phenomena (Benjamin, N.Y., 1976).

2) P. C. Hohenberg and B. I. Halperin, Rev. Mod. Phys. 49 (1977) 435.

3) A. Onuki and K. Kawasaki, submitted to Ann. Phys. (N.Y.).

4) A. Onuki and K. Kawasaki, presented at the Ōji Seminar on Nonlinear Nonequilibrium Statistical Mechanics (Kyoto, July 1978). A short summary will be published in the Supplement of Prog. Theor. Phys.

5) D. Beysens, M. Gbadamassi and L. Boyer, Saclay preprint. They have preformed the light scattering experiments by Aniline-cyclohexane critical mixture under shear flow and have observed the considerable decrease of the turbidity, the lowering of the critical temperature, and the anisotropy of the scattered intensity for $D\tau > 1$. Their results are in general accord with (2.9) and (2.13) $\sim$ (2.15).

6) A. Onuki and K. Kawasaki, submitted to Phys. Lett. Further details on the line shape will appear elsewhere.

7) A. Aharony, in Phase Transitions and Critical Phenomena, Vol. VI, C. Domb and M. S. Green, eds.(Academic Press, N.Y., 1976).

8) D. W. Oxtoby, J. chem. Phys. 62 (1975) 1463 ; D. W. Oxtoby and H. Metiu, Phys. Rev. Lett. 36 (1976) 1092.

9) M. Fixman, J. chem. Phys. 36 (1962) 310 ; W. Botch and M. Fixman, J. chem. Phys. 36 (1962) 3100 ; R. Sallavanti and M. Fixman, J. chem. Phys. 48 (1968) 5326.

10) T. Yamada and K. Kawasaki, Prog. Theor. Phys. 38 (1967) 1031 ; K. Kawasaki and J. D. Gunton, Phys. Rev. 8 (1973) 2048.

11) P. G. de Gennes, J. chem. Phys. 60 (1974) 5030. It is striking that there is a close analogy between the problems of polymer solutions and those of critical phenomena both in the static and dynamical aspects [de Gennes, Phys. Lett. 38 A (1972) 339 , and Macromol. 9 (1976) 587, 594]. Furthermore this analogy seems to hold to some extent even in various problems of steady states under flow field.

12) A. Onuki, Phys. Lett. 70 A (1979) 39. In two dimensional fluids linear transport coefficients diverge at zero frequency and wave number due to the so-called long time tail [B. J. Alder and T. E. Wainwright, Phys. Rev. A 1 (1970) 18]. In two dimensions large-scale hydrodynamic fluctuations are sensitively affected even by small external disturbances and steady states cannot be treated in the linear response scheme.

13) W. J. Skocpol and M. Tinkham, Rep. Prog. Phys. 38 (1975) 1049. A Cooper pair is accelerated and destroyed by electric field and the system becomes stable against infinitesimally small nuclei of the superconducting phase.

14) A. Schmid, Phys. Rev. 180 (1969) 527 ; T. Tsuzuki in Physics of Quantum Liquids, eds. R. Kubo and F. Takano, Syokabo, Tokyo (1971).

15) S. A. Wolf, D. U. Gubser and Y. Imry, Phys. Rev. Lett. 29 (1979) 324.

16) D. U. Gubser and S. A. Wolf, J. Phys. (Paris), Colloq. 39 (1978) C 6 - 579 ; M. K. Chien and R. E. Glover, III, in Low Temperature Physics, LT-13, edited by W. J. O'Sullivan, K. D. Timmerhaus, and E. F. Hammel (Plenum, New York, 1973), p. 653.

17) M. Suzuki, Int. J. Mag. 1 (1971) 123.

18) R. Kretschmer, K. Binder and D. Stauffer, J. Stat. Phys. 15 (1976) 267.

19) Z. Rácz, Phys. Rev. B 13 (1976) 263.

20) M. E. Fisher and Z. Rácz, Phys. Rev. B 13 (1976) 5039.

21) Z. Csépes and Z. Rácz, J. Phys. A 11 (1978) 575.

22) Z. Rácz and T. Tél, Phys. Lett. A 60 (1977) 3.

23) Z. Rácz and M. F. Collins, Phys. Rev. B 13 (1976) 3074; H. Ikeda, Prog. Theor. Phys. 55 (1976) 2027; H. Ikeda, Sci. Rep. Kanazawa Univ. 21 (1976) 19; N. J. White, J. Phys. C 9 (1976) L 187; M. Suzuki and H. Ikeda, Prog. Theor. Phys. 55 (1976) 2041.

24) E. Stoll, K. Binder and T. Schneider, Phys. Rev. B 8 (1973) 3266.

24') M. F. Collins and H. C. Teh, Phys. Rev. Lett. 30 (1973) 781; M. Sato and K. Hirakawa, J. Phys. Soc. (Japan) 42 (1977) 433.

25) R. Bausch and H. K. Janssen, Z. Physik B 25 (1976) 275; T. Yamada, T Ohta and K. Kawasaki, J. Stat. Phys. 16 (1977) 281.

26) M. Suzuki, Prog. Theor. Phys. 43 (1970) 882; T. Schneider and E. Stoll, Phys. Rev. B 10 (1974) 959.

27) H. Ikeda, Prog. Theor. Phys. 57 (1977) 687.

28) K. Binder, in Statistical Physics eds. L. Pál and P. Szépfalusy, Akademiai Kiadó, Budapest (1976).

29) J. W. Cahn, in Critical Phenomena in Alloys, Magnets and Superconductors eds. Mills, Ascher and Jaffee, McGraw-Hill, New York (1971).

30) Nucleation Phenomena, ed. A. C. Zettlemoyer, Elsevier, Amsterdam (1977).

31) B. E. Sundquist and R. A. Oriani, J. chem. Phys. 36 (1962) 2604; R. B. Heady and J. W. Cahn, J. chem. Phys. 58 (1973) 896.

32) J. S. Huang, W. I. Goldburg and M. R. Moldover, Phys. Rev. Lett. 34 (1975) 639.

33) K. W. Sarkies and N. E. Frankel, J. chem. Phys. 54 (1971) 433; Phys. Rev. A 11 (1975) 1724.

34) S. Fisk and B. Widom, J. chem. Phys. 50 (1969) 3219.

35) K. Binder and D. Stauffer, Advances in Phys. 25 (1976) 343.

36) C. Y. Mou and R. Lovett, J. chem. Phys. 62 (1975) 3298.

37) E. Brézin, J. C. Le Guillou and J. Zinn-Justin, in Phase Transitions and Crirical Phenomena Vol. 6, eds. C. Domb and M. S. Green, Academic Press, London (1976).

38) E. Brézin, Review Talk, European Particle Physics Conderence, Budapest (1977).

39) J. E. Hilliard, in Phase Transitions, ed. H. I. Aaronson, American Society of Metals, Park, Ohio (1970).

40) J. S. Huang, W. I. Goldburg, and A. W. Bjerkaas, Phys. Rev. Lett. 32 (1974) 921; A. J. Schwartz, J. S. Huang and W. I. Goldburg, J. chem. Phys. 62 (1975) 1847; ibid. 63 (1975) 599.

41) W. I. Goldburg, C. H. Shaw, J. S. Huang and M. S. Pilant, J. chem. Phys. 68 (1978) 484; M.-C. Wong and C. M. Knobler, J. chem. Phys. 66 (1977) 4707, and preprint

42) M. W. Kim, A. J. Schwartz and W. I. Goldburg, Phys. Rev. Lett. 41 (1978) 657; W. I. Goldburg, A. J. Schwartz and M. W. Kim, to be published in Suppl. Progr. Theor. Phys.

43) One of the authors (KK) is indebted to conversations with J. W. Cahn and J. S. Langer on this point.

44) J. S. Langer and M. Bar-on, Ann. Phys. (N.Y.) 78 (1973) 421, and the earlier references quoted therein.

45) J. S. Langer, M. Bar-on and H. D. Miller, Phys. Rev. A 11 (1975) 1417.

46) K. Binder, Phys. Rev. B 15 (1977) 4425; P. Mirold and K. Binder, Acta Met. 25 (1977) 1435; K. Binder, C. Billotet and P. Mirold, Z. Physik B 30 (1978) 183.

47) See e.g. K. Kawasaki, M. C. Yalabik and J. D. Gunton, Phys. Rev. A 17 (1978) 455, and the references quoted therein.

48) K. Binder, Z. Physik 267 (1974) 213 ; J. Marro, A. Bortz, M. Kalos and J. Lebowitz, Phys. Rev. B 12 (1975) 2000 ; M. Rao, M. Kalos, J. Lebowitz and J. Marro, Phys. Rev. B 13 (1976) 4328 ; A. Sur, J. Lebowitz, J. Marro and M. Kalos, Phys. Rev. B 15 (1977) 3014 ; J. Marro, J. Lebowitz and M. Kalos, preprint.

49) C. Billotet and K. Binder, preprint.

50) K. Kawasaki, in Synergetics, ed. H. Haken, B. G. Teubner, Stuttgart (1973); K. Kawasaki, Prog. Theor. Phys. 57 (1977) 826.

51) K. Kawasaki and T. Ohta, Prog. Theor. Phys. 59 (1978) 362.

52) J. S. Langer, Ann. Phys. (N.Y.) 65 (1971) 53 ; K. Kawasaki, J. Stat. Phys. 12 (1975) 365.

53) M. Gitterman, Rev. Mod. Phys. 50 (1978) 85.

54) V. L. Ginzburg and A. A. Sobyanin, Sov. Phys. Usp. 19 (1976) 773.

55) E. Levich and V. Yakhot, J. Phys. A 11 (1978) 2237.

56) H. Haken, Rev. Mod. Phys. 47 (1975) 67 ; R. Graham, in Fluctuations, Instabilities and Phase Transitions, ed. T. Riste, Plenum Press, New York (1976); H. Mori and K. J. McNeil, Prog. Theor. Phys. 57 (1977) 770 ; C. W. Gardiner and D. F. Walls, J. Phys. A 11 (1978) 161 ; A. Nitzan, Phys. Rev. A 17 (1978) 1513.

Note added

[1] After the present article was completed, we received a preprint from E. D. Siggia where he analyzes the late stage spinodal decomposition of critical liquid mixtures. For a quench at the critical concentration he predicts a crossover from the Lifshitz-Slyozov type law $k_m^{-1} \propto t^{1/3}$ to $k_m^{-1} \propto t$ when k_m^{-1} reaches $(k_B T/\sigma)^{1/2}$, σ being the surface tension. This crossover length, however, becomes roughly equal to ξ since $\sigma \sim k_B T/\xi^2$ for $d=3$.

[2] There is a new developing field of great interest which has not been touched upon. This is concerned with the dynamical behavior associated with phase transitions in two-dimensional systems without ordinary long range order. Here we only give some relevant recent references : V. Ambegaokar *et al*, Phys. Rev. Lett. 40 (1978) 783 ; B. A. Huberman *et al*, Phys. Rev. Lett. 40 (1978) 780 ; R. J. Meyerson, Phys. Rev. B 18 (1978) 3204 ; S. T. Chui and J. D. Weeks, Phys. Rev. Lett. 40 (1978) 733 ; B. I. Halperin and D. R. Nelson, Harvard University preprint (1979) and the references quoted therein.

EXPERIMENTAL ASPECTS OF TRANSITION PHENOMENA IN QUANTUM OPTICS

F.T. Arecchi

Università di Firenze and Istituto Nazionale di Ottica, Firenze
Italy

EXPERIMENTAL ASPECTS OF TRANSITION PHENOMENA IN QUANTUM OPTICS

F.T. Arecchi

Università di Firenze and Istituto Nazionale di Ottica, Firenze

ABSTRACT

Phase transitions in equilibrium systems are the result of a competition between the interparticle energy J and the thermal energy $k_B T$ which introduces disorder. In quantum optics, even when interparticle interactions are negligible as in a very dilute gas, there may be particle correlations due to the common radiation field. The transition from disorder to order consists in a passage from a regime where the atoms emit independently from one another, to a regime where the atoms emit in a strongly correlated way. It depends on a "cooperation number" C which is proportional to the atomic density. In a "pumped" system, as the density of active atoms is increased, the laser threshold is reached for C=1 and the coherent e.m. intensity is proportional to C-1.

The laser threshold and the optical bistability are discussed as examples, respectively, of 2nd and 1st order phase transitions in quantum optics.

By photon statistics methods the statistical features of these phenomena can be measured with high accuracy.

Furthermore, in nonequilibrium optical systems it has been possible to study for the first time the transient build-up of an ordered state by a rapid passage through an instability. These transients are characterized by large fluctuations which display a scaling behaviour.

1-INTRODUCTION

In the past fifteen years, the introduction of the photon statistics method (for the theory see Ref.1, for a description of experimental aspects see Ref.2) has allowed a careful investigation of quantum optical devices, such as the laser oscillators [3-7]. From the first experiments [2] the threshold point, where the gain due to the external excitation prevails over the internal losses, displayed the same features of a continuous phase transition in an equilibrium system (large increase in fluctuations, slowing down). While the Landau mean field model [8] was showing its weakness in giving the right critical exponents for most phase transitions [9], that model is sufficient to explain all statistical features of the laser threshold, hence motivating the development of a phase transition analogy, as already hinted for other nonequilibrium systems [10, 11], for the laser threshold [12, 13].
Similarly, evidence of discontinuous jumps and hysteresis effects in a laser with a saturable absorber [14, 15] suggested an analysis of the instability of such a device as a first order phase transition [16, 17]. Recently, injecting a laser field into an interferometer filled with absorbing atoms, saturations of the susceptibility has given rise to a discontinuous instability as in the previous case. Evidence of a hysteresis cycle suggests a region of coexistence of two stable points, hence the name of optical bistability [18]. The corresponding theory, either when the instability is due to the absorptive [19] or the dispersive [20] part of the atomic susceptibility, shows the characters of a 1st order phase transition.
Nonequilibrium systems can be driven through the instability point by a rapid passage, that is, at a rate larger than the local relaxation rate of steady state fluctuations. Such a transient situation was first observed for the laser instability [21]. A phenomenological theory [22] has shown the universality character of the anomalous transient fluctuations, which can be represented in terms of a scaling parameter. The general theory of these anomalous fluctuations has later been developed [23, 24]. Of course, since transient phenomena are not invariant for time translation, there is no equivalent in equilibrium systems. The superfluorescence, that is the spontaneous cooperative decay of N atoms all prepared in their excited state in the absence of a classical field to drive them [25, 26, 27], is a transient collective behaviour displaying a threshold. In such a case there is no stationary equivalent. The fluctuation enhancement here is only a theoretical conjecture [26, 24], since the experiments done so far give only single shot shapes [28, 29, 30] but not statistical fluctuations. The above introductory remarks are summarized in Table I.

Some recent investigations, theoretical [31, 32] and experimental [33] deal with the role of mode-mode coupling at the onset of the instability, either in steady state [31, 33] or in the transient [33]. A deeper investigation of these problems could give interesting analogies with the same phenomena in hydrodynamic instabilities.

This contribution is organized as follows.

In Sec. 2 we discuss cooperation in radiative interactions.

In Sec. 3 we give a statistical phenomenological description of quantum optical phase transitions [34].

In Sec. 4 we review photon statistics as a tool for measuring fluctuations in quantum optical systems.

In Sec. 5 we report the body of experimental data on the transition phenomena in quantum optics.

TABLE I

PHASE TRANSITIONS IN QUANTUM OPTICS

<table>
<tr><th colspan="2"></th><th>Thermal equilibrium</th><th>Quantum Optics</th></tr>
<tr><td rowspan="2">steady state</td><td>2nd ord.</td><td>- order/disorder
- para-ferromagnetic (H=0)</td><td>laser threshold</td></tr>
<tr><td>1st ord.</td><td>- para-ferromagnetic (H≠0)
- liquid gas</td><td>- laser plus saturable absorber
- optical bistability</td></tr>
<tr><td colspan="2">transient</td><td>= = = =</td><td>- laser transient
- superfluorescence</td></tr>
</table>

2-COOPERATION AND PHASE TRANSITION IN RADIATIVE INTERACTIONS

A gas of N non-interacting particles in contact with a thermal reservoir is a single-phase system. If however there is a nonzero interparticle energy J, such as the intermolecular potential in a real gas or the exchange energy in a magnetic spin system, as soon as J prevails over the fluctuation energy $k_B T$, there is a transition from a disordered phase characterized by almost free particles to an ordered phase characterized by a collective or cooperative behaviour.

A convenient classification of phase transitions is that of Ehrenfest [9] depending on which thermal derivative of the free energy F is discontinuous at the transition point.

In some instances, when the interparticle interactions are long-range compared to the interparticle distance, the system is well described by a free energy which is a series expansion of an order parameter q.

If, by symmetry arguments, the series has only even terms such as

$$F = U - TS = F_o + \frac{\alpha}{2} q^2 + \frac{\beta}{4} q^4 \qquad (2.1)$$

where F_o is the value at the transition temperature $T = T_c$,

$\alpha = a (T - T_c)$, with $a > 0$, and $\beta > 0$,

then it is an easy matter to show that at $T = T_c$, where $\alpha = 0$ the entropy is continuous (2nd order transition).

If an odd term has to be included [35]

$$F - F_o = \xi q + \frac{\alpha}{2} q^2 + \frac{\beta}{4} q^4 \qquad (2.2)$$

with the same α and β as above, then the transition is 1st order (entropy S discontinuous).

By Boltzmann's law, the probability P(q) of the order parameter q is given by (N = normalization factor)

$$P(q) = N e^{-F/k_B T} \qquad (2.3)$$

Both free energy F and probability P are plotted in Figs. 1 and 2.

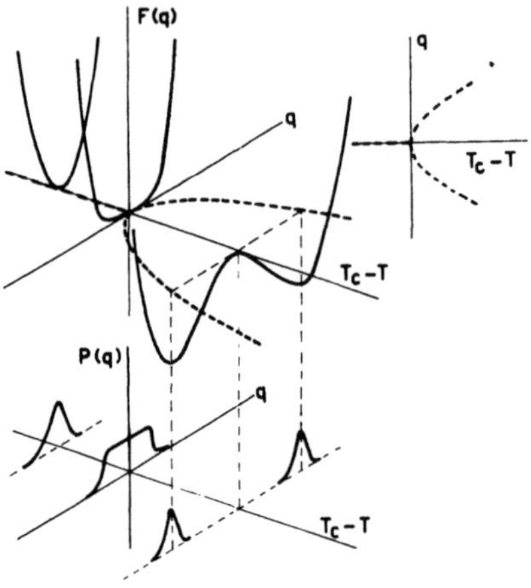

Fig. 1 - Second order phase transition.Free energy F(q) and probability density P(q) versus the order parameter q at different temperatures T.The locus of equilibrium points is displayed separately in a q - T plane.

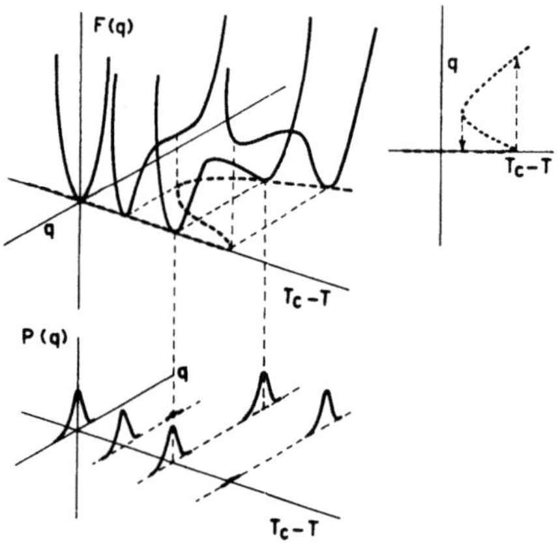

Fig. 2 - First order phase transition. Plots as in fig. 1.

In optics, even when interparticle interactions are negligible as in a very dilute gas, there may be particle correlations due to the common radiation field. Even though such situations could be described classically, they require the large radiation fields provided by the laser to be seen experimentally.

Generally, a quantum optical device is an open system which is fed by a source of energy and radiates electromagnetic energy toward a sink. Since we are talking of radiating systems, the relevant parameter is the field E. We shall show situations such as

i) single mode laser threshold,

ii) threshold of a laser with a saturable absorber,

iii) optical bistability,

where the E field is the order parameter in the Landau sense.

There are however transition phenomena as

iv) superfluorescence

which are also characterized by a threshold but where the order parameter is an atomic variable, and the field E appears only as a transient pulse.

In the first case we can plot probability distributions of E which are topologically equivalent of fig.1 and 2, hence we can draw analogies with 2nd and 1st order phase transitions.

The onset of a quantum optical transition is related to the value of an adimensional cooperation parameter

$$C = \frac{\gamma_c^2}{\gamma K} = (\text{constant}) \cdot \rho \tag{2.4}$$

where γ_c^2 is the square of the cooperative emission rate [36] and is proportional to the density ρ of radiating atoms, γ is the atomic relaxation rate and K is the loss rate of the electromagnetic field. In the case of a cavity made of two facing mirrors with reflectivity R at a distance L, K is

$$K = \frac{c}{L}(1 - R) \tag{2.5}$$

whereas for a lump of material of length L without mirrors, it is

$$K = c/L \tag{2.5'}$$

Cooperation in quantum optics was first considered by Dicke [25],who introduced the idea of superradiance.

In order to cooperate, however, the atoms must exchange information, that is, putting them on a line, the last one should "see" the radiation of the first one before decaying. By this argument, we find for the cooperation rate [36]

$$\gamma_c = (\rho \lambda^2 c \gamma)^{1/2}. \tag{2.6}$$

Putting for γ the usual expression for the spontaneous emission rate, we have

$$\gamma_c^2 = \frac{\omega \mu^2}{2\hbar \varepsilon_0} \rho \qquad \text{(S.I. units)} \tag{2.7}$$

where ω is the emission frequency, ε_0 the vacuum dielectric constant and μ the electric dipole moment of the radiative transition.

The cooperative rate comes from radiation damping considerations but we must account that the two partners, atoms and field, are also each coupled with its own reservoir which introduces its own loss rate. A simplified picture can be given in terms of coupled Maxwell and Schrödinger equations written for a field coupled with N two level atoms by a resonant transition at frequency ω . Writing field E, polarization P and population inversion density D in terms of adimensional slowly varying variables (V is the interaction volume)

$$E = \sqrt{\frac{\hbar\omega}{2\varepsilon_0 V}} \left[a(x,t)\, e^{-i(\omega t - kx)} + c.c. \right]$$

$$P = \frac{\mu}{V} \left[S(x,t)\, e^{-i(\omega t - kx)} + c.c. \right] \qquad (2.8)$$

$$D = \frac{1}{V} \Delta(x,t) ,$$

the coupled equations reduce to [37] (at resonance a and S are real quantities)

$$\left(\frac{\partial}{\partial t} + c\frac{\partial}{\partial x}\right) a = g S - k a$$

$$\frac{\partial S}{\partial t} = 2 g a \Delta - \gamma_\perp S \qquad (2.9)$$

$$\frac{\partial \Delta}{\partial t} = -2 g a S - \gamma_\parallel (\Delta - N/2)$$

where

$$g = \left(\frac{\omega \mu^2}{2\hbar\varepsilon_0 V}\right)^{1/2} \qquad (2.10)$$

is the coupling constant and K , $\gamma_\perp, \gamma_\parallel$ are loss rates. The normalization is such that $a^2 = n$ is the photon number and Δ the number of inverted atoms in the volume V. N is a source term. For an absorbing medium in the ground state (no pumping) it must be changed into $-N_{tot}$. We neglect here the space derivative $\partial / \partial x$.
Care can be taken by transforming to a suitable moving frame both in the cases of an amplifying [37] and of an unexcited [38, 39] medium.
To give the order of magnitude, for a dilute gas of atoms with allowed transitions

in the visible and for $V \simeq 1\ cm^3$, it is

$$g \sim 10^4\ s^{-1}\ , \qquad \gamma_\perp \sim \gamma_\parallel \sim \gamma \sim 10^8\ s^{-1}\ ,$$

and (2.11)

$$K \sim 10^7\ s^{-1} \quad \text{or} \quad \sim 10^{10}\ s^{-1}\ ,$$

depending on whether the gas is in a laser cavity or distributed over a length of some centimeters, without mirrors at the ends.

Relations (2.11) suggest two distinct time regimes depending on whether $K \ll \gamma$ or $K \gg \gamma$. In the first case the fast atomic variables S, Δ relax toward their equilibrium values, while the slow variable a changes little. Hence a is the order parameter. In the second case the field escapes rapidly from the interaction volume and the atomic variables are the slow ones.

It will be shown that

i) neglecting all losses, the field-atoms coupling gives rise to collective pulses with a characteristic time given by γ_c^{-1} ,

ii) for $\gamma \gg K$ (adiabatic elimination of atomic variables) the field evolves with a rate γ_c^2/γ which must be larger than K to have a laser-like transition. This condition amounts to $C > 1$.

iii) for $K \gg \gamma$ (adiabatic elimination of field variables) the atomic quantities evolve with a rate γ_c^2/k which must be larger than γ to have a superfluorescent transition. This again is equivalent to $C > 1$.

2.1-THE LOSSLESS CASE

If $\gamma = 0$, $S^2 + \Delta^2$ is a conserved quantity , hence one can introduce a new variable ϑ such that

$$S = N/2 \ \sin\vartheta$$
$$\Delta = N/2 \ \cos\vartheta \tag{2.12}$$

It follows immediately that

$$\dot{\vartheta} = 2\,g\,a\ ; \tag{2.13}$$

replacing this in the first equation (2.9) for k=0 one obtains

$$\ddot{\vartheta} = \gamma_c^2 \sin\vartheta \tag{2.14}$$

which is a nonlinear pendulum equation, giving periodic solutions in terms of Jacobi functions [39], with pulse durations of the order of γ_c^{-1}. Hence γ_c is the <u>natural</u> coupling constant of an atom-field system, <u>independently</u> from boundary conditions or interactions with thermal reservoirs.

2.2-SINGLE MODE LASER

When $\gamma \gg \kappa$ the last two equations can be solved at steady state in terms of the field amplitude giving

$$S = \frac{gN}{\gamma} \frac{a}{1+x^2} .$$

Replacing S in the field equation

$$\dot{a} = \frac{g^2 N}{\gamma} \frac{a}{1+x^2} - \kappa a = \frac{\gamma_c^2}{\gamma} \frac{a}{1+x^2} - \kappa a . \tag{2.15}$$

Here we have introduced

$$x = 2ga/\gamma . \tag{2.16}$$

Since $g/\gamma \sim 10^{-4}$ it takes $a^2 \sim 10^8$ photons to have $x \sim 1$. Hence for $n < 10^8$ photons the saturation term in the polarization can be approximated as a cubic correction and the field equation becomes

$$\dot{x} = \kappa\left[(C-1)x - Cx^3\right] \tag{2.17}$$

where C is the cooperation parameter introduced in (2.4).

The steady stable solution x^2 goes smoothly from zero to a non-zero value as C goes from below to above 1 (fig. 3). For x one has a $(c-1)^{\frac{1}{2}}$ power law as for a 2nd order phase transition in the Landau model.

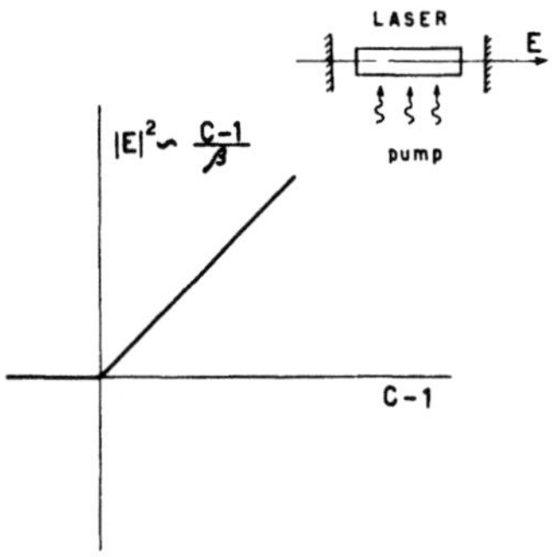

Fig.3 - Plot of intensity of the laser field versus the cooperation number (C-1 is proportional to difference between gain and losses).

2.3-OPTICAL BISTABILITY

Take the same configuration (atoms within a cavity) leaving the atoms in the ground state and injecting an external field y. The steady solution is [19]

$$y = x + C\frac{x}{1+x^2} \tag{2.18}$$

Three plots of transmitted versus injected field are given in fig. 4 for increasing atomic densities, that is, for increasing C values. For $C \simeq 0$ the cavity is tuned to transmission 1, hence transmitted and incident fields are equal. For $C < 1$ the system is in the linear absorption regime and there is very little transmitted field. For $C > 1$ the system jumps discontinuously into the saturated regime where the atoms become transparent, hence returning to full transmission.
By decreasing the impinging field one gets a hysteresis cycle peculiar of the 1st order phase transition. Since there is a region with two stable points the phenomenon is called optical bistability.

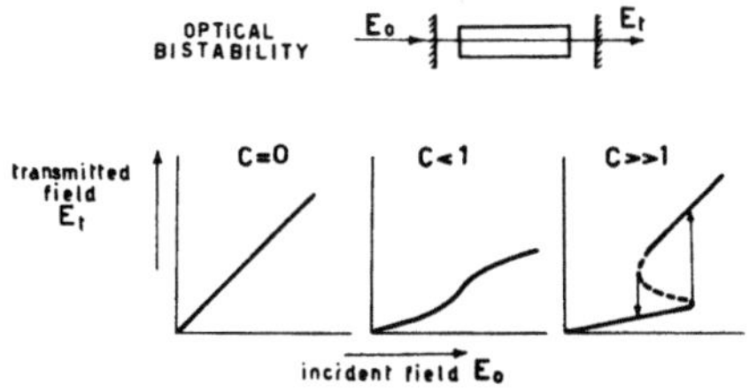

Fig.4 - Plots of transmitted field versus impinging field in an optical bistable device,for three different values of the cooperation number C.

2.4- π PULSES [37] AND SUPERFLUORESCENCE [26]

When $K \gg \gamma$ the field can be adiabatically eliminated giving

$$a = \frac{g}{K} S , \tag{2.19}$$

For times shorter than γ^{-1}, one can neglect atomic losses and make use of the ϑ- representation.
By combining (2.12) and (2.13) with (2.19) one obtains

$$\dot{\vartheta} = \frac{\gamma_c^2}{K} \sin\vartheta \equiv \frac{1}{\tau_R} \sin\vartheta . \tag{2.20}$$

The solution is [37] (so called π pulses)

$$a \sim \operatorname{sech} (t - t_D) / \tau_R . \tag{2.21}$$

It shows a duration determined by the competition between γ_c^2 and the loss rate K. In order to have the cooperation, this duration must be shorter than the single atom decay. The threshold condition is then

$$1/\tau_R = \gamma_c^2/\chi > \gamma$$

that is $C > 1$, as in the above cases.

The delay time t_D depends on the initial conditions. If the system is prepared in the fully excited state, it can decay only by spontaneous emission (superfluorescence) thus requiring a quantum theory [26] for the evaluation of t_D .

3-STATISTICAL DESCRIPTION OF QUANTUM OPTICAL TRANSITIONS

Let us consider a nonlinear stochastic system described by a macroscopic variable q and driven by a dissipation term k(q) and a stochastic force f(t) which is Gaussian and δ -correlated in time with correlation amplitude 2 D(q).
The Langevin equation describing the motion is

$$\dot{q} = k(q) + f(t) \tag{3.1}$$

and the Fokker-Planck equation for the time dependent probability density P(q,t) is[7]

$$\frac{\partial P}{\partial t} = -\frac{\partial}{\partial q}[K(q) P] + \frac{\partial^2}{\partial q^2}[D(q) P] \tag{3.2}$$

If, as in many quantum optical cases, the diffusion term D does not depend on q and k(q) admits a potential

$$V(q) = -\int k(q)\, dq \tag{3.3}$$

then the stationary solution of (3.2) is immediately written as

$$P(q) = N\, e^{-V(q)/D} \tag{3.4}$$

where N is a normalization constant.
By comparison with distribution (2.3) for an equilibrium system, the pseudo-potential V(q) plays the same role as the free energy, and situations as in fig. 1 and 2 can be foreseen.
Indeed, "ad hoc" addition of stochastic forces accounting for noise processes to the laser equation transforms it into a nonlinear Langevin equation with a pseudo-potential

$$V(x) = -\frac{k}{2}(C - 1)\, x^2 + \frac{kC}{4} x^4 \, . \tag{3.5}$$

This indicates that the laser threshold is as a 2nd order phase transition in an e-quilibrium system. The role of the temperature as control parameter is here taken by the density of excited atoms which changes C.
Some extra care must be taken since a radiation field is a complex parameter. Plotting each of the curves of fig. 1 versus a complex parameter on a complex plane $x = |x|\, e^{i\varphi}$ rather than on an axis x, potential and probability curves become surfaces having rotational symmetry [6, 35] . While the modulus has the constraint already

discussed, the phase φ is free to diffuse. Hence this will be the ultimate noise for a laser highly above threshold. If the rotational symmetry is broken by the injection of an external signal [40], this symmetry break leaves a single minimum [41], forbidding the coexistence of two stable points and hence a discontinuous jump. Things are different if the external signal is injected into a passive medium (optical bistability). From Sec. 3 one can immediately evaluate the following pseudo-potential [42]

$$V(x) = \frac{|x|^2}{2} + \ln|x| + 2C\ln(y - |x|) \tag{3.6}$$

which under suitable power expansion, gives curves like fig. 2, showing the analogy with a 1st order phase transition.

The time dependent solution $P(q_0 / q, t)$ is simple in the case of a linear force $k(q) = -\beta q$.

In such a case the time dependent probability is a Gaussian with its center value decaying exponentially to zero at a rate β and its variance given by D/β,

$$P(q_0|q,t) = \frac{1}{\sqrt{2\pi\frac{D}{\beta}(1-e^{-2\beta t})}}\, e^{-\frac{(q - q_0 e^{-\beta t})^2}{\frac{D}{\beta}(1-e^{-2\beta t})}}\,. \tag{3.7}$$

Linearizing the laser equation one sees that $\beta = C - 1$. Hence at threshold ($\beta = 0$) we expect a divergence in the fluctuations and an infinite relaxation time (critical slowing down). In fact the infinities are smoothed by the nonlinearities.

The slowing down at threshold shows that even in a many-mode system one can make a description in terms of the critical field amplitude (order parameter) since this becomes the slowest variable.

In fig. 5 we report the stationary intensity distributions for the laser [6] after eqs. (3.4) and (3.5).

To make comparisons with experimental data, we must account for the change in statistics due to the photon process.

For the superfluorescent case, eq. (2.21) describes the motion of a point on a sphere Indeed, if for a moment we leave our oversimplified outlook and consider the Fokker-Planck equation as a suitable phase-space density evolution, arising from a quantum Master equation, it is very convenient to choose a diagonal representation for the density operator of the system, so that the weight-function of the diagonal projection is the quantum equivalent of the classical probability $P(q)$. In the field case this is done in the coherent state representation [1] where states are mapped as points on a complex plane.

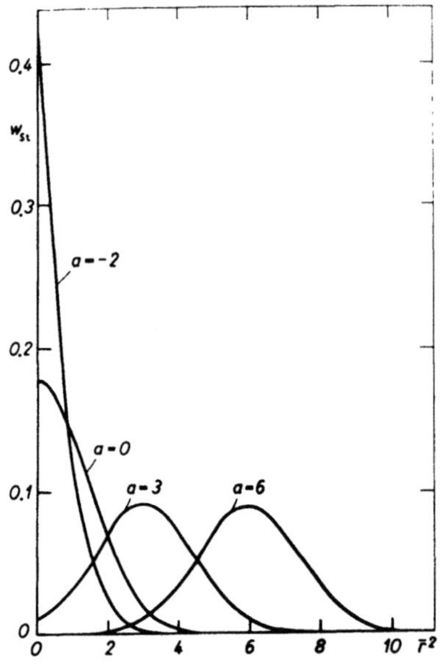

Fig.5 - The stationary laser distribution as a function of the normalized intensity, for different pump parameters (Ref.6)

In the atomic case this is done in the atomic coherent state representation [43], where states are mapped as points on a sphere. Projecting on a plane tangent on a pole of the sphere one obtains a single Fokker-Planck equation in a plane [44a]. Otherwise one can carry the calculation on the sphere [44b]. In both cases one obtains results equivalent to the Master equation treatment [26].
The Haake-Glauber plane equation [44a] looks like the equation for a forced system,

$$\dot{q} = + |\beta| \cdot q + f(t)$$

which, starting from an initial point q_o diverges. A nonlinear transformation back on the sphere gives the hyperbolic sechant behaviour for the field plus the anomalous fluctuations discussed in Sec. 5.

4-PHOTON STATISTICS AND FLUCTUATIONS IN QUANTUM OPTICS

Consider a photodetector illuminated by a light beam. By an electronic gate lasting for a time T, the number n of photons annihilated at the photosurface in T is counted. The random variable n has a statistical distribution p(n) that can be determined by iterating the above procedure for a large number of samples.

In fig. 6 we show an experimental plot of the statistical distribution of photocounts p(n) versus the number of counts [45, 46]

$$n = \eta \langle I \rangle T \, . \qquad (4.1)$$

$\langle I \rangle$ is the average intensity, T the gating time, η a constant accounting for the quantum efficiency of the detector plus other instrumental factors. The three curves refer to three physical cases which are indistinguishable from the point of view of classical optics: same average photon number $\langle n \rangle$, same diffraction-limited plane wave, same line-width $\Delta\omega$ filtered out in such a way that

$$\tau = 1/\Delta\omega >> T .$$

From the point of view of PS, the three radiation fields are dramatically different, as seen from the figure.

The three fields L, G, S correspond to the following cases: L comes from a stabilized single mode well above threshold. A moment analysis shows that it is well approximated by a Poisson distribution with a variance

$$\langle \Delta n^2 \rangle = \langle n \rangle \, . \qquad (4.2)$$

G is obtained by scattering L over a collection of microparticles in Brownian motion, being sure that the correlation time be longer than T, and then putting up a diffraction-limited plane wave at a given angle. A moment analysis shows that it is a Bose-Einstein distribution with a variance

$$\langle \Delta n^2 \rangle = \langle n \rangle (1 + \langle n \rangle) \, . \qquad (4.3)$$

S is the superposition of L and G over the same spatial mode.

A heuristic view of the photodetection process explains the above results, without recurring to the theory [1]. If the field is uniform as we expect for a stabilized laser the photons, being particles with zero mass, can not be localized; hence there is no a-priori correlation between two annihilation events at two different points either in space or in time. The photocounts from a single detector whose average number is proportional to the square field and the measuring time T,

$$\langle n \rangle = |E|^2 \, T \, \eta \qquad (4.4)$$

(η = quantum efficiency of the detector), must then be distributed as a Poissonian, i.e. ,

$$p(n) \doteq K\ (E,\ T\ |\ n) = \frac{\langle n \rangle}{n!}\ e^{-\langle n \rangle} . \qquad (4.5)$$

This is shown in fig. 7a.

If now the complex field is randomly distributed with a statistics P_1 (E, t) and each measurement lasts for a time T much smaller than the coherence time τ_c (in order to have a constant field within each sample) then we must average the detector statistics (4.5) over the field statistics

$$p(n,\ T,\ t) = \int K\ (E,\ T\ |\ n)\ P_1\ (E,\ t)\ d^2E \qquad (4.6)$$

In fig. 7 the results are shown pictorially in the three cases of fig.6.

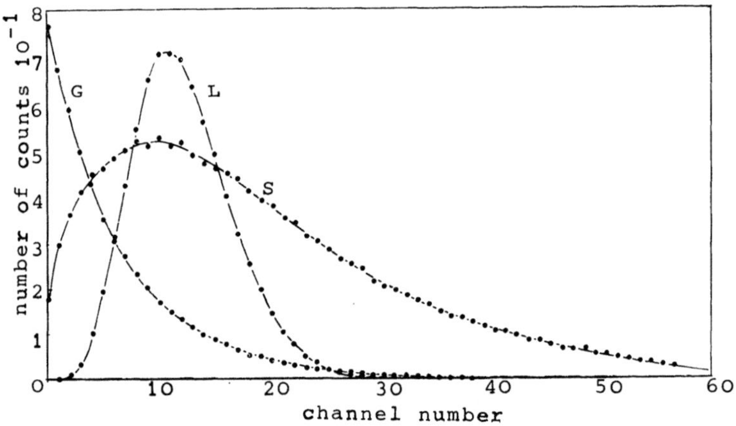

Fig. 6 - Photocounts distributions obtained by sampling at equal time intervals three types of radiation fields
L = laser field from a stabilized single-mode laser
G = Gaussian field
S = linear superposition of L and G onto the same space mode (after Ref.46)

A single p(n,t) gives only an integrated information on the time evolution of the field. By a suitable set-up one can measure the joint distribution $p(n_1,\ t_1\ ;\ n_2\ ,\ t_2)$ and then evaluate all cross moments or correlation functions

$$G^{(2k\ ,\ 2\ell)}\ (t_1,\ t_2) = \langle n_1^{\ k}\ n_2^{\ \ell} \rangle \qquad (4.7)$$

and fully characterize a Markov field. However, often one is interested only in the first cross moment (intensity correlation)

$$G^{(2,2)}(t_1, t_2) = \langle n_1 \; n_2 \rangle \; . \qquad (4.8)$$

This can be obtained by a single correlator.
For more details, see Ref. 47 , 48.

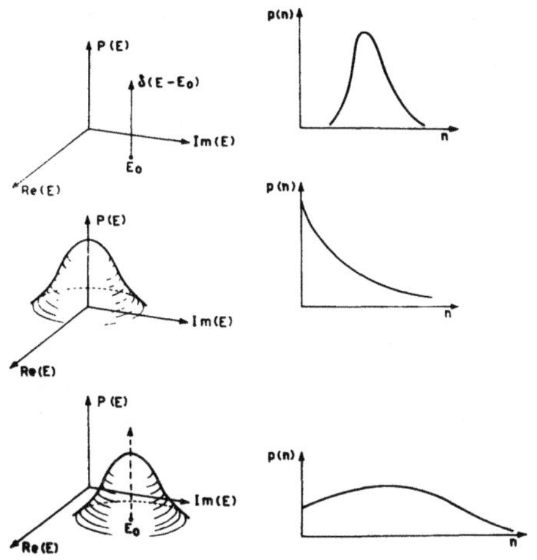

Fig. 7 - Qualitative plots of the field and photon statistical distributions in the cases of an ideal coherent field of a thermal equilibrium field (Gaussian with zero average) and the superposition of the two.

5-STATISTICAL EXPERIMENTS IN QUANTUM OPTICS

5.1 - WHAT WE EXPECT FOR LASERS

From the phenomenological description of Sec. 3 we expect at threshold the appearance of a nonzero order parameter, a large increase in fluctuations and a critical slowing down. To correct that picture we must add that

i) the laser equation is nonlinear, hence there is neither divergence in the fluctuations, nor zero line-width. The infinities are smoothed by the nonlinearity and the theoretical expectations are plotted together with the experimental points in the coming figures 8-11;

ii) the laser field is a complex parameter which should be described in modulus and phase:

$$E = |E| \; e^{i\varphi}.$$

The photon statistics destroys phase information because $n \sim |E|^2$. Phase information is recovered by performing an interference experiment with two independent lasers [50], so that the output intensity is as

$$n \sim |E_1 + E_2|^2 = I_1 + I_2 + 2\,\mathrm{Re}\left(E_1^{*} \cdot E_2\right).$$

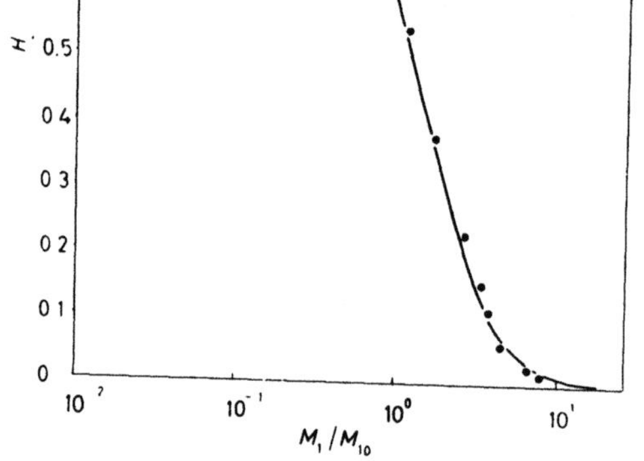

Fig. 8 - Measured and theoretical values of the reduced second order factorial moment of the photon distribution $H_2 = \langle n\,(n-1)\rangle / \langle n\rangle^2 - 1$ versus the intensity M_1 normalized to the threshold value M_{10} (Ref. 52)

In order to interfere the laser field with itself at a later time, one has to build a very long Michelson interferometer. This was done indeed [51] and the experiments agree with the nonlinear theory at threshold. For phase fluctuations above threshold, we expect a phase diffusion process which yields a stationary probability distribution, with a very definite modulus as in fig. 1 when q becomes very large and the spread of P(q) much smaller than q but with a phase completely uncertain. The lack of monochromaticity of a laser above threshold is then due purely to phase fluctuations. These can be shown to be inversely proportional to the emitted power $|E|^2$, that is, the coherence time of the laser for phase fluctuations becomes longer as the output field increases (Townes formula)

$$\tau_\varphi \sim |E|^2 .$$

iii) The description of Sec. 3 refers to stationary statistical processes. In Sec. 5.4 we present some time-dependent statistical features without equivalent in thermodynamic phase transitions and whose relevance in nonequilibrium systems is being recognized in these years [24,34].

5.2-INTENSITY FLUCTUATIONS OF LASERS

In this section we describe the experimental results obtained by means of the PS method in the study of the statistics of the e.m. field of a stabilized laser operating in different conditions. The measurements were done on a 6328 A He-Ne laser, single mode, with one mirror supported by a piezoceramic disc in order to stabilize against fluctuations and to move the mode position with respect to the atomic line (for details see Ref. 2).

The measurement of p(n) was performed as described in Sec. 4. For comparing experiments and theory we use the second reduced factorial moment of the photocount distribution

$$H_2 = \frac{\langle n(n-1)\rangle}{\langle n\rangle^2} - 1 = F_2 / F_1^{\,2} - 1 \qquad (5.1)$$

which goes from 1 (Gaussian field distribution, well below threshold) to 0 for an amplitude-stabilized field (well above threshold), and the third one

$$H_3 = \frac{\langle n(n-1)(n-2)\rangle}{\langle n\rangle^3} - 1 = F_3 / F_1^{\,3} - 1 \qquad (5.2)$$

which goes from 5 (Gaussian distribution) to 0 (amplitude-stabilized field) (Fig. 8, 9) [52].

From the stationary solution of the dynamic laser equations the distribution of photocounts and the associated factorial moments can be derived [4-6]. One can see from the figures that the agreement between experiments and theory is very good.

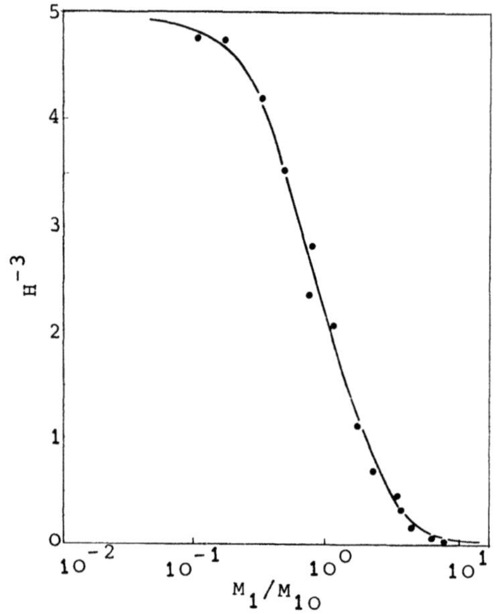

Fig.9-Measured and theoretical values of the reduced third order factorial moment of the photon distribution

$H_3 = \langle n(n-1)(n-2) \rangle / \langle n \rangle^3 - 1$

(Ref. 2)

Measurements of the power spectrum of the intensity fluctuations for a He-Ne laser both below and above threshold were first reported by Freed and Haus [54]. Since those measurements left about 2 decades of intensity around the threshold unexplored, it was possible to interpret them with accuracy by a linear model, hence they left open the possibility of zero width at threshold. In a later experiment [55] the threshold region was thoroughly explored. Here a comparison with the results of the nonlinear theory was required [6].

We report in fig. 10 the measurements which show good agreement with theoretical predictions.

In later measurements exploiting a nonlinear correlation technique it was possible to attribute separate values to the main eigenvalues in the series of exponentials describing the decay of intensity fluctuations at threshold (fig. 11) [57].

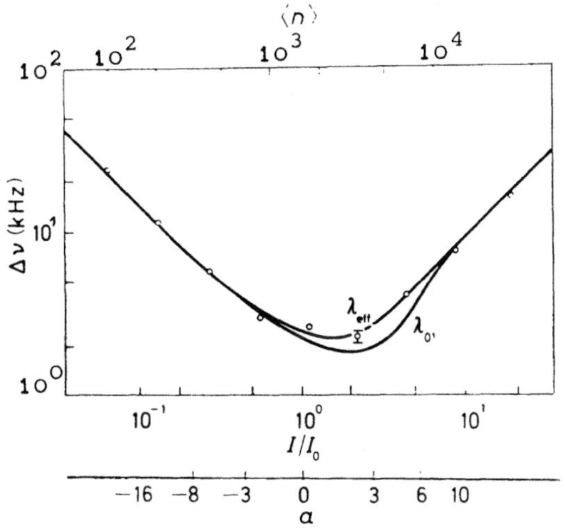

Fig.10-"Effective"linewidth $\lambda_{eff.}$ of the laser intensity fluctuations versus the intensity I normalized to the threshold value I_o .

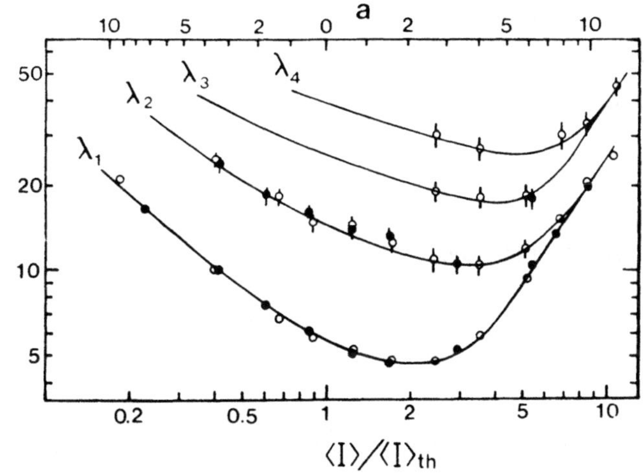

Fig.11- Relaxation rates of the laser intensity fluctuations. The full curves represent the theoretically predicted results.

5.3-PHASE FLUCTUATIONS IN LASERS

Calling by x(t) the complex amplitude of the laser field, the first-order correlation function

$$G^{(1)}(t) = \langle x^*(t)\; x(0)\rangle$$

has a decay depending on both phase and amplitude fluctuations. The decay rate would have a divergence in asymptotic (below and above threshold) theories which is smoothed by nonlinearity [6]. Well above threshold, where the amplitude is stabilized, the residual linewidth is mainly due to phase fluctuation. A measurement well above threshold would then give the "quantum phase noise".

A measurement in the threshold region was possible by building a very long folded Michelson interferometer [51]. Fig. 12 shows the good agreement with the theory.

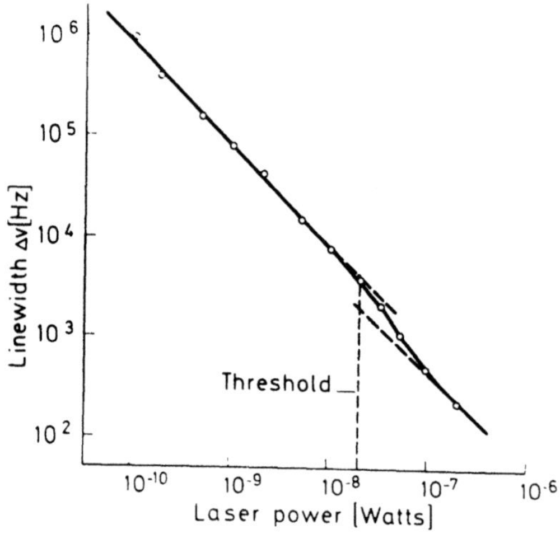

Fig.12-Experimental linewidth versus laser power (circles). The solid curve represents the theoretical prediction (Ref. 51)

5.4-TRANSIENT LASER FLUCTUATIONS AROUND THRESHOLD [21, 22]

By the joint use of a Q-switched gas laser and of PS a non-stationary statistical ensemble can be studied, measuring the time evolution of a laser field during a fast build-up.

We put a Kerr cell within a single-mode laser cavity. Starting with some pre-set pump and cavity parameters, but with the optical shutter closed, the Kerr cell is switched "on" at the instant t=0. The laser field undergoes a transient build-up, from an initial statistical distribution corresponding to the equilibrium between gain and losses far below threshold, up to an asymptotic condition above threshold. At some later time τ we perform photocount measurements for a measuring interval T, very small compared to the build-up time which is in our case of the order of some microseconds.

Once a steady-state condition has been reached, an amplitude-stabilizing operation is performed by sampling the laser output and comparing this with a standard reference signal. This is equivalent to 'preparing' an identical initial state for a suc-

cessive measuring cycle.
After the sampling, the shutter is switched off for about 10 ms. At the end of this interval the shutter is again switched on and the above described cycle of operation is repeated. In this way we collect an ensemble of macroscopically identical events. By successively varying τ we obtain the time evolution of photocounts distribution $p(n,T,\tau)$. A set of experimental results is shown in Fig. 13 [21]. The average photocount number $\langle n \rangle$ and the associated variance $\langle \Delta n^2 \rangle = \langle n^2 \rangle - \langle n \rangle^2$ are reported as a function of the time delay in figures 18 and 19 [22]. One can see a transition from a Bose-Einstein type distribution for initial times to a Poisson-like distribution for long delays. But, at variance with the stationary field distributions of fig. 5 the intermediate distributions are much wider, showing large transient fluctuations.

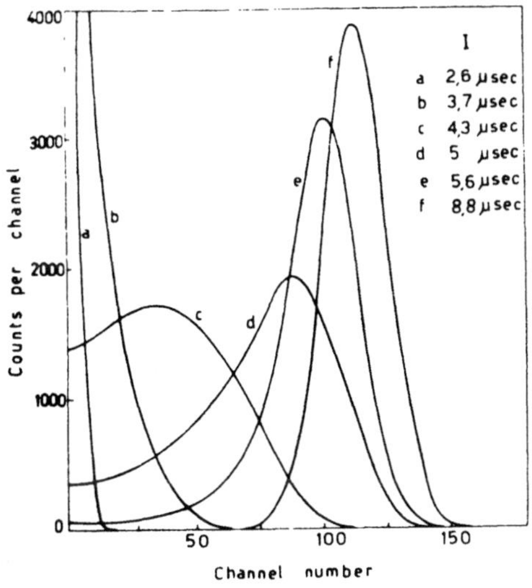

Fig.13- Experimental statistical distributions with different time delays obtained on a laser transient. The solid lines connect the experimental points which are not shown to make the figure clearer. All distributions are normalized to the same area. (Ref. 21).

Solid lines in fig. 14 and 15 represent theoretical results computed with a suitable choice of parameters for best fit with the experimental points.

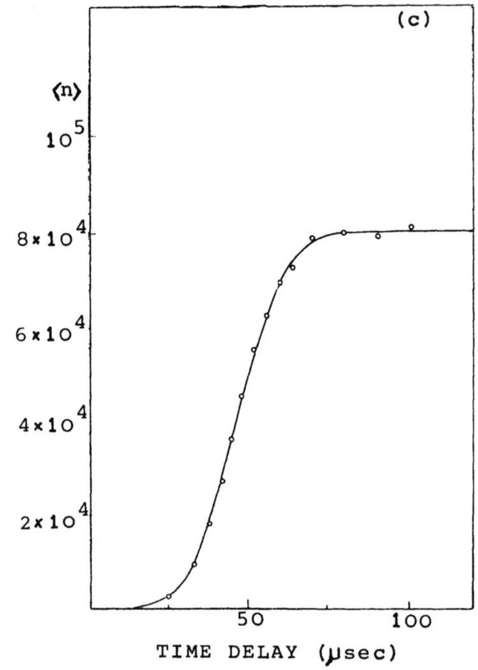

Fig.14-Evolution of the average photon number ⟨n⟩ inside the cavity as a function of the time delay . Solid lines represent best-fit results computed from the theory of Scully, Lamb and Sargent. (Ref.22)

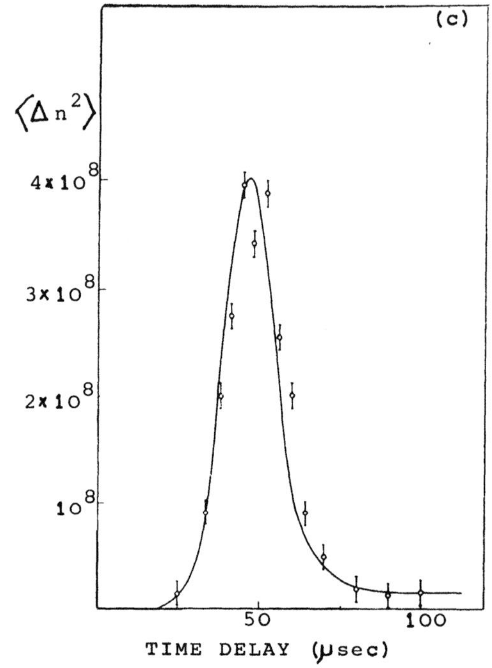

Fig.15-Evolution of the variance $\langle\Delta n^2\rangle$ of the statistical distribution of photons inside the cavity, as a function of the time delay. Solid lines represent theoretical results. (Ref.22).

Similar measurements were later reported by other groups [61]. The latter experiment was performed closer to threshold by means of a Pockels-cell shutter. The results agree with the computation of Risken [6]. It was possible to fit the data with a phenomenological theory in terms of the semi-classical evolution $n(\tau; n_0)$ depending on the initial photon number n_0,

$$n(\tau, n_0) = \frac{n_0 d}{d\, e^{-\beta d \tau} + n_0 (1 - e^{-\beta d \tau})} \tag{5.3}$$

then evaluating the various moments by using as a weight function the initial statistical distribution $p(n_o)$ of the photons below threshold. By introducing a scaled quantity [22]

$$Z = \frac{d\, e^{-\beta d \tau}}{\bar{n}_o (1 - e^{\beta d \tau})} \qquad (\bar{n}_o = \langle n_o \rangle) \tag{5.4}$$

it is possible to show that the ratio between variance and mean photon number squared is only a function of z.

$$r \equiv \frac{\langle \Delta n^2 \rangle}{\langle n \rangle^2} = \frac{z(1-H) - H^2}{(1-H)^2} \tag{5.5}$$

where $H = H(z)$ depends only on z.

Eq. (5.5) gives therefore a universal function of the laser transient. The ratio r is plotted in fig. 16.

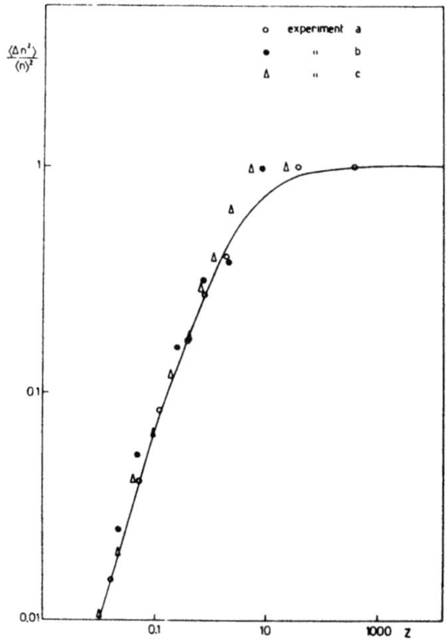

Fig.16-Relative variance in the experiments of fig.14,15 plotted versus the scaling parameter z. Larger z values correspond to small delays.The solid line comes from the phenomenological model of Ref. 22.

These scaling considerations were later [23, 24] extended to any nonlinear stochastic system crossing an instability at a rate so fast that initial conditions play an essential role.

5.5 - EXPERIMENTS ON OTHER QUANTUM OPTICAL TRANSITIONS

We show briefly some data on superfluorescence and optical bistability, noting that in both cases there are only preliminary experiments, and it is impossible to make any statistical inference either on average behaviour or fluctuations.

The cleanest experiment on superfluorescence is that of Ref. 30 where care has been put to satisfy the threshold conditions and the cooperativity requirements discussed in Sec. 2, that is,

$$K > \gamma_c > \gamma_c^2 / k > 1/t_D > \gamma \tag{5.6}$$

or, in words: escape time < cooperative rate < pulse duration < delay time between preparation and occurrence of the maximum < atomic decay rates (both homogeneous and Doppler).

Furthermore, the duration of the preparation pulse was chosen consistently shorter than the delay time. All parameters beside t_D have been given in Sec. 2. Starting from the fully excited state, t_D requires quantum considerations [27] and is given by

$$t_D \sim \tau_R \ln N \tag{5.7}$$

where $\tau_R = k / \gamma_c^2$ is the pulse width.

In fig. 17 some superfluorescent shapes are given for different atomic densities.

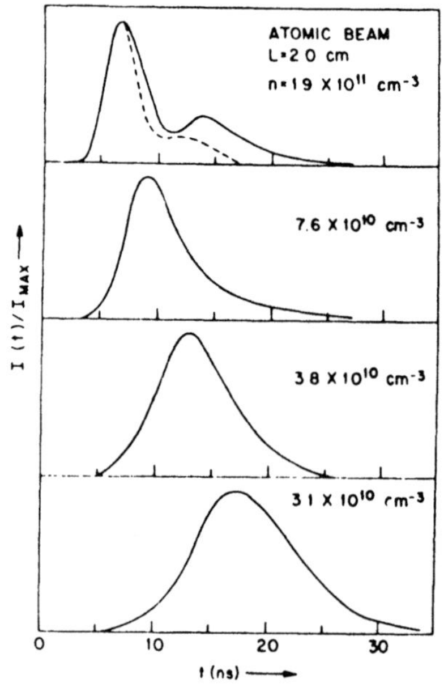

Fig.17-Superfluorecence pulse at 2.9 μm in the 7 $P_{3/2}$ to 7 $S_{1/2}$ transition in Cesium atoms. L is the sample length. Normalized single-shot shapes for several densities n (Ref.30)

Fig. 18 is a plot of t_D versus τ_R.
To decide how much of the phenomenon is classical and the relevant role of the initial quantum fluctuation one should measure the variance in the photon statistics.
The optical bistability in Na atoms was measured in Ref. 18 and some data are given in fig. 19.
The phenomenon depends sensitively on the mutual frequency positions of injected field, cavity and atomic resonance.
Here also there is a lack of statistical experiments to make comparisons with the theoretical previsions on spectrum and photon statistics[42].

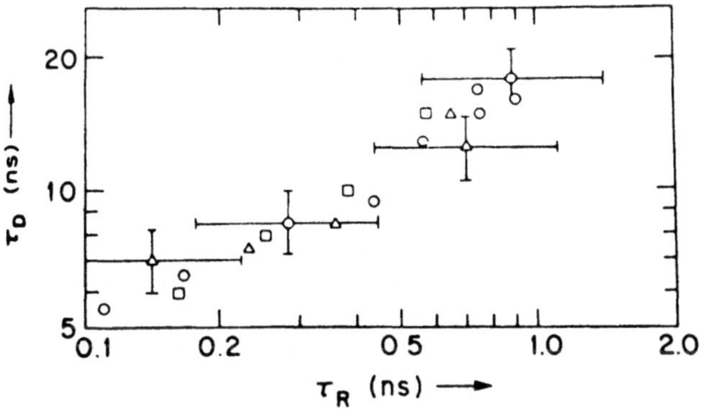

Fig.18 - Delay time t_D of the superfluorescent pulse with respect to the preparation time as a function of the superfluorescent characteristic time τ_R. (Ref. 30)

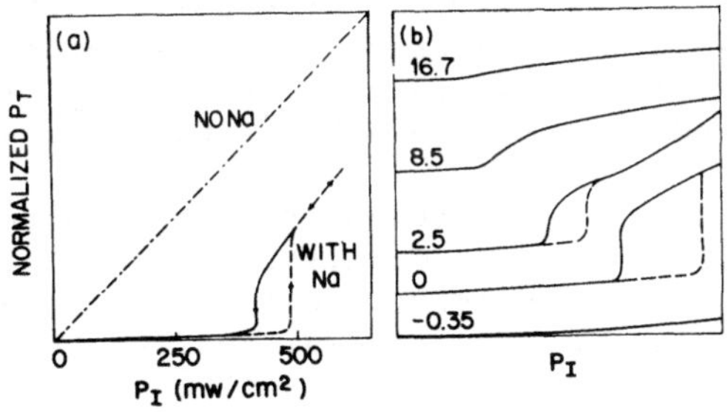

Fig.19-(a)Optical bistability in Na vapor.The oscilloscope trace is dashed for increasing input intensities and solid for decreasing. (b)Characteristic curve,dependence on Fabry-Perot plate separation is given in MHz. (Ref.18)

REFERENCES

1)-R.J.Glauber in Quantum Optics and Electronics (Proc.of 1964 Les Houches School), McGraw-Hill 1965, 65-185.
2)-F.T.Arecchi in Quantum Optics (Proc.of 1967 Varenna School),Academic Press 1969, 57-110.
3)-Since the first works on the subject
(W.E.Lamb,Oral communication at the Third International Conference on Quantum Electronics,Paris Feb.11-15,1963,and Phys.Rev.134,A 1429 (1964);
H.Haken and H.Sauermann, Z.Phys.,173,47 (1963)), a large body of papers has appeared on the subjects dealing with the quantum and statistical aspects of the laser oscillator.Complete reviews can now be found in two books,namely:
H.Haken,Laser Theory, Handbuch der Phys.ed.by S.Flügge,vol.XXV/2c, Springer 1970; and
M.Sargent III, M.O.Scully, W.E.Lamb,Laser Physics ,Addison-Wesley 1974.
The laser around threshold was the first system in which to exploit the photon statistical methods introduced after the theory of Glauber (Ref.1).A review is given in Ref.2. For interpreting the experiments it has been sufficient to use the phenomenological approaches introduced in the next Refs.4,5 and reviewed in Ref.6. Those phenomenological approaches exploit the statistical treatment of a non-linear oscillator as given, for example, in Ref.7.
4)-H.Risken, Z.Phys.,186,85 (1965) and 191,302 (1966).
5)-R.D.Heampstead and M.Lax, Phys.Rev.161,350 (1967).
6)-H.Risken in Progress in Optics, ed.by E.Wolf 8,239-294 (1970).
7)-R.L.Stratonovich, Topics in the theory of random noise, Gordon and Breach,vol.I (1963) and vol.II (1967).
8)-L.Landau, Phys.Z.Sovietunion 11,26 (1937); 11,545 (1937).
9)-H.E.Stanley, Introduction to phase transitions ,Oxford Press 1971.
10)-R.Landauer,J.Appl.Phys.33,2209 (1962).
11)-E.Pytte, H.Thomas,Phys.Rev.Letters 20,1167 (1968).
12)-R.Graham, H.Haken,Z.Physik 237,31 (1970).
13)-M.O.Scully, V.Degiorgio,Phys.Rev.A2,1170 (1970).
14)-V.N.Lisitsyn, V.Chebotayev, JEPT Letters 7,1 (1968).
15)-T.H.Lee, P.B.Schaefer, W.B.Barker,Appl.Phys.Letters 13,373 (1968).
16)-J.F.Scott, M.Sargent III, C.Cantrell,Opt.Comm.15,13 (1975).
17)-L.A.Lugiato, P.Mandel, S.T.Dembinski, A.Kossakowski,Phys.Rev. A18, 238 (1978).
18)-H.M.Gibbs, S.L.McCall, T.Venkatesan,Phys.Rev.Letters 36,1135 (1976).
19)-R.Bonifacio, L.A.Lugiato,Optc.Comm.19,172 (1976).
20)-R.Bonifacio, L.A.Lugiato,Lett.N.Cimento 21,517 (1978).
21)-F.T.Arecchi, V.Degiorgio, B.Querzola,Phys.Rev.Letters 19,1168(1967).
22)-F.T.Arecchi, V.Degiorgio, Phys.Rev.A3,1108 (1971).
23)-R.Kubo in Synergetics,ed.by H.Haken, B.G.Teubner 1973.
24)-M.Suzuki,Contribution to the XVII Solvay Conference on Physics,1978.
25)-R.H.Dicke,Phys.Rev.93,99 (1954).
26)-R.Bonifacio, P.Schwendimann, F.Haake, Phys.Rev.A4,302 and 804 (1971).
27)-R.Bonifacio, L.A.Lugiato,Phys.Rev.A11,1507 and A12,587 (1975).
28)-N.Sckribanowitz,I.Herman,J.MacGillivary,M.Feld,Phys.Rev.Lett.30,309

28)- (1973).
29)-M.Gross, C.Fabre, P.Pillet, S.Haroche,Phys.Rev.Letters 36, 1035 (1976).
30)-H.M.Gibbs,Q.Vrehen,H.Hikspoors,Phys.Rev.Letters 39,547 (1978).
31)-S.Grossmann, P.H.Richter,Z.Phys.249,43 (1971) and 255,59 (1972).
32)-F.T.Arecchi, A.M. Ricca,Phys.Rev.A15,308 (1977).
33)-M.M.Tehrani, L.Mandel,Phys.Rev.A17,694 (1978).
34)-For a complete discussion including a microscopic approach, see R.Graham,contribution to the XVII Solvay Conference on Physics, 1978.
35)-For a detailed treatment see H.Haken, Synergetics, Springer Verlag, 1977.
36)-F.T.Arecchi, E.Courtens, Phys.Rev.A2,1730 (1970).
37)-F.T.Arecchi, R.Bonifacio, IEEE J.Quantum El.1,169 (1965).
38)-S.McCall, E.L.Hahn, Phys.Rev.Letters 18,908 (1967).
39)-F.T.Arecchi, V.Degiorgio, C.Someda,Phys. Letters A27,588 (1968).
40)-W.Chow, M.Scully, E.van Stryland,Opt.Comm.15,6 (1975).
41)-L.A.Lugiato, Lett.Nuovo Cimento (to be published).
42)-R.Bonifacio, L.A.Lugiato,Phys.Rev.Letters 40,1023 (1978).
43)-F.T.Arecchi, E.Courtens, R.Gilmore, H.Thomas in Proc.of Esfahan Symposium, 1971 and Phys.Rev.A6,2211 (1972).
44)-a) F.Haake, R.J.Glauber,Phys.Rev.A5,1457 (1972);
b) L.M.Narducci, C.Coulter, C.Bowden,Phys.Rev.A9,829 (1974).
45)-F.T.Arecchi, Phys.Rev.Letters 15,912 (1965).
46)-F.T.Arecchi, A.Berné, P.Burlamacchi,Phys.Rev.Letters 16,32 (1966).
47)-F.T.Arecchi, A.Berné, A.Sona,Phys.Rev.Letters 17,260 (1966).
48)-F.T.Arecchi, V.Degiorgio in Laser Handbook, North Holland,1972, vol.I,pp.191-264.
49)-F.T.Arecchi, V.Degiorgio,Phys.Letters 27A,429 (1968).
50)-A.E.Siegman, R.Arrathoon,Phys.Rev.Letters 20,901 (1968); K.R.Manes, A.E.Siegman,Phys.Rev.A4,373 (1971).
51)-H.Gerhardt, H.Welling, A.Gütner,Phys.Letters 40A,191 (1973).
52)-F.T.Arecchi, G.Rodari, A.Sona,Phys.Letters 25A,59 (1967).
53)-M.O.Scully, W.E.Lamb,Phys.Rev.159,208 (1967).
54)-C.Freed, H.Haus,Phys.Rev.141,287 (1966).
55)-F.T.Arecchi, M.Giglio, A.Sona,Phys.Letters 25A, 341 (1967).
56)-J.A.Armstrong, A.W.Smith, Phys.Rev.140,A 55 (1965).
57)-M.Corti, V.Degiorgio, F.T.Arecchi,Opt.Comm.8,329 (1973).
58)-A.Schawlow, C.H.Townes,Phys.Rev.112,1940 (1958).
59)-M.Sargent, M.O.Scully, W.E.Lamb,Appl.Optics 9,2423 (1970).
60)-E.Baklanov, S.Rautian, B.Troshin, V.Chebotayev, Sov.Phys.JEPT 29, 601 (1969).
61)-D.Meltzer, L.Mandel, Phys.Rev.Letters 25, 1151 (1970).

ADDRESS LIST OF PARTICIPANTS

ARECCHI F.T., Istituto Nazionale di Ottica
Largo Enrico Fermi, 6 - 50125 ARCETRI-FIRENZE (Italy)

ATTEN P., Laboratoire d'Electrostatique 166 X
38042 GRENOBLE-CEDEX (France)

AUSLOOS M., Institut de Physique B5 - Université de Liège
Sart Tilman - 4000 LIEGE 1 (Belgium)

BAERISWYL D., Laboratories RCA Ltd.
Badenerstrasse 569 - 8048 ZÜRICH (Switzerland)

BAUSCH R., Institut für Festkörperforschung
KFA, Postfach 1913 - 5170 JÜLICH (Germany)

BERGE P., Service de Physique du Solide et Résonance Magnétique
B.P. 2 - 91190 GIF-SUR-YVETTE (France)

BEYSENS D., Commissariat à l'Energie Atomique - C.E.N. - SPSRM
B.P. 2 - 91190 GIF-SUR-YVETTE (France)

BHATTACHARJEE J.K., Department of Physics - University of Maryland
COLLEGE PARK, Md. 20742 (U.S.A.)

BLINC R., Institute J. Stefan - University of Ljubljana
LJUBLJANA (Jugoslavia)

BRENIG L., Service de Chimie-Physique II - C.P. 231 - ULB
Campus Plaine, Bd du Triomphe - 1050 BRUXELLES (Belgium)

BREUER N., Institut für Theoretische Physik - Universität Düsseldorf
Universitätsstrasse 1 - 4000 DÜSSELDORF (Germany)

BURKHARDT R., Institut für Physik - Universität Basel
Klingelbergstrasse 82 - 4058 BASEL (Switzerland)

CALMETTES P., Commissariat à l'Energie Atomique - C.E.N. - ORME des Merisiers
D.Ph. - SRM
B.P. 2 - 91190 GIF-SUR-YVETTE (France)

CHOQUARD Ph., Laboratoire de Physique Théorique - EPFL
14, av. de l'Eglise Anglaise - 1006 LAUSANNE (Switzerland)

DEKER U., Institut für Physik - Universität Basel
Klingelbergstrasse 82 - 4058 BASEL (Switzerland)

DOHM V., Institut für Festkörperforschung
KFA, Postfach 1913 - 5170 JÜLICH (Germany)

DE DOMINICIS C., Service de Physique Théorique - C.E.N. Saclay
B.P. 2 - 91190 GIF-SUR-YVETTE (France)

DROZ M., Département de Physique Théorique - Université de Genève
24, quai Ernest Ansermet - 1211 GENEVE 4 (Switzerland)

DUBOIS M., Service de Physique du Solide et Résonance Magnétique
B.P. 2 - 91190 GIF-SUR-YVETTE (France)

EISENRIEGLER E., Institut für Festkörperforschung
KFA, Postfach 1913 - 5170 JÜLICH (Germany)

ENZ C.P., Département de Physique Théorique - Université de Genève
24, quai Ernest Ansermet - 1211 GENEVE 4 (Switzerland)

FERRELL R.A., Department of Physics - University of Maryland
COLLEGE PARK, Md. 20742 (U.S.A.)

FOURNIER J.D., Observatoire de Nice
B.P. 252 - 06007 NICE CEDEX (France)

GALASIEWICZ Z., Institute of Theoretical Physics - Wroclaw University
Ul. Cybulskiego 36 - 50205 WROCLAW (Poland)

GBADAMASSI M., Commissariat à l'Energie Atomique - C.E.N. - D.Ph.G. - SRM
B.P. 2 - 91190 GIF-SUR-YVETTE (France)

GOERTZ R., Universität Düsseldor
Universitätsstrasse 1 - 4000 DÜSSELDORF (Germany)

GRAHAM R., Fachbereich 7, Physik - Universität Essen - GHS
Postfach 6843 - 4300 ESSEN (Germany)

GREYTAK T.J., Massachusetts Institute of Technology - Room 13-2074
CAMBRIDGE, Ma. 02139 (U.S.A.)

GUNTON J.D., Physics Department - Temple University
PHILADELPHIA, Pa. 19122 (U.S.A.)

HONGLER M.-O., Département de Physique Théorique - Université de Genève
24, quai Ernest Ansermet - 1211 GENEVE 4 (Switzerland)

HORNER H., Institut für Theoretische Physik - Universität Heidelberg
Philosophenweg 19 - 6900 HEIDELBERG (Germany)

HORSTHEMKE W., Service de Chimie Physique II - ULB
C.P. 231 - Campus Plaine, Boulevard du Triomphe
1050 BRUXELLES (Belgium)

JANSSEN H.K., Institut für Theoretische Physik III - Universität Düsseldorf
Universitätsstrasse 1 - 4000 DÜSSELDORF (Germany)

KAMENSKI W., Landau Institute for Theoretical Physics - Academy of Science USSR
Vorobjevskoe Shosse 2 - 117334 MOSCOW (USSR)

KAWASAKI K., Department of Physics - Faculty of Science - Kyushu University
FUKUOKA 812 (Japan)

KHMEL'NITZKII D.E., Landau Institute for Theoretical Physics
CHERNOGOLOVKA, Moscow Region (USSR)

KINZEL W., Institut für Festkörperforschung
KFA, Postfach 1913 - 5170 JÜLICH (Germany)

KIRKHAM J., Department of Physics - University
SOUTHHAMPTON (U.K.)

KONDOR I., Institute for Theoretical Physics - R. Eötvös University
Puskin u. 5-7 - 1088 BUDAPEST VIII (Hungary)

KRAGLER R., Laboratories RCA Ltd.
Badenerstrasse 569 - 8048 ZÜRICH (Switzerland)

KUNZ H., Lab. de Physique Théorique - EPFL
14, av. de l'Eglise Anglaise - 1006 LAUSANNE (Switzerland)

LARKIN A.I., Landau Institute for Theoretical Physics
CHERNOGOLOVKA, Moscow Region (USSR)

LIPOWSKY R., LS Prof. Wagner - Universität München
Theresienstrasse 37/III - 8000 MÜNCHEN 2 (Germany)

MALASPINAS A., Department of Physics - University of Crete
CRETE (Greece)

MARTIN Ph., Laboratoire de Physique Théorique - EPFL
14, av. de l'Eglise Anglaise - 1006 LAUSANNE (Switzerland)

MAZENKO G.F., James Franck Institute of Physics - University of Chicago
CHICAGO, Ill. 60637 (U.S.A.)

MEISSNER G., Fachbereich Physik - Universität des Saarlandes
6600 SAARBRÜCKEN 11 (Germany)

MEYER H., Department of Physics - Duke University
DURHAM, N.C. 27706 (U.S.A.)

MÜLLER K.A., IBM Research Center
P.O. Box 218 - YORKTOWN HEIGHTS, N.Y. 10598 (U.S.A.)

NOLAN M., Department of Applied Physics - Stanford University
STANFORD, Calif. 94305 (U.S.A.)

DE PASQUALE F., Istituto di Fisica - Fac. Ingegneria - Università
Piazzale delle Scienze 5 - 00185 ROMA (Italiy)

PEKALSKI A., Institute of Theoretical Physics - University of Wroclaw
Ul. Cybulskiego 36 - 50205 WROCLAW (Poland)

PELITI R., Istituto di Fisica "G. Marconi" - Università di Roma
00185 ROMA (Italy)

PIRC R., Institut "Jozef Stefan" - University of Ljubljana
P.O.B. 199 - LJUBLJANA (Jugoslavia)

REITER G., 510 A - Physics Department - Brookhaven National Laboratory
UPTON, L.I., N.Y. 11973 (U.S.A.)

RUHLAND J., Sektion Physik der Universität München - LS Prof. Wagner
Theresienstrasse 37/III - 8000 MÜNCHEN 2 (Germany)

RYTER D., Institut für Physik - Universität Basel
Klingelbergstrasse 82 - 4056 BASEL (Switzerland)

SANCHO J.M., Dept. de Fisica Teorica - Universidad de Barcelona
Diagonal 647 - BARCELONA 28 (Spain)

SAN MIGUEL M., Dept. de Fisica Teorica - Universidad de Barcelona
Diagonal 647 - BARCELONA 28 (Spain)

SASVARI L., Institute for Theoretical Physics - R. Eötvös University
Puskin u. 5-7 - 1088 BUDAPEST (Hungary)

SHEPHERD T.J., Royal Signals and Radar Establishment
St. Andrews Road - Great Malvern - WORCS WR14 3PS (U.K.)

SJÖLANDER A., Institute of Theoretical Physics - Chalmers University of Techn.
FACK - 40220 GÖTEBORG 5 (Sweden)

SOUSA J.B., Fac. de Ciencias Fisica - Universidad do Porto
PORTO (Portugal)

SULEM P.L., Observatoire de Nice
B.P. 252 - 06007 NICE CEDEX (France)

SUZUKI M., Department of Physics - University of Tokyo
7-3-1 Hongo, Bunkyo-Ku - TOKYO (Japan)

SZEPFALUSY N., Fachbereich Physik - Universität des Saarlandes
6600 SAARBRÜCKEN 11 (Germany)

SZEPFALUSY P., Fachbereich Physik - Universität des Saarlandes
6600 SAARBRÜCKEN 11 (Germany)

SCHNEIDER T., IBM Zürich Research Laboratory
Säumerstrasse 4 - 8803 RÜSCHLIKON (Switzerland)

SCHUSTER H.G., Institut für Theoretische Physik - Universität Frankfurt
Robert-Mayerstrasse 8-10 - 6000 FRANKFURT/MAIN (Germany)

SCHWABL F., Institut für Physik - Universität Linz
4045 LINZ (Austria)

STOLL E., IBM Zürich Research Laboratory
Säumerstrasse 4 - 8803 RÜSCHLIKON (Switzerland)

TEL T., Institute for Theoretical Physics - R. Eötvös University
Puskin u. 5-7 - 1088 BUDAPEST (Hungary)

THOMAS H., Institut für Physik - Universität Basel
Klingelbergstrasse 82 - 4056 BASEL (Switzerland)

TOMBESI P., Istituto di Fisica "G. Marconi"
Piazzale delle Scienze 5 - 00185 ROMA (Italy)

TOYODA T., Freie Universität Berlin - FB 20, WE 5
Arnimallee 3 - 1000 BERLIN 33 (Germany)

TURSKI L.A., Instytut Fizyki Teoretycznej - Uniwersytet Warszawki
Ul. Hoza 69 - 00681 WARSAW (Poland)

VAN DEN BROECK C., Vrye Universiteit Brussel
Pleinlaan 2 - 1050 BRUSSEL (Belgium)

VELARDE M.G., Dept. Fisica C-3 - Universidad Autonoma de Madrid
CANTOBLANCO, MADRID (Spain)

VELASCO BELMONT R.M., Av. Michoacān y La Purisima
Iztapalapa Apartado Postal 55-532 - MEXICO 13, D.F. (Mexico)

WESFREID J.E., C.E.N. Saclay - D.Ph.G. - SRM
B.P. 2 - 91190 GIF-SUR-YVETTE (France)

YAMAZAKI Y., Theoretische Physik - Universität des Saarlandes
6600 SAARBRÜCKEN (Germany)

YASUE K., Département de Physique Théorique - Université de Genève
24, quai Ernest Ansermet - 1211 GENEVE 4 (Switzerland)

ZALCZER G., Commissariat à l'Energie Atomique - S.P.S.R.M.
B.P. 2 - 91190 GIF-SUR-YVETTE (France)

ZAMBRINI J.-C., Département de Physique Théorique - Université de Genève
24, quai Ernest Ansermet - 1211 GENEVE 4 (Switzerland)

Selected Issues from

Lecture Notes in Mathematics

Vol. 561: Function Theoretic Methods for Partial Differential Equations. Darmstadt 1976. Proceedings. Edited by V. E. Meister, N. Weck and W. L. Wendland. XVIII, 520 pages. 1976.

Vol. 564: Ordinary and Partial Differential Equations, Dundee 1976. Proceedings. Edited by W. N. Everitt and B. D. Sleeman. XVIII, 551 pages. 1976.

Vol. 565: Turbulence and Navier Stokes Equations. Proceedings 1975. Edited by R. Temam. IX, 194 pages. 1976.

Vol. 566: Empirical Distributions and Processes. Oberwolfach 1976. Proceedings. Edited by P. Gaenssler and P. Révész. VII, 146 pages. 1976.

Vol. 570: Differential Geometrical Methods in Mathematical Physics, Bonn 1975. Proceedings. Edited by K. Bleuler and A. Reetz. VIII, 576 pages. 1977.

Vol. 572: Sparse Matrix Techniques, Copenhagen 1976. Edited by V. A. Barker. V, 184 pages. 1977.

Vol. 579: Combinatoire et Représentation du Groupe Symétrique, Strasbourg 1976. Proceedings 1976. Edité par D. Foata. IV, 339 pages. 1977.

Vol. 587: Non-Commutative Harmonic Analysis. Proceedings 1976. Edited by J. Carmona and M. Vergne. IV, 240 pages. 1977.

Vol. 592: D. Voigt, Induzierte Darstellungen in der Theorie der endlichen, algebraischen Gruppen. V, 413 Seiten. 1977.

Vol. 594: Singular Perturbations and Boundary Layer Theory, Lyon 1976. Edited by C. M. Brauner, B. Gay, and J. Mathieu. VIII, 539 pages. 1977.

Vol. 596: K. Deimling, Ordinary Differential Equations in Banach Spaces. VI, 137 pages. 1977.

Vol. 605: Sario et al., Classification Theory of Riemannian Manifolds. XX, 498 pages. 1977.

Vol. 606: Mathematical Aspects of Finite Element Methods. Proceedings 1975. Edited by I. Galligani and E. Magenes. VI, 362 pages. 1977.

Vol. 607: M. Métivier, Reelle und Vektorwertige Quasimartingale und die Theorie der Stochastischen Integration. X, 310 Seiten. 1977.

Vol. 615: Turbulence Seminar, Proceedings 1976/77. Edited by P. Bernard and T. Ratiu. VI, 155 pages. 1977.

Vol. 618: I. I. Hirschman, Jr. and D. E. Hughes, Extreme Eigen Values of Toeplitz Operators. VI, 145 pages. 1977.

Vol. 623: I. Erdelyi and R. Lange, Spectral Decompositions on Banach Spaces. VIII, 122 pages. 1977.

Vol. 628: H. J. Baues, Obstruction Theory on the Homotopy Classification of Maps. XII, 387 pages. 1977.

Vol. 629: W. A. Coppel, Dichotomies in Stability Theory. VI, 98 pages. 1978.

Vol. 630: Numerical Analysis, Proceedings, Biennial Conference, Dundee 1977. Edited by G. A. Watson. XII, 199 pages. 1978.

Vol. 636: Journées de Statistique des Processus Stochastiques, Grenoble 1977, Proceedings. Edité par Didier Dacunha-Castelle et Bernard Van Cutsem. VII, 202 pages. 1978.

Vol. 638: P. Shanahan, The Atiyah-Singer Index Theorem, An Introduction. V, 224 pages. 1978.

Vol. 648: Nonlinear Partial Differential Equations and Applications, Proceedings, Indiana 1976–1977. Edited by J. M. Chadam. VI, 206 pages. 1978.

Vol. 650: C*-Algebras and Applications to Physics. Proceedings 1977. Edited by R. V. Kadison. V, 192 pages. 1978.

Vol. 656: Probability Theory on Vector Spaces. Proceedings, 1977. Edited by A. Weron. VIII, 274 pages. 1978.

Vol. 662: Akin, The Metric Theory of Banach Manifolds. XIX, 306 pages. 1978.

Vol. 665: Journées d'Analyse Non Linéaire. Proceedings, 1977. Edité par P. Bénilan et J. Robert. VIII, 256 pages. 1978.

Vol. 667: J. Gilewicz, Approximants de Padé. XIV, 511 pages. 1978.

Vol. 668: The Structure of Attractors in Dynamical Systems. Proceedings, 1977. Edited by J. C. Martin, N. G. Markley and W. Perrizo. VI, 264 pages. 1978.

Vol. 675: J. Galambos and S. Kotz, Characterizations of Probability Distributions. VIII, 169 pages. 1978.

Vol. 676: Differential Geometrical Methods in Mathematical Physics II, Proceedings, 1977. Edited by K. Bleuler, H. R. Petry and A. Reetz. VI, 626 pages. 1978.

Vol. 678: D. Dacunha-Castelle, H. Heyer et B. Roynette. Ecole d'Eté de Probabilités de Saint-Flour. VII-1977. Edité par P. L. Hennequin. IX, 379 pages. 1978.

Vol. 679: Numerical Treatment of Differential Equations in Applications, Proceedings, 1977. Edited by R. Ansorge and W. Törnig. IX, 163 pages. 1978.

Vol. 681: Séminaire de Théorie du Potentiel Paris, No. 3, Directeurs: M. Brelot, G. Choquet et J. Deny. Rédacteurs: F. Hirsch et G. Mokobodzki. VII, 294 pages. 1978.

Vol. 682: G. D. James, The Representation Theory of the Symmetric Groups. V, 156 pages. 1978.

Vol. 684: E. E. Rosinger, Distributions and Nonlinear Partial Differential Equations. XI, 146 pages. 1978.

Vol. 690: W. J. J. Rey, Robust Statistical Methods. VI, 128 pages. 1978.

Vol. 691: G. Viennot, Algèbres de Lie Libres et Monoïdes Libres. III, 124 pages. 1978.

Vol. 693: Hilbert Space Operators, Proceedings, 1977. Edited by J. M. Bachar Jr. and D. W. Hadwin. VIII, 184 pages. 1978.

Vol. 696: P. J. Feinsilver, Special Functions, Probability Semigroups, and Hamiltonian Flows. VI, 112 pages. 1978.

Vol. 702: Yuri N. Bibikov, Local Theory of Nonlinear Analytic Ordinary Differential Equations. IX, 147 pages. 1979.

Vol. 704: Computing Methods in Applied Sciences and Engineering, 1977, I. Proceedings, 1977. Edited by R. Glowinski and J. L. Lions. VI, 391 pages. 1979.

Vol. 710: Séminaire Bourbaki vol. 1977/78, Exposés 507–524. IV, 328 pages. 1979.

Vol. 711: Asymptotic Analysis. Edited by F. Verhulst. V, 240 pages. 1979.

Lecture Notes in Physics

For information about Vols. 1–23, please contact your bookseller or Springer-Verlag.

Vol. 24: R. F. Snipes, Statistical Mechanical Theory of the Electrolytic Transport of Nonelectrolytes. V, 210 pages. 1973.

Vol. 25: Constructive Quantum Field Theory. The 1973 "Ettore Majorana" International School of Mathematical Physics. Edited by G. Velo and A. Wightman. III, 331 pages. 1973.

Vol. 26: A. Hubert, Theorie der Domänenwände in geordneten Medien. XII, 377 Seiten. 1974.

Vol. 27: R. K. Zeytounian, Notes sur les Ecoulements Rotationnels de Fluides Parfaits. XIII, 407 pages. 1974.

Vol. 28: Lectures in Statistical Physics. Edited by W. C. Schieve and J. S. Turner. V, 342 pages. 1974.

Vol. 29: Foundations of Quantum Mechanics and Ordered Linear Spaces. Advanced Study Institute, Marburg 1973. Edited by A. Hartkämper and H. Neumann. VI, 355 pages. 1974.

Vol. 30: Polarization Nuclear Physics. Proceedings 1973. Edited by D. Fick. IX, 292 pages. 1974.

Vol. 31: Transport Phenomena. Sitges International Schools of Statistical Mechanics, June 1974. Edited by G. Kirczenow and J. Marro. XIV, 517 pages. 1974.

Vol. 32: Particles, Quantum Fields and Statistical Mechanics. Proceedings 1973. Edited by M. Alexanian and A. Zepeda. V, 132 pages. 1975.

Vol. 33: Classical and Quantum Mechanical Aspects of Heavy Ion Collisions. Proceedings 1974. Edited by H. L. Harney, P. Braun-Munzinger, and C. K. Gelbke. VII, 311 pages. 1975.

Vol. 34: One-Dimensional Conductors GPS Summer School Proceedings, 1974. Edited by H. G. Schuster. VII, 371 pages. 1975.

Vol. 35: Proceedings of the Fourth International Conference on Numerical Methods in Fluid Dynamics, 1974. Edited by R. D. Richtmyer. V, 457 pages. 1975.

Vol. 36: R. Gatignol, Théorie Cinétique des Gaz à Répartition Discrète de Vitesses. II, 219 pages. 1975.

Vol. 37: Trends in Elementary Particle Theory. Proceedings 1974. Edited by H. Rollnik and K. Dietz. V, 472 pages. 1975.

Vol. 38: Dynamical Systems, Theory and Applications. Proceedings 1974. Edited by J. Moser. VI, 624 pages. 1975.

Vol. 39: International Symposium on Mathematical Problems in Theoretical Physics. Proceedings 1975. Edited by H. Araki. XII, 562 pages. 1975.

Vol. 40: Effective Interactions and Operators in Nuclei. Proceedings 1975. Edited by B. R. Barrett. XII, 339 pages. 1975.

Vol. 41: Progress in Numerical Fluid Dynamics. Proceedings 1974. Edited by H. J. Wirz. V, 471 pages. 1975.

Vol. 42: H II Regions and Related Topics. Proceedings 1975. Edited by D. Downes and T. L. Wilson. XII, 488 pages. 1975.

Vol. 43: Laser Spectroscopy. Proceedings 1975. Edited by S. Haroche, J. C. Pebay-Peyroula, T. W. Hänsch, and S. E. Harris. X, 466 pages. 1975.

Vol. 44: R. A. Breuer, Gravitational Perturbation Theory and Synchrotron Radiation. VI, 196 pages. 1975.

Vol. 45: Dynamical Concepts on Scaling Violation and the New Resonances in e^+e^- Annihilation. Edited by B. Humpert. VII, 248 pages. 1976.

Vol. 46: E. J. Flaherty, Hermitian and Kählerian Geometry in Relativity. VIII, 365 pages. 1976.

Vol. 47: Padé Approximants Method and Its Applications to Mechanics. Edited by H. Cabannes. XV, 267 pages. 1976.

Vol. 48: Interplanetary Dust and Zodiacal Light. Proceedings 1975. Edited by H. Elsässer and H. Fechtig. XII, 496 pages. 1976.

Vol. 49: W. G. Harter and C. W. Patterson, A Unitary Calculus for Electronic Orbitals. XII, 144 pages. 1976.

Vol. 50: Group Theoretical Methods in Physics. 4th International Colloquium. Nijmegen 1975. Edited by A. Janner, T. Janssen, and M. Boon. XIII, 629 pages. 1976.

Vol. 51: W. Nörenberg und H. A. Weidenmüller. Introduction to the Theory of Heavy-Ion Collisions. IX, 273 pages. 1976.

Vol. 52: M. Mladjenović, Development of Magnetic β-Ray Spectroscopy. X, 282 pages. 1976.

Vol. 53: D. J. Simms and N. M. J. Woodhouse, Lectures on Geometric Quantization. V, 166 pages. 1976.

Vol. 54: Critical Phenomena. Sitges International School on Statistical Mechanics, June 1976. Edited by J. Brey and R. B. Jones. XI, 383 pages. 1976.

Vol. 55: Nuclear Optical Model Potential. Proceedings 1976. Edited by S. Boffi and G. Passatore. VI, 221 pages. 1976.

Vol. 56: Current Induced Reactions. International Summer Institute, Hamburg 1975. Edited by J. G. Körner, G. Kramer, and D. Schildknecht. V, 553 pages. 1976.

Vol. 57: Physics of Highly Excited States in Solids. Proceedings 1975. Edited by M. Ueta and Y. Nishina. IX, 391 pages. 1976.

Vol. 58: Computing Methods in Applied Sciences. Proceedings 1975. Edited by R. Glowinski and J. L. Lions. VIII, 593 pages. 1976.

Vol. 59: Proceedings of the Fifth International Conference on Numerical Methods in Fluid Dynamics. 1976. Edited by A. I. van de Vooren and P. J. Zandbergen. VII, 459 pages. 1976.

Vol. 60: C. Gruber, A. Hintermann, and D. Merlini, Group Analysis of Classical Lattice Systems. XIV, 326 pages. 1977.

Vol. 61: International School on Electro and Photonuclear Reactions I. Edited by C. Schaerf. VIII, 650 pages. 1977.

Vol. 62: International School on Electro and Photonuclear Reactions II. Edited by C. Schaerf. VIII, 301 pages. 1977.

Vol. 63: V. K. Dobrev et al., Harmonic Analysis on the n-Dimensional Lorentz Group and Its Application to Conformal Quantum Field Theory. X, 280 pages. 1977.

Vol. 64: Waves on Water of Variable Depth. Edited by D. G. Provis and R. Radok. 231 pages. 1977.

Vol. 65: Organic Conductors and Semiconductors. Proceedings 1976. Edited by L. Pál, G. Grüner, A. Jánossy and J. Sólyom. 654 pages. 1977.

Vol. 66: A. H. Völkel, Fields, Particles and Currents. VI, 354 pages. 1977.

Vol. 67: W. Drechsler and M. E. Mayer, Fiber Bundle Techniques in Gauge Theories. X, 248 pages. 1977.

GPSR Compliance

The European Union's (EU) General Product Safety Regulation (GPSR) is a set of rules that requires consumer products to be safe and our obligations to ensure this.

If you have any concerns about our products, you can contact us on ProductSafety@springernature.com

In case Publisher is established outside the EU, the EU authorized representative is:

Springer Nature Customer Service Center GmbH
Europaplatz 3
69115 Heidelberg, Germany

Batch number: 10838062

Printed by Printforce, the Netherlands